FOOD SCIENCE AND NUTRITION

Food Science and Nutrition

[According to UGC Syllabus for UG, PG and Diploma Courses]

M.N. Ahmed

ANMOL PUBLICATIONS PVT. LTD.
NEW DELHI - 110 002 (INDIA)

ANMOL PUBLICATIONS PVT. LTD.
4374/4B, Ansari Road, Daryaganj
New Delhi - 110 002
Ph.: 23261597, 23278000
Visit us at: www.anmolpublications.com

Food Science and Nutrition

First Published, 2005

PRINTED IN INDIA

Published by J.L. Kumar for Anmol Publications Pvt. Ltd., New Delhi - 110 002 and Printed at Mehra Offset Press, Delhi.

Contents

Preface

Tourism and Hotel Industry are the two significant sectors of modern day economy and are complimentary to each other. The new world order has turned the whole mankind into one family or society. Today's world is a global village, where distances have no meaning. Although, tourism and travelling are the two phenomena, as old as the civilized world, yet, tourism evolved into a regular and well-organised industry, during the last century only.

Hotel industry is part and parcel of tourism sector. The tourists are honoured guests and the hotels offer them the demanded hospitality. Hence, the two form a correlated industrial sector, which is growing and flourishing fast, with promises for future.

Extension of education and expansion of knowledge have created a new anxiety, among the members of the new generation, regarding the above said two areas. In fact, the two sectors offer high profile career avenues and good opportunities for aspirants and job-seekers. However, like other disciplines, Tourism and Hotel Management also require proper training for professionals. Hence, a large room for exhaustive and exclusive books on the subject, is there.

This series of comprehensive books has been outlined within the parameters of the syllabi of various universities and institutes, in a broad-based manner, in order to cover all streams of the discipline and for the benefit of students at graduate and post-graduate levels.

The undersigned is confident of a due recognition and proper acknowledgement from concerned circles. Further expectation is that this series of useful books would prove to be practically helpful to the professionals and students equally.

— **Editor**

1

Introduction

In India, several dietary and clinical surveys carried out in different parts of our country have revealed that malnutrition is rampant. Nutrition education assumes special significance in the Indian context because the problem of malnutrition in India is mainly due to ignorance, poverty and lack of knowledge regarding the value of foods. Dietary practices, especially in children and pregnant and lactating mothers, are often governed by social taboos based on food fads. Nutrition education is the foundation for improvement in the dietary habits of his vulnerable set of population. Rigid dietary habits need correction and only systematic nutrition education programmes can bring changes in dietary habits. Nutrition awareness is essential to bring changes in dietary habits and creating nutrition awareness entirely depends on education and training.

Nutrition education requires identification of needs and collection of baseline data enables to identify the needs. It is a popular belief that young mothers or adolescents must be educated about the basic principles of health and nutrition. Studies on nutrition education in our country have proved that even elderly women need education as they are the ones who mislead the youngsters. Various topics of nutrition can be selected for nutrition education. Selection of foods and preparation of foods, feeding practices of various groups, special diets during pregnancy and lactation, infancy, pre-school age, supplementary feeding, weaning foods, low cost recipes, food sanitation, food-borne infections, food adulteration, nutrition and health, therapeutic diets, food fads and fallacies and deficiency diseases and their prevention are some common topics used for nutrition education.

Extension education methods can be utilised for nutrition education. Collection of background data, analysis of problems, plan of action, calendar of activities, selection of audio-visual aids, methods of teaching and evaluation are the various steps involved in nutrition education programme.

The various approaches suitable for nutrition education are (1) Individual contacts through home visits, consulting individual training, (2) Group contacts through simple talks, demonstrations and discussions, (3) Group demonstrations, (4) Hospital visits, and (5) Nutrition rehabilitation training. These are the methods suitable for rural groups.

Nutrition education and training can be implemented through schools and colleges. Good nutrition education, if followed effectively, can be a useful tool to bring economic development in the country as it provides good health. Better health enables people to improve their productive efficiency and it enhances the purchasing capacity and the consumption patterns of the population. Food and nutrition is thus intimately linked with production and economic development of a country.

Training in nutrition at various levels—schools, colleges, adult education centres, village level workers—is essential. Housewives are ignorant about the principles of nutrition as they are not trained anywhere. Thus the researches on these lines are wasted as the messages are not passed on to the people. Various feeding programmes and nutrition programmes which are implemented through various organisations can be used as nutrition education channels. Feeding programmes under the family and child welfare departments, school feeding programmes, the composite programme for women and pre-school children, the applied nutrition programmes, world food programmes and other state programmes can be utilised for education purposes.

Programmes of Nutrition

A number of nutrition programmes were initiated by the Government of India. Most of them are functioning under the guided supervision of development department, social welfare department and voluntary agencies in various states.

Applied Nutrition Programmes : The applied nutrition programme (ANP) is evolved on the basis of the fundamental premise that though malnutrition is primarily due to mass poverty it is also due to ignorance and that appropriate choice of foods and feeding practices without involving extra cost from the family can improve the nutritional status of the people, especially young children.

Aims and Objectives

It is well known that many young children in India are undernourished. This undernourishment is dangerous. If children do not receive adequate nourishment when they are young, they will never grow to be strong, healthy and productive adults. Furthermore, undernourished children are more likely to become ill. Undernourishment is directly or indirectly responsible for illness and death.

The goal of ANP is to prevent the bad effects of undernourishment before they can occur. The ANP works at the village and family levels to improve nutrition, particularly of young children. To bring about such improvement the programme supports the local production of good and nutritious foods, consumption of such foods, especially by young children, pregnant women and nursing mothers, and education about good foods and proper feeding of young children and women who are pregnant or nursing (nutrition education).

Improvement in nutrition will not be effective unless there is also improvement in health. The ANP should be implemented with local health workers to provide health check-ups, immunisation, safe drinking water and proper sanitation. The ANP does not have enough funds to assist all the village in a block. Therefore, it is necessary to choose those villages which have a reasonable chance of success. As many ANP activities as possible should be started in the village selected.

The ANP also does not have enough funds to cover all the families in the village selected to participate in the programme. Care should be taken to reach the most disadvantaged families such as the families of the Harijans, tribal people, small cultivators and landless labourers, which have young children, pregnant women and nursing mothers.

By concentrating on these selected villages and families it should be possible to demonstrate clearly the results of the programme. This should lead to the voluntary use of beneficial practices in other villages of the blocks which is called the radiation effect

Education of Nutrition

People will not change their beliefs and habits concerning food unless they are convinced that changes will benefit them and their children. Therefore, the goal of ANP is to educate village families about the benefits of nutritious foods, good child feeding practices and improved preparation and preservation of foods.

Many traditional practices such as prolonged breastfeeding and eating roti and rice with dal are basically good. Nutrition education should emphasise such traditional practices and at the same time should teach new practices where necessary. By using nutritious local foods and good feeding practices it is possible to improve the health of undernourished children at little or no extra expense to the family.

Nutrition education should be provided in cooperation with schools, Balwadis, PHCS, sub centres, Maniia Mandals and Yuvak Mandals. It can be given by people such as school teachers, Balsevikas, Mukhya Sevikas, Gram Sevikas and medical staff to the PHC and sub-centres. Efforts should be made to bring this eduction to the poorest villagers, who may not ordinarily participate in village organisations. The education should stress on basic messages about nutritious foods, child feed and child care.

The method of practical demonstration with the participation of the villagers ought to be used as much as possible. In the demonstrations, only local food and easily available equipment should be used; charts, posters and film shows should also be used to make the information clear and interesting.

Main Messages : Breastfeed as long as possible. Introduce some solid food 5 to 6 times a day. Do not reduce food in illness. Use health services available. Get children immunised. Keep yourself and your surroundings clean. Drink clean water. Have no more than 2 or 3 children 2 to 3 years apart. Try to eat some green leafy vegetables every day.

Activities in Production

The ANP gives many kinds of assistance for production of good nutritious foods depending on local conditions. The programme now emphasises assistance to home gardens, school gardens and community gardens. These gardens are used to grow dals, leafy vegetables and other nutritious vegetables and fruits. The products are then used to demonstrate the preparation and use of the foods.

Gardens should become as self-sufficient as possible. After the first season, seeds should be obtained either from the garden itself or from the sale of surplus products. Local manure should be used as much as possible. When available, waste water from the village should be used for irrigation.

Home gardens should be emphasized where possible, because of their importance in making nutritious foods available to the individual.

School gardens will be supported if the following conditions are met—suitable land is available close to the school and water for irrigation is available at a reasonable cost. The garden is made part of the regular school programme of the students and the students must look after the garden themselves.

Community gardens can be used both to demonstrate the proper cultivation of nutritious food and to produce food for feeding purposes. Cash crops may be grown to make the gardens self-supporting but the entire net profit of the garden must be used for demonstration feeding projects at health centres or balwadis. The garden should be run by a village organisation such as a Yuvak Mandal or Mahila Mandal.

Foods of Animal Origin : The ANP now emphasises the cultivation of nutritious dals, vegetables and fruits in home, school and community gardens rather than the production of foods of animal origin, which are much more expensive. Moreover, if local conditions are suitable, activities such as poultry, fisheries, piggeries and goat keeping can be supported.

Activities in Consumptions

Feeding Projects : Feeding projects under ANP will be chiefly for the purpose of practical demonstrations for nutrition education.

Agents for Change : The ANP cannot succeed unless villagers themselves accept the need for new practices. Therefore, the cooperation of village leaders is essential in planning and running the programme in each village. The support of women and youth is also important.

2

Programme Organisation

The Department of Rural Development (Ministry of Agriculture and Irrigation), Government of India has overall charge of the programme. The State Department of Community Development or its equivalent is in charge of the ANP in the State, the officer responsible is the State ANP coordinator. At the district level, the District Collector will have overall charge of the programme. He will be assisted by other district development officials.

At the block level, the BDO will have overall charge of the programme. He will be assisted by a coordination committee including the head of the Panchayat Samiti, the Block Agricultural Extension Officer, the Mukhya Sevika, the Doctor in charge of the PHC and other block officials.

At the village level, the gram sevika and the gram sevak will be chiefly responsible for the implementation of the programme.

Initiation of the Programme : As soon as possible, after the start of the financial years on first April, the BDO and other officials of the new ANP blocks will receive an orientation to the ANP and training in the preparation of the Block Plan. The BDO will then begin the work as the block plan is completed. He will submit copies to the State ANP coordinator and to the UNICEF field representatives, if necessary. The State Government may be asked by the Government of India to submit a modified plan or provide additional information. As soon as the block plan is approved by the Government of India, the State will ensure that all posts in the blocks including those of gram sevikas, are filled. Preparatory activities such as the digging of wells will begin and the training of block and village staff will proceed.

Block Plans : Experience has shown that the condition and needs of blocks all over India and even of villages in the same block can be quite different. It is important to the ANP to plan realistically in terms of the needs and resources of each block and of the villages of that block. The BDO and his staff will be responsible for preparing a realistic block plan based on their local knowledge of the block. The BDO and other block officials will receive special training in the preparation of a block plan. The BDO and the medical officer under the supervision of the State Nutrition Officer will conduct a survey to identify local nutritional problems. Then after consulting block and village workers as well as the villagers, the BDO will select the villages for ANP activities that will be suitable in these villages. He will write a detailed plan of action for the block. The importance of making realistic plans before ANP activities cannot be overemphasised.

Selection of Villages : To make the programme effective, it is important to introduce as many ANP activities as practical in the village simultaneously. Villages with facilities that will improve the chances for successful implementation of the programme should be given preference. Such facilities may include water for drinking and agricultural purposes, a road, bus services, proximity of a health centre or sub centre, availability of land for garden. Other government programmes or activities in the villages in a block should be selected for intensive integrated development under the ANP. However, if other villages in the block show interest, they can also be assisted to develop particular activities.

The ANP activities must be planned so that they continue even after the 5-year programme of assistance is finished. If activities are introduced in a village only in the third and fourth year of the programme, it is not likely that they will become established before the end of the programme. Therefore, it is better to choose all of the village in the first two years of the programme and start all activities in these villages as soon as possible.

Suitability to Local Conditions : If it is essential the ANP should be tailored to the special needs and conditions of each block and each village. In the past the practice has been to introduce ANP activities according to a standard pattern and a fixed financial limit. This is not realistic and is usually not successful. Now the needs of each village and the activities to be carried out will be used to decide what assistance and how much will be given to each block.

Nutrition education should be carried on in every village selected for ANP. Production activities must emphasise the growing of nutritious dals, vegetables and fruits. However, these activities can be carried out in many ways, depending upon the local conditions and resources. For instance, if a community garden is not possible because village houses are scattered or because land or water is not available, it may be possible to plan on funds to increase the number of home gardens in that village. Activities such as poultry or fisheries should be started only if local conditions are suitable and such activities can be maintained. Local officials and villagers must consider carefully the local conditions, resources and needs and then decide on the best plan.

Participation of the Community

Unless the community itself is concerned about the usefulness of the activities of the ANP and unless the community is involved in planning and carrying them out, the programme will not succeed.

In making the block plan, the BDO should consult the villagers. He will then be able to assess the problems and resources of the villages and, together with the villagers, decide the activities to be started. The village organisation such as the Mahila Mandals and Yuvak Mandals play an important role in the ANP. Members of these organisations will receive special training. They will then be able to organise such activities as common gardens, nutrition education and balwadi training.

The ANP will not succeed unless the block staff and the

villagers understand the purpose of the programme and the rules in maintaining the work. The ANP training that they will receive should explain to them their work and also that of others so that all can work together to make the ANP a success. As soon as possible, after the selection of new blocks, the BDO agricultural officer, education officers, medical officers, Mukhya Sevikas and other important block officials will receive an orientation to the programme and training in the preparation of the Block Plan.

Later, training will be given to gram sevikas, bal sevikas, teachers, panchayat officials and members of Mahila and Yuvak Mandals. The training will emphasise nutrition education. Mukhya and Gram Sevikas will receive practical training in teaching and demonstration. School teachers will be given material on nutrition so that they can make nutrition a part of their regular lessons.

The Government of India contributed Rs. 30,000 per year for each block. The state government contributed Rs.51,000 per year for each block. The United Nations Children's Emergency Fund (UNICEF) will contribute supplies of equipment, cash grants and assistance for training to a maximum value of Rs. 1,60,000 per block, spread over the operational period of the programme. The figure of Rs.20,000 is a ceiling the UNICEF supports and does not necessarily represent the amounts that will be provided to every block. Actual expenditure in each block will be determined by the needs and capacities of the individual blocks. The central and state funds will be placed in a common pool and disbursements will be accounted for by the state government on block by block basis.

Composite Programme

The pre-school age is the most crucial period in a child's life. This is the period when the child increases the activity and is exposed to the environment outside the home. Providing a nutritious and wholesome diet and immunising the child against diseases common during childhood are of utmost importance. Equally important is providing education to this age group through balwadis. Mere distribution of food to the children will not have a substantial effect unless the economic status of the mother is also simultaneously improved. For the successful implementation of programmes aimed at the overall development of the child, associate organisations like Mahila Samajams have to be fully involved in such programmes. After taking all the above aspects into consideration the Composite Programme for Women and Pre-school (CPWP) children was evolved and put into operation as part of the Applied Nutrition Programme in all the blocks in the state from the year 1975-76 in a phased manner with the following objectives:

1. To improve the health of the children (both physical and mental).
2. To inculcate a sense of discipline and equality among the children.
3. To promote healthy living habits.
4. To create a desire among the children to go to school and acquire knowledge.
5. To impart nutrition and health education to mothers.
6. To provide additional income, for the mother for giving special attention to the needs of the children.

   The main features of the regular ANP are found in the CPWP also. The CPWP has the following important aspects additionally:

   (i) Education to the children.

   (ii) Health check-up and immunisation of pre-school and pregnant and nursing mothers.
7. The medical officer of the Public Health Centre is expected to visit the balwadi at least once a month and the Auxiliary Nurse- Midwife once a week, for health check-up and immunisation. Growth charts for the children are maintained in duplicate—one kept by the balwadi teacher and the other by the mother. Growth charts to record the health status of the children and one DDS (Diet Drug Supplement) set costing Rs.400 is given by the UNICEF to every balwadi.
8. The target group consists of economically, socially and nutritionally backward children within the age group of

0 to 5 years and pregnant and nursing mothers. In each regular ANP centre 70 children and 30 mothers are enrolled. In a CPWP centre 100 children and 30 mothers are enrolled. The beneficiaries are selected by a committee consisting of the Block Development Officer or his representative, the Medical Officer of the Public Health Centre or his representative, the president of the Panchayat and the president of the Mahila Samajam. The socio-economic and nutritional status of the family will be the basic criterion for selection.

9. The feeding programme under both ANP and CPWP is implemented through organised Mahila Samajams. In the case of CPWP, Samajams having at least five cents of land with a building are chosen as implementing agencies. While sanctioning centres it is ensured that there is no other feeding centre within a radius of one kilometre. In the regular ANP feeding centre feeding is conducted only once a day. In the case of the CPWP, children attending balwadi classes are fed a second time. In specially recommended cases of severe malnutrition, there is provision for one more feeding. In the regular ANP centres feeding is done 25 days in a month. In the CPWP centres feeding is conducted 26 days a month.
10. The nutritive value of the feed is 14 grammes protein and about 280 calories. The composition of the diet is as follows:

    (i) Regular ANP

    (ii) CPWP first feeding:

CPWP first feeding:	grain	80 gms
	grain	80 gms
	oil	07 gms
	milk powder	30 gms
	Second feeding: grain	30 gms

11. The cost of the feed, at subsidised rate, works out to 40 paise and 50 paise on an average in the regular ANP and CPWP centres, respectively.
12. In the CPWP centres the balwadi teachers are given an

allowance of rupees 100 per month. In all other centres the conveners are paid a monthly allowance of 25. No separate preparation charges are paid. The fuel requirements are met out of voluntary contributions in kind collected by the Mahila Samajams. In fact, the whole programme revolves around the Mahila Samajams.

13. The programme is guided and supervised by the personnel of the Development Department. At the village level, lady Village Extension Officers are in charge of the programme. Their work is supervised by the Block Development Officer and the Extension Officer (women's welfare) at the block level. At the district level, the Assistant Development Commissioner and the District Women's Welfare Officer, under the guidance of the Collector, control and supervise the implementation of the programme.

14. Apart from feeding, nutrition education, health check-up and immunisation, there is also the production component of the programme consisting of financial assistance and technical advice for goat rearing, backyard poultry, kitchen gardening, a building for balwadi, environmental sanitation and training of balwadi teachers. This is a critical component of the ANP, in so far as the objective is to include and inculcate among the vulnerable sections in the rural areas, the habit of producing and consuming with very little additional strain, foods of high nutritional value. The basic concept behind the package of services and activities under the ANP is itself one of extension, that is, to train and educate the rural poor about the various aspects of nutrition and production of essential food items of nutritive value so that in course of time the people themselves would look after their own nutrition requirements balanced.

One-meal-a-day Programme

15. In 1979, the International Year of the Child, the government decided that supplementary nutrition programme should cover all undernourished pre-school children in the state. Towards the realisation of this objective, the one-meal-a-day programme was introduced in 1979-80. During that

year the programme covered 4.13 lakhs of pre-school chidrcn through 4,825 feeding centres. The programme is being contributed with the same number of feeding centres. The target group under this programme is children under five years and feeding is conducted 300 days in a year, that is, an average of 25 days in a month.

16. The quantity of food materials supplied under this programme to each beneficiary is as follows:

1.	CSM/Bulgar Wheat	40gms
2.	Pulses	10 gms
3.	Rice	40 gms
4.	Oil	10 gms

This meal costs 12.5 paise at subsidised rates,

17. In rural areas, the District Collector is empowered to administer this scheme through the staff of the Development Department. In urban areas this task is entrusted to the Social Welfare Department. The feeding programme is actually carried out by voluntary organisations. In the rural areas the beneficiary children are selected by a committee consisting of the Block Development Officer, the Medical Officer of the Primary Health Centre, the president of the Panchayat and the president of the local Mahila Samaj. In urban areas the Committee consists of the municipal commissioner, the municipal health officer or medical officer and the president of the voluntary agency concerned.

Special Nutrition Programme (SNP)

18. The Special Nutrition Programme was launched as a nationwide programme in 1971, to combat the incidence of malnutrition among the vulnerable groups like children below 6 years, pregnant women and lactating mothers. It was first introduced in the urban slums and tribal areas. Subsequently, it was extended to other areas like poverty-stricken nutritionally-deficient pockets.

19. The objective of the scheme is to supplement the diet of selected beneficiaries with additional protein and calories

to make up the deficiency in their daily food intake. Under this programme special sweet bread from Modern Bakeries, special milk bread from private bakeries, milk from Milma or cooperative societies, locally available food consisting of rice and pulses, are supplied. The quantity of food supplied contains an average of 5 to 6 gms of protein and 250 to 300 kcals. In addition Adexeline capsules containing vitamins A and D are also being supplied.

20. Originally the number of beneficaries in most of the feeding centres sanctioned during 1979 and 1980 was fixed at 100 per centre. Ten per cent of the beneficiaries in the SNP feeding centres will be pregnant women and lactating mothers and the rest will be children below the age of six years.

21. While introducing the scheme the per capita rate per beneficiary per day was fixed at 23.5 paise consisting of 18 paise for food, 2 paise for transportation of food materials and 3.5 paise for administrative charges. During 1974 the per capita unit cost of food was enhanced from 18 paise to 20 paise per beneficiary per day. Subsequently, the per capita rate was enhanced to 30 paise per head per day (i.e., food charges 24.5 paise, transportation 2 paise and administrative expenses 3.5 paise).

Integrated Child Development Services (ICDS)

22. The latest approach to nutrition is to provide a package of services comprising nutrition, medical check-up, immunisation, referral services, day-care services, education for mothers and functional literacy for adult women. These services are extended with the Anganwadi as the focal point. There will be one Anganwadi for every thousand population. The entire population in a block will be covered by setting up the requisite number of Anganwadis. This scheme is known as the ICDS and is evolved along the lines of the national pattern. As soon as the ICDS starts functioning in a block, the Special Nutrition Centres are merged into it.

23. The ICDS, as a centrally sponsored scheme, has been introduced through blocks in states. Central assistance is

obtained for administrative charges. Equipment is donated by the UNICEF. Only the nutrition component of the scheme is financed out of the state plan outlay. Besides the above, there are ICDS blocks sponsored and financed by the state governments.

24. The food usually being supplied to the beneficiaries under the ICDS are gruel and green gram. The diet costs 30 paise per child below 3 years and 50 paise per child above 3 years. The target group under the ICDS is the same as in the case of the Special Nutrition Programme. The number of feedings in a year is also the same. Of the beneficiaries 10 per cent will be pregnant women and lactating mothers and the rest will be pre-school children.

World Food Programme (WFP)

25. During the Fifth Plan, the Government of India introduced a new scheme, the "World Food Programme" (WFP), as a part of the Special Nutrition Programme. During this period 2,000 feeding centres started functioning under this scheme. During 1979-80 another 300 feeding centres were also started. In each centre, there are 100 beneficiaries out of which 12.5 per cent will be pregnant women and lactating mothers and the rest will be children below the age of 6 years. At present WFP covers nearly 2,30,000 beneficiaries spread over 2,300 feeding centres.

26. Uppuma made out of soya fortified Bulgar wheat and butter oil/salad oil are being distributed to the beneficiaries under WFP as detailed below:

    Soya-fortified Bulgar wheat is given to pre-school children (0-6 years) 80 gms per day per head and to pregnant women and lactating mothers 125 gms per day per head.

    Butter oil/salad oil is given 10 gms each to all beneficiaries.

Organisational Arrangement—Social Welfare Department

27. The feeding centres under SNP and WFP in the State are run by the local bodies like municipalities, corporations, panchayats and voluntary organisations like Bharat Sevak Samaj, Gandhi Smarak Nidhi and Mahila Samajams. In the case of SNP feeding centres there is one organiser and

one helper to distribute the food materials to the beneficiaries. An amount of Rs. 20 and Rs.10 per month respectively is being paid to them as honorarium. The transportation of food materials to the SNP feeding centres is being arranged by the Government through the contractors. Under the ICDS the Anganwadi worker gets an honorarium of Rs. 175 per month and the helper gets Rs.50 per month. In the case of WFP feeding centres, an allowance of Rs.40 per month is being paid to the voluntary organisations as grant. In addition to this the local bodies like panchayats are given financial assistance for the purchase of cooking vessels to the feeding centres. The transportation of food materials from the Warehousing Corporation Depot to the feeding centres is being done by the voluntary organisations themselves. They are also permitted to sell the empty containers and to utilise the amount for meeting the transportation cost.

28. The Health Services Department operates with this department to provide health services in the nutrition programme. Medical check- up and immunisation to the beneficiaries in the centres in urban areas are being conducted by the health staff of the corporations and municipalities and those in the rural and tribal areas by the Medical Officers in Primary Health Centres. Government has accorded sanction to meet the additional expenses towards petrol charges in connection with the visit of Medical Officers of the Primary Health Centre to the nearby feeding centres of WFP for the conduct of Health Education, Nutrition Survey and Immunisation.

29. The organisers and helpers are the nominees of the voluntary organisation responsible for implementing the nutrition programme. In every centre, there is an organiser and a helper. The organisers are responsible for the proper maintenance of printed records such as admission register, attendance register, stock register and medical check-up register. They are also vested with the responsibility of receiving the food supplies, their proper preservation,

storage and distribution to the beneficiaries. The helpers are responsible for the preparation of food articles for distribution. They also render assistance to the organiser for spot feeding of children.

30. There are two officers, one in the gazetted cadre (District Social Welfare Officer) and the other in the non-gazetted cadre (Inspector) for each district for the supervision of the feeding centres. Four inspectors in the non-gazetted cadre are appointed in the directorate for attending to the scrutiny of the bills of the SNP. The supervisors are entrusted with the work of supervision of feeding centres, arrangement of Nutrition Education Camps, selection of location of feeding centres, payment of honorarium to the organisers and helpers of feeding centres, allotment of food materials consistent with the number of beneficiaries in each feeding centre, allocation of cost of food materials and transportation charges.
31. One-day training camps at the rate of two camps in each district are being conducted by the department for giving nutrition education and training to the organisers of the feeding centres and to the mothers of the beneficiaries. The importance of nutrition, child care and family planning are being taught by the Medical Officer and Block Development Officer, in the camp in the feeding centres given to the supervisor-inspectors by the professors of medical colleges.
32. Some feeding centres in the States have now become defunct for various reasons such as non-availability of special bread, practical difficulties in serving cooked food made out of rice and pulse and irregularities noticed in the supply of food articles.

    The recommendation of the District Committee is therefore necessary for allotting and cancelling and redistributing the centres to voluntary agencies. These committees are not being convened at regular intervals. Hence, much difficulty is being experienced in the matter of starting, cancelling and redistributing the centres.

Health-Based Nutrition Programme

33. The Department of Health Services implements mainly three nutrition programmes of which two are meant for prophylaxis against nutritional anaemia among mothers and children and against blindness in children caused by vitamin A deficiency. Under these two programmes only iron and vitamins are supplied to the beneficiaries. Feeding is conducted only under the Integrated Health Package Programme. A brief description of the above programme follows:

Prophylaxis against Nutritional Anaemia among Mothers and Children

34. This programme was started in the year 1970-71. The family welfare programme supplies the cost of drugs as grant to the state government. Expectant and breastfeeding mothers, acceptors of family planning methods like IUD, tubectomy and pre-school children receive the benefit of this scheme. Tablets containing iron and folic acid are distributed through AN clinics, PH clinics and paediatric clinics of hospitals, PHCs and FW centres under the guidance of Medical Officers.

35. Haemoglobin percentage is obtained before starting the programme and at 3 months interval wherever facilities are available. The drugs are issued for a week at a time. Each beneficiary gets 100 tablets.

    Adult: 60 gms iron 0.5 mgs folic acid

    Child: 20 gms iron 0.1 mgs folic acid.

    Difficulties experienced are irregular supply and lack of motivation to come to the centre. The most deserving beneficiaries often do not come to the centre or do not come regularly.

Prophylaxis against Blindness in Children by Vitamin A Deficiency

36. Blindness due to vitamin A deficiency is more prevalent in the pre-school age group. Studies have shown that vitamin A in oil given orally is readily absorbed and stored in the liver from where it is gradually released for

utilisation in the tissues. Vitamin A is procured by the Department of Family Planning and distributed to the State Health Department

37. The vitamin A capsules are administered to children in the age group of 1 to 5 years in homes, personally by peripheral workers like AIM and EPHA. Two lakhs of IU/ 60,000 microgrammes of vitamin A is given at an interval of 6 months till the children attain 5 years of age. In view of the limited financial resources the programme is restricted to areas where there is maximum prevalence of vitamin A deficiency.

3

The Energy

Energy is defined as the capacity for doing work. From sunlight plants store energy as carbohydrate with the help of chlorophyl. Animals and men avail their energy through foods. When food is metabolised in the body, energy is released. Body requires this energy for all its activities. For all muscular activities and for the function of vital organs like heart, lungs, alimentary canal, nervous tissues and glandular tissues energy is required.

The Requirement

The energy requirement of a person depends on various factors. The most important factor in assessing the energy requirement is the basal metabolic rate of the person.

Basal Metabolic Rate (BMR) : This is defined as the amount of heat or energy required by the body to do the involuntary work of the body. Functions of brain, heart, liver, kidneys, lungs, secretary activities of the glands and intestinal movements are the basal activities in our body.

Factors Affecting the BMR : Basal metabolic rate of a person depends on the size, shape, composition, body weight, age and physiological condition of the person. Nutritional status and endocrine system of the body also exert their influence on basal metabolic rate of a person. Climatic condition of the surroundings also affects the basal metabolic rate.

1. *Size:* Since the heat loss in body is proportional to the skin surface, a tall and thin person has greater surface and thus higher basal metabolic rate compared to the short person. The body composition shows varitations in energy use. With little fat deposition in the body the basal metabolic rate increases and thus a tall, thin person has higher rates of basal metabolism compared to a short, fat man. Constant muscular activities of an athlete demand about 5 per cent more basal metabolic rate.
2. *Sex:* Sex also makes a variation in energy requirement. The metabolic rate of women is 6-10 per cent lower than that of men.
3. *Growth period:* During the growth period the basal metabolic rate and thus the energy requirement are increased. The highest BMR is during the first two years. After the growth period, especially after 25 years of age, there is a decline in energy requirement.
4. *Endocrine glands:* The thyroid gland exerts influence over energy requirements. Thyroid hyper-activity will speed up basal metabolism. The pituitary gland also increases the metabolic fate if it has disturbance. Adrenaline increases the BMR.
5. *Nutritional status:* During undernourishment the basal metabolic rate also declines to an extent of thirty per cent.
6. *Pregnancy condition:* During pregnancy and lactation the basal metabolic rate is increased by about 5 per cent during the first and second trimester and 12 per cent during the third trimester.
7. *Sleep:* During sleep the BMR is less than in the waking state.
8. *Climate:* Climatic conditions of the environment also affect the energy requirement. If the temperature falls below 14°C the energy requirement increases.
9. *Body temperature:* Fever increases the BMR.
10. *Disease conditions* : Diseases like typhoid fever, medullary diseases and lymphatic leukaemia show an increase in the BMR.

11. *Physical activities:* Physical activities half an hour before BMR measurement show high rates of BMR.

Total energy requirement of 4 person varies according to the basal metabolic rate, effect of food, activity involved, age, sex, physiological conditions and climatic conditions of the surroundings.

Energy requirements of an adult in India was reviewed by 'The Expert Group' of the FAO/WHO in terms of a reference man and reference woman. The reference man is in the age group of 20-39 years, with a weight of 55 kgs, without any disease and with a capacity to perform 8 hours of moderate activity. When not engaged in work a reference man spends 8 hours in bed and 4-6 hours in moving around or in a sitting position and 2 hours either walking or doing household activities.

In the case of a reference woman, the difference is only in her body weight that is, as against 55 kg of a man she weighs 45 kg. Instead of the physical activity of the occupation, the woman does household duties. Other conditions are the same in the case of a reference woman.

The energy expenditure for man and woman is calculated considering their internal and external activities. The FAO/WHO expert group (1983) suggested some recommendations as given in Table.

Energy Allowance for Various Groups

Category	*Reference body weight*	*Activity*	*Energy allowance kcals.*
Man	55 kg	light	2400
		moderate	2800
		heavy	3900
Woman	45 kg	light	1900
		moderate	2200
		heavy	3000
	Pregnancy	2nd and 3rd trimester	+300
	Lactation	first six months	+550
		6-12 months	+400

Additional requirement of energy is needed for the growth of the foetus, placenta and maternal tissues during pregnanoy. The BMR is also increased due to increased internal activities. Daily 150 kcals during the first trimester and 300 kcals during the rest of pregnancy is recommended. The energy cost during the term of pregnancy is 62,500 kcals. Additional energy requirement during lactation is for the secretion of milk. For a normal output of 850 ml of milk during the first 6 months 550 kcals/day is recommended.

During infancy the energy requirement is high. For 0-3 months 120 kcalsAg, 3-5 months 115 kcals/kg, 6-8 months 110 kcal/kg, 9-11 months 105 kcal/kg and at one year 112 kcals are recommended.

For children, the energy requirement varies according to their body weight The ideal weight of the children according to their age must be considered for assessing energy requirements. Thus in the latest recommendations, some suggestions were made as shown in Table.

Energy Allowance for Children

	Age group	*Body weight*	*Energy allowance kg. kcal.*
Children	1-3 years	12.03 kg	1220 kcal
	4-6 years	18.87 kg	1720 kcal
	7-9 years	26.37 kg	2050 kcal
Boys	10-12 years	34.30 kg	2420 kcal
	13-15 years	47.03 kg	2660 kcal
	16-18 yrars	56.50 kg	2820 kcal
Girls	10-12 years	36.47 kg	2260 kcal
	13-15 years	45.53 kg	2300 kcal
	16-18 years	50.00 kg	2200 kcal

Specific Dynamic Action : Energy requirement is affected by the type of food ingested. The extra heat which is produced after taking food is known as the specific dynamic action (SDA). The stimulating effect of carbohydrates, fats and protein on energy metabolism is different. Protein foods produce the highest per cent increase in energy metabolism. The SDA of protein is about 30 per cent while carbohydrates and fats exert 6 per cent and 4 per cent, respectively. The average SDA of a mixed diet is about 8-10 percent.

The activities which demand maximum energy are in the following order:

> Walking very fast, severe exercises, running, swimming, sawing wood, labourer's work, carpentry, metal and industrial work, walking slowly, laundry work, typing and ironing.

Measuring Units

Energy is measured in terms of Kcalorie. A Kcalorie is defined as the amount of heat required to raise the temperature of 1 kilogramme (1 litre) of water by 1°C. But joule is the new term used to measure energy. A joule is defined as the amount of work done or heal generated by an electric current of 1 ampere acting for 1 second against a resistance of 1 ohm. This unit is very small for measuring energy in nutrition and so 1 kilojoule (KJ) or megajoule (MJ) is used for practical use.

1 Kilojoule	=	1000 joules
1 Megajoule	=	10,00,000 joules
1 Calorie	=	4.184 joules (J)
1 Kcalorie	=	4.184 Kjoules
1,000 Calories	=	4.184 Mj (Megajoules)

Energy Value of Foods : Energy value of foods is otherwise known as the caloric value of foods. To measure the caloric value of food, an apparatus known as Bomb Calorimeter is used. In this apparatus a known amount of food is burnt and the rise in temperature of water in a container is recorded. The heat liberated by 1 gramme of carbohydrate is recorded as 4.1 kcals, protein 5.6 kcals and fat 9.4 kcals.

The energy value of food is otherwise known as the fuel value of foods. This is determined by means of an instrument known as Bomb calorimeter (Fig.).

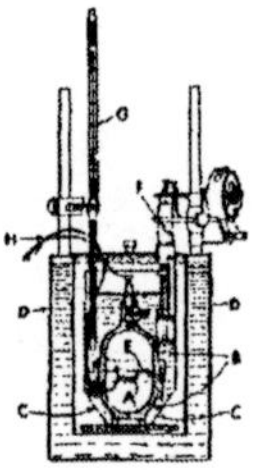

Fig. : Bomb Calorimeter

A weighed sample of dried food in pellet form is placed in a neavy steel container called a bomb. The bomb is held in place in a well insulated vessel. This is surrounded by a known volume of water.

The bomb is charged with oxygen to a pressure of about 300 pounds to the square inch. After the temperature of the water has been determined the sample is ignited by means of an electric fuse. The heat from the sample is dissipated into the water. By noting the change in the temperature of the water the energy value of the food can be calculated.

Physiological Fuel Value of Foods : Small losses occur in digestion when carbohydrates, fats and proteins are utilised by the body. Digestability coefficiency of western diets are available and based on these data:

the physiological fuel value of carbohydrate 4kcals

the physiological proteins 4kcals

the physiological fats 9kcals

Among these nutrients the body loses maximum energy from protein. This is due to the loss during digestion. The digestibility of various foods varies which affects the physiological fuel value.

Among the various sources of energy, foods rich in fats and dairy produce are the best. Dried fruits, cereals and other starchy foods and legumes are fair sources of energy.

Determination of Energy Requirements of the Body : There are two methods generally applied for measuring the energy requirements of human body. They are direct and indirect calorimetry methods.

Direct Calorimetry : In this method the amount of heat produced by the body is measured directly. At Water-Rosa, Benedict perfected the human respiration calorimeter. To measure the heat production in an individual a specially constructed chamber, called a respiration calorimeter, is used. It is determined through the relation between energy output and oxygen consumed in a person in this calorimeter.

Respiration calorimeter consists of an airtight copper chamber insulated by wooden wall with air spaces in between. Enough provision for sitting or to lie down is made through a folding bed, chair and table. Food is provided through the opening at one end and the excreta is removed through another opening at the other end. The chamber is very well arranged for the individual to relax, read or write. It is insulated to prevent the entry and exit of heat through the walls. The chamber is ventilated by a current of air, the CO_2 and water given off are removed by soda lime and sulphuric acid respectively. For oxygen supply, a known amount of oxygen is introduced to the chamber through a gas metre with an air current. The oxygen consumption and CO_2 production are calculated using a formula. The heat produced is also measured by circulating a current of water through copper pipes and measuring the quantity of water that has been circulated through the chamber and also the difference between the temperature of the water entering and leaving the chamber.

Respiratory Quotient: Respiratory Quotient is the ratio between the volume of CO_2 given out and the volume of O_2 consumed by the human subjects.

$$\text{Respiratory Quotient (RQ)} — \frac{\text{Volume of } CO_2 \text{product}}{\text{Volume of } O_2 \text{ consumed}}$$

The RQ varies with the type of food being oxidised. When glucose is oxidised RQ is 1, for a fatty acid it is 0.7 and for protein it is 0.8. Under resting condition with no food for 12-14 hours the RQ is 0.82. In this condition carbohydrate, fats and protein present in the body are utilised.

Indirect Calorimetry **:** In this method oxygen consumed and carbon dioxide excreted in a given time under basal condition is measured. Many experiments on people of different ages have shown that 1 litre of oxygen is equal to 4.825 kcals when the person is in Basal Metabolic condition. The basal metabolic rate is measured by indirect calorimetry. The person has to observe certain conditions before measuring the basal metabolic rate. They are:

1. In a post-absorptive state, i.e., 12-16 hours after the last meal- preferably in the morning.
2. In a reclining state but awake.
3. Before the test the person must take one hour rest if he is engaged in any work.
4. Relaxed and free from emotional upset, excitement or fear.
5. Body temperature must be normal.
6. Room temperature must be comfortable.

In these favourable conditions the BMR is measured. For an adult man the BMR rate is 34.2-36.7 kcals/sqm/hr. and for an adult woman 30.9-35.1/kcals/sqm/hr.; for children it is more. For 7-8 year boys it is 49.1/kcals/sqm/hr. and for girls it is 45.2/ kcals/sqm/hr.

4

Planning for Diet

Planning a balanced diet within the income level needs precise knowledge about the sources of nutrients, their requirements for various groups, seasonal availability of various foodstuffs and dietary habits of the group.

Blending these theoretical knowledge into a day's menu needs skill which has to be acquired through practice. There are certain basic principles which has to be observed while planning a balanced diet. As the name indicates a planned diet should be balanced. A diet is called balanced when all the nutrients required by the body are present in correct proportion. A judicious selection from the food groups will supply adequate nutrients for maintenance, repair and growth. Other factors to be considered are compositon of the family and the physiological phases of family members, dietary habits of the family, food budget of the family, variety in meal patterns, appearance, texture and taste of meals, selection of food for seasons (cold and hot climate condition) and nutritional balance.

While planning a diet, the composition of the family has to be considered. Infants, children, teenagers, pregnant and lactating mothers and geriaric people need food in different amounts. There are variation in their nutritional requirements and their digestive capacity. Quality and quantity and often the preparation need certain changes. A careful planning gives provisions for all these without making great changes in the preparation pattern of foods.

Dietary habits of the family also have to be considered while meal planning. Food habits are influenced by earlier experiences in life: sensory, aesthetic, economic, geographic, social and cultural factors. Often food habits change according to the new environment, new values and availability of items. When a child is exposed to group feeding practices slight changes in likes and dislikes of foods are common. Age and sex also influence food choices. The best example is an adolescent's dietary habits. There are vegetarian, non-vegetarian, lacto-aro-vegctarian and pesco-vegetarian dietary habits (plant foods and fish but no milk, egg, meat or poultry).

Food budget is a very important factor in meal planning. Money available to buy food items, time available for food preparation, job pattern, life-style of the family members, location of residence, and family values decide how much has to be spent on food. First of all to reduce food expenditure, one must have an idea about low cost nourishing foods for selection. Intelligent selection, good marketing, just enough bulk purchase, seasonal buying, food processing and preservation, better food preparation methods, avoiding more indulgence in ready-to-eat foods, fast foods, readymade food and hotel catering reduce food budget.

Variety in meal makes eating a pasure. The same meal, if repeated, creates monotonous taste and reduces appetite. Variety in choice of food items, methods of preparation in colour, texture and flavour induce appetite and provide satiety value to the meal. When food is prepared in a proper form with characteristic taste and flavour, it derives satiety value.

Above all these factors, a nutritional balance in each meal is essential. Successful meal planning principles suggest that each meal must be nutritionally balanced. One-third of the total nutrients must be in the lunch, one-third in the dinner and the remaining one-third in breakfast and teatime. Skipping a meal is not good. Many people often skip breakfast which is very bad as far as the body is concerned. Studies have showed that the performance of a person after mid-morning is very poor if they skip their breakfast

Based on these principles, if a diet is planned it will be a balanced one. The Indian Council of Medical Research has suggested the amounts of various food items to make a balanced diet for various groups.

Some Clues for Different Balanced Diet —Low Cost BD : Millets, cereals, roots and tubers, pulses and leafy vegetables can be included liberally. Cheap seasonal fruits and salad vegetables lend attraction and freshness to the diets.

Balanced diets of middle income groups can include nodente amounts of protective and proteins-rich food. Milk, egg, fish and meat can be included in moderate amounts while pulses, nuts, fruits and other foods can be used in liberal amounts.

Balanced Diets for High Cost : In a high cost diet protecthe foods and protein-rich foods can be included liberally. Wet surveys in India have shown that malnutrition is prevalent among high class people also. B vitamins and certain minerals are not included m enough quantities in their dietaries. Over-nutrition due to high consumption of milk, meat, fats, ghee and sweets are common.

Food Groups and Guidelines

The ultimate aim of consuming food is to maintain health. There is no single foodstuff which can contribute all the nutrients needed by the body. Only a judicious selection can provide all the nutrients in required amount. In order to select the sources of all nutrients and that also in correct proportion one must know about the basic principles of food selection. The nutritionists and food scientists all over the world have put in a lot of efforts to formulate certain guidelines in these matters. Basic food groups suitable to different countries were evolved by the nutritionists.

The US Department of Agriculture suggested a number of food group plans like the Basic-4 food group.

The Basic-7 Food Group and the Basic-11 Group Plan : The Nutrition Expert Group of L.C.M.R. (India) suggested a five-food group plan.

Based on their nutritive values foods are grouped into 11. They are:

1. Cereals and millets
2. Pulses (legumes)
3. Nuts and oil seeds
4. Vegetables
5. Fruits
6. Milk and milk products
7. Egg
8. Meat, fish and other fleshy foods
9. Fats and oils
10. Sugar and jaggery
11. Spices and condiments.

In India this classification has great significance.

Cereals and Millets : They supply the major portion of calories (70%), more than 50 per cent of B Vitamins. In a poor vegetarian diet cereals supply more than 70 per cent of protein.

Rice and wheat are abundantly used as staple cereals in the world. In India rice is used mainly in the diet of South India, whereas wheat is the main staple in all other parts.

The major millets used are jowar, ragi and bajra. Though they are cheaper they are superior in their mineral supply and protein content Ragi is a very good source of calcium.

Pulses and legumes are fair sources of proteins and they are good sources of B vitamins, especially thiamine and riboflavin. Quantitatively their protein content is high but qualitatively they are not as superior as animal proteins. But in an Indian diet cereals and pulses supplement mutually in their amino acid content and improve the quality of protein. Pulses and legumes are comparatively cheap sources of protein. Cowgram, horsegram, soyabean, dry field beans and dried peas are some of the cheap pulses.

Nuts and oilseeds are fair sources of protein. Coconut is a good source of energy. It has also fair sources of B vitamin. Vegetables are unique in their vitamins and mineral content. Vegetables are again grouped into yellow vegetables, leafy vegetables and as other vegetables. Yellow vegetables and leafy vegetables are rich sources of vitamin 'A' (carotene), calcium, iron, folic acid, riboflavin and vitamin 'C. But the iron content in vegetables is comparatively poorly utilised by the body. Other vegetables are good sources of minerals like calcium, trace elements and vitamin 'C.

Apart from these, vegetables have a high percentage of water in them. They contribute the alkaline ash and the bulk or cellulose to the body. Beetroot, carrot and ladyfinger are high in sodium content. Fruits, especially citrus fruits, are good sources of ascorbic acid. Some starchy fruits are rich in fructose and dextrose; yellow fruits contribute carotene, the provitamin A. Fruits as a group supply water and electrolytes like potassium. Most of the fruits are rich sources of potassium and poor sources of sodium. Dates, dry raisins, grapes, peaches, lime and ripe papaya are very high in their potassium content Bananas are rich in starch content in the raw form and in sugar content in the ripe form.

Milk and milk products are the best sources of calcium and ribo-flavin and second to the meat groups in protein content. Except a iron and vitamin 'C milk is good in all other nutrients. Milk products are high in biological value and milk cream is excellent in vitamin A.

Egg has biologically very good protein. Egg white is rich in protein and egg yolk is rich in fat, iron, sodium, potassium, calcium and magnesium. Vitamins like thiamine, riboflavin, vitamins A and D are also present in egg. It plays very many roles in cookery.

Meat, fish and other flesh foods rank first in protein content. They are also good sources of phosphorous, magnesium, iron, thiamine, niacin, B 6, and B 12. The biological value of meat protein is high. Fat is present in a good quantity in it.

Fats and oils contribute mainly calories and they are good sources of essential fatty acids and vitamins A and E.

Sugar and jaggery are the common sweetening agents in our diet. They are rich sources of carbohydrate.

Spices and condiments are necessary foodstuffs used for flavouring foods. Apart from these functions some of them contribute certain nutrients. Red chillies and coriander supply carotene; green chillies vitamin 'C turmeric and tamarind iron; and garlic stimulates synthesis of certain vitamins.

Then food is classified into 11 groups for convenience in selection.

But in different countries different food group plans are followed by the nutritionists to formulate a balanced diet.

In India the nutrition expert group of ICMR suggested the five-food group plans.

They are given below:

	Group	*Nutrients*
1.	*Milk group*	
	This consists of milk and milk products and protein-rich foods like pulses, nuts, meat, fish and egg.	This group supplies proteins, minerals and vitamins.
2.	*Fruits and green leafy vegetables*	
	Fruits like papaya, orange, mango, Indian gooseberry, guava and green leafy vegetables like spinach, amaranthus, drumstick leaves.	Rice sources of vitamin 'C' vitamin A (carotene), minerals and roughage.
3.	*Other vegetables*	
	Peas, beans, drumsticks, ladyfingers, bringals, cucumber, bittergourd, ashgourd, plantain etc.	They are fair to good sources of minerals, vitamins and roughage and water content.
4.	*Cereals, roots and tubers*	
	Rice, wheat, maize, ragi corn, tapioca, potato, sweet potato, yam, colocacia etc.	Rich sources of starch, fair to good sources of proteins and B vitamins. In an Indian diet this group supplies about 75-80 per cent of calories and 50-60 per cent of proteins.

Contd...

Group	Nutrients
5. *Fats and oils and pure carbohydrate foods*	
Butter, ghee, vegetable oils, sugar, jaggery, honey, starch and egg powders, custard powders etc.	They are rich sources of energy, sential fatty acids, vitamins A and E and cholesterol.

When these food groups were suggested they were known as Basic groups. Basic-7 food group plan, Basic-4 food group plan, and Basic-11 food group plan of US Department of Agriculture was the first in these series. Basic dietary pattern and daily food guides were suggested by different nutrition boards. In India the National Institute of Nutrition of ICMR has done remarkable work in this field. Guidelines for a balanced diet, composition of balanced diet and balanced diet for different regions of India, viz., balanced diets for the eastern region, northern region, southern region and western region were suggested by the NIN and it was periodically revised based on their advanced research in this field. Guidelines for low-cost balanced diets were prepared by them for a student of nutrition. In our Indian background this has great significance.

Choice of Food

Most of us have to plan either our own or other people's diets if only to the extent of selecting from a hotel or cafe menu what they well eat.

The housewife has to plan meals for her family, the members of which may be of different ages and have very different likes and dislikes. Meals have to be planned for the school dining-room, the institutional canteen etc.

Everyone should therefore have some knowledge of the values of foods and of the factors that influence choice, recognizing that the more costly foods are not necessarily those of the highest nutritional value. Grapes, for example, contain only traces of vitamins compared with the amount present in an orange or cabbage.

During and since the war, the science of nutrition has been applied to home economics as never before and in 1952 the Ministry of Food (U.K.) published a Report on Domestic Food Consumption and Expenditure in which data are given on nutrients consumed by different sections of the community and the extent to which these satisfy nutritional needs.

In 1950, the *mean* intake of the chief vitamins per head of the population was as follows:

Vitamin A	3,536 i.u. (a)
Vitamin B_1	1.51 mg.
Riboflavin	1.69 mg.
Nicotinic acid	13.0 mg.
Vitamin C	84 mg. (a)
Vitamin D	172 i.u. (a)

Including Welfare Foods : Except for vitamin D which is definitely low and riboflavin which seldom reaches the estimated needs, all these are at the level of adequacy. However, the mean figure is obtained by averaging some above and some below and that in the case of the vitamins of the B complex, all of which are near the requirement level, there must be as many individuals receiving less that this as there are consuming more.

There was not, in 1950, the marked difference between the intake of people in different income groups as was found by the pre-war survey. It has therefore been possible to make estimates of the food consumption of the whole population with regard to the supply of the principal nutrients.

Vitamin A	20.7	per cent obtained from vegetables
	17.4	per cent obtained from fats, milk and cheese
	14.5	per cent obtained from meat (incl. bacon)
	7.1	per cent obtained from eggs
Vitamin B_1	43.8	per cent obtained from bread and cereals
	22.7	per cent obtained from vegetables
	14.8	per cent obtained from meats
	12.5	per cent obtained from milk
Riboflavin	40.0	per cent obtained from milk and cheese

	17.6	per cent obtained from bread and cereals
	16.5	per cent obtained from meat
Nicotinic acid	34.8	per cent obtained from meat
	31.9	per cent obtained from bread and cereals
	20.1	per cent obtained from vegetables
Vitamin C	37.2	per cent obtained from potatoes
	33.0	per cent obtained from fruit
	10.0	per cent obtained from green vegetables
Vitamin D	36.0	per cent obtained from fats
	22.7	per cent obtained from fish

The report points out that after August 1950 when the extraction rate of national flour was lowered from 85 per cent to 81 per cent, there was a corresponding drop in vitamin B intake in all groups of the population. Since bread and cereals supply over 40 per cent of the vitamin B_1, this change was to be expected.

Average figures are no guide to individual differences. Everyone can do something, by the application of a knowledge of food values, to improve his or her own food intake and can do so without in any way becoming a food-faddist. It should be a natural habit to choose the better food when several choices are presented and such good habits can be developed easily in children.

If a mother or a school matron has enough knowledge to provide children with a nutritious diet then they are likely to grow up liking the nutritious foods to which they have become accustomed. Habits about food tastes are very pronounced and it is sometimes difficult training children (or adults) to like something new.

We would like to study various ways in which meals can be improved by various additions.

One of the simplest breakfasts taken in Great Britain is toast with butter and marmalade, accompanied by tea, but many people add to this either porridge or a cereal with milk or a cooked dish such as bacon or fish, or fruit, sometimes raw, sometimes cooked; and some take coffee in stead of tea.

The accompanying table shows the effect of a number of different additions to what is taken as a 'basal' breakfast.

The Effects of Various Additions to a 'Basal' Breakfast

Basal breakfast		*Vit. A or equiv. in carotene* i.u.	*Vit. B_1* µg.	*Ribo-flavin* µg.	*Niacin* µg.	*Vit. C* mg.	*Vit. D* i.u.
Toast (81 percent edtraction)	2. oz	—	98	34	660	—	—
Butter	¼ oz.	284	—	—	—	—	—
Milk	1 oz.	30	13	40	—	—	4
Tea							
Suger	—	—	—	—	—	—	—
		314	111 =.111 mg.	74 =.074 mg.	660 =.66 mg.	—	4.3
Additions							
(1) 5 oz. milk for cereal or coffee	150	65	200	—	—	1.5	
(2) Cornflakes	2 oz.	—	—	—	—	—	—
(3) Porridge	2 oz.	—	250	80	600	—	—
(4) Bemax	1 oz.	—	450	200	1,700	—	—
(5) I egg	2 oz.	500	74	340	—	—	3.4
(6) 2 x 2 oz. tomatoes	4oz.	500 Approx	80	40	400	20	—
(7) Rasher bacon	1 oz..	—	160	125	1,200	—	
(8) Herring	6 oz.	250	—	—	6,000	—	1,500
(9) Haddock	4 oz.	70	48	300	1,500	—	—
(10) Grape-fruit pulp	2 oz.	3	20	14	100	12	—
(11) Stewed prunes	4 oz.	300-400	32	68	—	1	—

The same kind of analysis can be made of the snacks obtainable at nearly every snack bar.

Basal breakfast		*Vit. A or equiv. in carotene* i.u.	*Vit. B_1* µg.	*Ribo-flavin* µg.	*Niacin* µg.	*Vit. C* mg.	*Vit. D* i.u.
1. *Baked beans on toast*							
		Baked beans		3 oz.	70	60	30
600	—	—					
Nat. bread	1 oz.			49	17	330	
Margarine	1/8 oz.	71	—	—	—	—	7
		141	109	47	930	—	7
2. *Spaghetti on toast*							
Spaghetti (uncooked)	1 oz.	—	20	—	600	—	—
Nat. bread	1 oz.	—	49	17	330	—	—
Margarine	1/8 oz.	71	—	—	—	—	7
		71	69	17	930	—	7
3. *Sardines on toast*							
Sardines	2 oz.	150	20	160	1,300	—	560
Nat. bread	1 oz.	—	49	17	330	—	—
Margarine	1/8 oz	71	—	—	—	—	7
		221	69	177	1,630	—	567
4. *Grilled tomato on toast*							
Tomatoes	3 oz.	420	60	30	300	6	—
Nat. bread	1 oz.	—	49	17	330	—	—
Margarine	1/8 oz	71	—	—	—	—	7
		491	109	47	630	6	7
5. *Welsh rarebit*							
Cheese	1 oz.	369	9	140	100	—	4
Nat. bread	1 oz.	—	49	17	330	—	—
Margarine	1/8	71	—	—	—	—	7
		440	58	157	430	—	11

Basal breakfast		*Vit. A or equiv. in carotene i.u.*	*Vit. B_1* µg.	*Ribo-flavin* µg.	*Niacin* µg.	*Vit. C* mg.	*Vit. D* i.u.
6. *Poached egg on toast*							
		1 egg	2 oz.	500	74	340	—
—	34						
Nat. bread	1 oz.	—	49	17	330	—	—
Margarine	1/8	71	—	—	—	—	7
		571	123	357	330	—	41

The value as carotene is halved to give an approximate equivalent in terms of i.u. vitamin A.

It is worth nothing that all of these (except tomatoes) are completely lacking in vitamin C, that sardines make the only appreciable vitamin D contribution and provide the only large quantity of niacin and that, except for numbers 3 and 6, not one contains as much as one-tenth of the day's requirement of vitamins of the B complex.

While these meals are useful, people consuming them regularly must remember to make up for these defects at other meals.

It is sometimes assumed that any fruit will supply vitamin C. This is erroneous as can be seen by reference to the table of fruit contents and it is worth repeating the values of a few.

One orange with about 3 oz. of edible pulp will yield 45-50 mg.

One apple with about 3 oz. of edible portion yields only 10-15 mg.

the same weight of raw cherries
grapes about
pear } 3 mg.

Fruit drinks taken with a snack lunch supply vitamin C only if they do, in fact, contain the juice of the fruit. Some brands do and it is worth examining the label on the bottle. Some drinks, e.g. Ribena and Rosena have a declared vitamin-C potency.

An attempt was made during the war years to popularize the salad lunch, based on the Oslo breakfast of Norway and this met with a measure of success although the general preference in Great Britain is still for 'meat and two veg.' However, the following example illustrates the value of the salad meal especially when it is combined with such highly protective foods as egg, cheese and one or other of the fat fish.

The value of these cold salad foods can be turned to practical use in the preparation of picnic meals and sandwiches of which a few examples cab be given.

From these examples it will be clear that certain foods are mentioned over and over again. These are the ones with a high vitamin content which should find a place in the diet of everyone.

Typical salad lunch	*Amount o.z.*	*Vit. A or equiv. in carotene i.u.*	*Vit. B_1 µg.*	*Ribo-flavin µg.*	*Niacin µg.*	*Vit. C mg.*	*Vit D i.u.*
Canned salmon	4	280	40	240	7,200	—	680
Hard-boiled egg	1	225	18	85	—	—	8.5
Cheese	1/3	123	3	47	33	—	1
Lettuce	1	363*	17	20	100	3	—
Tomato	2	140*	40	20	200	14	—
Beetroot	1	—	5	20	30	1.7	—
Wholemeal bread	2	-180	100	1,200	—	—	
Butter	¼	281	—	—	—	—	4.25
Orange	3	21*	60	30	300	48	—
Coffee (= Milk)	5	150	60	180		1	1.5
		1,583 i.u.	423 µg. = .423 mg.	742 µg. = .742 mg.	9,063 µg. = 9.063 mg.	67.7 mg.	695.25 i.u.
Proportion of day's approx. needs		½	1/3	½-½	3/4	all	all

Sandwich Fillings of Good Vitamin Content

	A	*B_1*	*Ribo-flavin*	*Niacin*	*C*	*D*	*E*
Grated cheese with lettuce & tomato	x		x		x		
Grated cheese with black currant jam	x		x		x		
Grated cheese with Marmite & watercress	x		x	x	x		
Grated cheese with Bemax & tomato	x	x	x	x	x	x	x
Buttered egg with parsley	x		x	x			x
Buttered egg with tomato	x	x	x		x		x
Buttered egg with lettuce & Bemax	x	x	x	x	x	x	x
Sardine with lettuce	x		x	x		x	
Salmon with tomato	x		x	x	x	x	
Ham with black currant jam.		x		x	x		
Ham with grated carrot & lettuce		x	x		x	x	

No hard and fast rules need be made about what must be included in the day's meals and what is better omitted but the examples given show how a knowledge of vitamin values can be applied to the planning of diets for a single individual or for groups of any size.

If vitamin values are forgotten it is easy to omit essentials, but the necessary information is available and no one today need eat a poor diet through ignorance. It is one function of teachers and others who have an opportunity to influence the tastes and habits of the population in dietary matters, to make sure that proper guidance is given on this subject.

The science of nutrition is a development of the present century. This has been a period during which there have been many advances in science and in medicine advances which are so closely interconnected that it is difficult to distinguish the effects of one alone.

There are, however, certain pointers indicating where nutrition has played a part.

The effect of improved diet on the heights and weights of boys at school can be compared with the increases in height and weight of London schoolboys during the half-century.

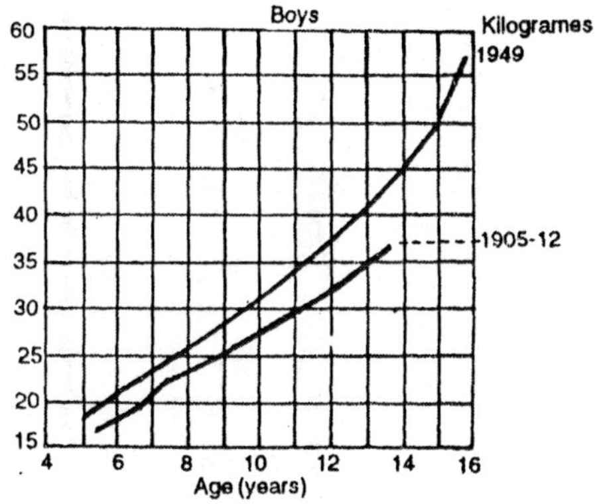

Fig. County of London. Average weights of London school children for the years 1905-12 and 1949.

Other effects known to be because of nutritional factors are the fall in the incidence of rickets and in the death rate of children from this cause.

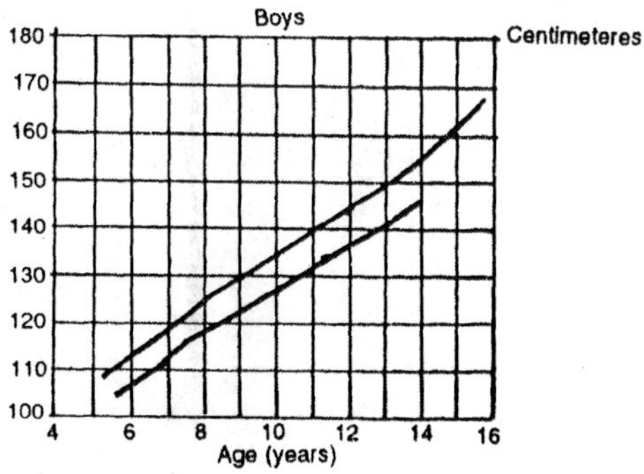

Fig. County of London. Average heights of London school children for the years 1905-12 and 1949.

Another improvement for which a better knowledge of nutrition is responsible is the reduced occurrence of complications of pregnancy, particularly the form of illness called toxaemia. Now this is partly due to improved nutrition because during 1940-42 the addition of various vitamins and minerals to the intake of over, 2, 000 pregnant women was tried out and the effects noted. It was found that the number of cases of toxaemia was reduced one-third.

The practice of giving extra vitamins has become more general and is no doubt one reason among others for the increased safety of motherhood today.

We now know something of the chemistry of vitamins, we can detect and measure their concentration with a fair degree of accuracy; some have been synthesized. We know something of their action in the body and of the effects of introducing into it, either by mouth or by injection, quantities far in excess of what could be taken in any food and this has led to the use of vitamins in massive doses in the treatment of disease. But there are still gapsin our knowledge, gaps which are gradually being filled as one discovery after another is made.

Chemists and biologists continue to work on obscure problems connected with the behaviour of vitamins. Isolated results of their works may seem sometimes to be of little importance. It is only when a result obtained in one laboratory is linked with another discovery made somewhere else that ultimately a fact of practical value to ordinary people emerges.

Such knowledge has wide applications. It influences agricultural policy, international economics, the world supply of food. These are problems too vast and complicated, but every individual can learn enough to be able to select the most nutritious food from what is available and to play a part in influencing public opinion in favour of producing and eating the foods which are best for nutrition and health.

5

Water and Milk

About sixty to seventy per cent of the total body weight of a person is water. Water is present inside the cells of tissues (the intracellular water) and outside the tissue cells (the extracellular water). The extracellular water is present in the blood plasma, water in the space between the tissues is known as interstitial water and in lymph. Water and electrolytes are essential constituents of cellular metabolism. It is not only a component of our body but also acts as a vehicle for chemical transport. Water and electrolytes must be supplied to the body regularly to maintain the cellular function and to regulate the excretion through kidneys, lungs and skin.

The Water

Water intake and loss must be balanced. Water is in jested through food and also as drinking water. Apart from this, a considerable amount of water is formed in the tissues through the oxidation of carbohydrates, fats and protein in their metabolism.

Water Intake and Loss from the Body

Water intake	*(Ml)*	*Daily excretion*	*(Ml)*
As fluids	1,500-1,750	Urine	1,200-1,500
Water in solid food	600-900	Perspiration	700-900
Water from oxidation of carbohydrate, protein, fat metabolism	300-350	Faeces	100-200
Total	2,400-3,000		2,400-3,000

Water is lost through urine, skin, lungs and faecal matter. In a normal healthy person the- water intake and water loss is more or less equal. The excretion through the kidneys and skin changes according to the climatic conditions. In a dry climate more perspiration and in the cold weather more urination takes place. If the fluid intake is less thirst occurs. When two per cent of the body weight is lost dehydration starts.

Water is absorbed quickly and it is circulated to the tissues. From the tissues water is passed on to the capillaries. The capillary walls and cell membranes separate the water compartments. Water can move in and out of all cells by osmosis. The osmotic pressure in the plasma proteins regulates the exchange of water from tissues to capillaries. Antidiuretic hormone of posterior pituitary, hypothalamus, electrolytes like Na and K and adrenal cortex hormone aldosterone regulate the water metabolism in the body.

Antidiuretic hormone of the posterior pituitary increases the reabsorption of water in kidney tubules by increasing the permeability of distal tubules and collecting tubules. This mechanism enables them to retain more water in the body.

Hypothalamus is having the thirst centre. In dehydration plasma is concentrated and osmoconcentration stiumulates the thirst centre of hypothalamus. This induces the person to be very thirsty. By taking water the water balance in the body is regulated.

Various solutes present in the body fluids also help to regulate water metabolism. Plasma proteins exert low osmotic pressure and this helps to maintain the water in the blood vessels. Aldosterone increases reabsorption of Na by the kidney.

About 170 litres of fluid pass through the glomeruli of the kidneys daily. A healthy person's kidneys excrete about 1.5 litres of urine from the above fluids. The urine formation is reduced if the intake of fluids is less. Then the urine becomes more concentrated with urea which forms the main bulk in urine. The normal output of urea is 30 mg for which 750 ml of urine must be formed. If urine formation is less urea is retained in the body which is harmful to the body; when less urine is formed in summer season more sweat is formed. Excessive sweating occurs with increased external temperature and humidity, fever and exercise. Water is also lost through insensible perspiration.

Apart from maintaining the intracellular and extracellular water metabolism, water also takes part in a number of activities. All the vital functions of the body depend on the presence of proper water content. Digestion depends on its juices and all digestive juices are watery in nature. Urine, the excretory medium of our body, contains 97 per cent water. Water also helps in regulating body temperature. By evaporation of water, by respiration and through sweating the body temperature is regulated.

Requirement of Water : Water requirement of a person varies with climate, age, activity, dietetic habits and body build. Activity is an important factor which affects the water requirements. When activity is more intense more heat is produced and more water is needed to dissipate the heat produced. The type of diet also influences the water requirement. A high mineral content increases the water requirement. Consumption of sweets demands extra water for their dilution in the stomach. Beverages like tea, coffee and cocoa are diuretics and they stimulate excretion of water. If excess water is injested the kidney responds to it after 13-50 minutes and the extra water is excreted within three hours. The water balance will be maintained by the excess urine output. The sensation of thirst is one means of meeting the body's need for water.

Dehydration : Water depletion occurs when water loss is not met by water intake. On water depletion changes occur in the body fluids. Thirst, weakness, loss of weight, dryness of the skin, mucous membranes, tachycardia, disorientation and delerium are the symptoms manifested by dehydration. If not corrected coma and death occur. The urine changes in its specific gravity and concentration. The salt level is diminished in the urine and serum levels of sodium, urea and chlorides are elevated. Intercellular water is depleted to compensate the water content of the secretions. This results in cellular dehydration and vital organs fail to function. Plasma volume and blood volume are reduced which leads to reduced output and further cardiac failure.

Acid-base Balance : Acid-base balance refers to the regulation of the hydrogen ion concentration of body fluids. From metabolic processes acids are continuously produced and have to be eliminated. Lungs and kidneys are the major agents in carrying out this function in our body. Normally the body is maintained in neutrality. In some pathological conditions like diabetes mellitus or fasting for weight reduction and in renal failure the acid-base balance is affected. Acidosis occurs in this conditon.

Carbonic acid is the main form of acid formed in the body from metabolism. Carbohydrates, protein and fats yields Co_2 and water during their metabolism. The hydrated form of carbon dioxide is carbonic acid. Intermediary products like lactic acid, pyruvic acid, keto acids, uric acid, sulphuric acid, phosphoric acid from phosphoprotein are acidic in nature.

Certain foods in the body produce alkaline ash as the cations (Na^+. K^+) will allow the body to retain more bicarbonate ions thus producing alkaline effect. The neutrality of the body is maintained through the body fluids and their circulation, the acid buffer system like bicarbonate carbonic acid through haemoglobin and its capacity to form carbamale, regulation in respiration and the final effort from kidney on emergencies.

Acidosis : It is a condition in which the hydrogen ion concentration is increased or there is an excessive loss of base. The ratio of bicarbonate to carbonic acid is less than 20:1 and the pH is below 7.35.

Alkaksis : This is a condition in which the hydrogen ion concentration is decreased or the base is increased. The ratio of bicarbonate to carbonic acid is greater than 20:1 and the pH is 7.45.

The Milk

Milk is commonly considered as a complete food as it contains all six of the essential foodstuffs. Milk protein is of excellent quality and it promotes growth and maintenance of body tissues. Therefore, milk is an ideal food for infants and children. But milk is very low in iron and ascorbic acid content Calcium and phosphorus levels in milk are very high. The ratio of calcium and phosphorus and the presence of vitamin A help its maximum utilisation. Milk is an excellent source of vitamin A. In it the cream is rich in vitamin A and if the same is removed as in slammed milk, its vitamin A content becomes low. Riboflavin, a B vitamin, is present in significant quantities in milk. If the milk is exposed, to light riboflavin is lost. Fats present in milk are readily digested. The carbohydrate of milk is lactose and it is hydrolysed into glucose and galactose. Whole milk contains 4.7 per cent carbohydrate, 4 per cent fat, 3.3 per cent protein and 88 per cent water. Fats in the form of oleic, palmitic and stearic acids is present in milk. Milk protein is casein and whey protein has lactalbumin and lactoglobulin.

The Average Composition of Milk and Milk Products (gm\100 ml)

Milk product	*Protein (gm)*	*Fat (gm)*	*Calcium (mg)*	*VitA. (I.U.)*	*Vit.C. (mg.)*	*Calories (kcal.)*
Buffalo's milk	4.3	8.8	210	160	3	118
Cow's milk	3.2	4.1	149	150	2	67
Buttermilk	0.8	1.1	30	...	...	15
Skimmed milk	2.5	0.1	120	...	...	29
Condensed milk	9.1	8.4	280	430	...	328
Cheese	24.1	25.1	790	273	...	348
Goat's milk	3.7	5.6	170	60	2	84
Human milk	1.0	3.9	20	70	3	67
Curd	2.9	2.9	120	40	4	51

In developing countries, the per capita availability of milk is poor whereas in developed countries it is more.

Among these buffalo's milk has a higher proportion of total solids and fats than cow's milk. Casein coagulation is also high.

Goat's milk is rich in cream and poor in protein compared to buffalo's milk. Human milk is easily digestible because it becomes solid on reaching the stomach. Casein is the chief protein of milk. It is a phosphoprotein which contains protein and phosphoric acid. An enzyme rennin aggregates and precipitates casein and it is coagulated or curdled. Bacterial action also curdles milk. Casein is present in milk in the form of calcium caseinate. Rennin converts calcium caseinate into calcium and paracasein. Again it forms a smooth, soft clot of calcium paracaseinate which is non-dispersible. If milk is overheated it forms a weaker clot. Dried milk also forms a weak clot. Milk also contains lactalbumin, a simple protein of the albumin class. Prolonged and direct heating of milk also forms a weak clot. Milk also contains lactalbumin, a simple protein of the albumin class. Prolonged and direct heating of milk helps it to settle down at the bottom and sides of the container in which milk is heated. Use of double cooker rectifies this. The souring of milk is due to the presence of lactose, the carbohydrate. Lactic acid-forming bacteria use lactose as food to produce lactic acid in milk. Fat is present in emulsified form. Creamy layer of milk consists of fats. Milk contains potassium, sodium, calcium and magnesium chlorides and phosphates. On heating milk calcium precipitates on the sides of the container. Of all food materials milk is the best source of calcium and careless heating reduces its availability. Even though iron is present in minute quantities it is highly nutritional.

Processing of Milk

Pasteurisation of Milk : Milk is a favourable medium for bacterial growth. Louis Pasteur, the French scientist developed the heat treatment of milk for inactivating disease-causing micro-organisms.

Milk-borne diseases are tuberculosis, typhoid, hoof and mouth diseases, undulant fever, scarlet fever, septic, sore throat gastroenteritis and diphtheria. But pasteurisation destroys certain advantageous bacteria also. There are two methods of pasteurisation—the flash method and the holding method. In the flash method the milk is heated at a temperature of 160°F for 30 to 60 seconds and in the holding method the milk is heated to temperatures of 140° to 145° F and held there for at least 30 minutes. In both the methods, after the heat treatment, the milk is cooled quickly to about 43° F. Pasteurisation reduces the thiamine and ascorbic acid content whereas the nutritive content of other vitamins is not much affected.

Evaporated Milk : Evaporated milk is produced by evaporating 60 per cent of water from milk and the product is then sterilised and sealed in cans. To evaporate, the milk is heated at a high vacuum from 130° to 150* F. The concentrated product is homogenised to reduce the fat globules and then placed in cans, heated and sterilised at temperatures from 240° to 245° F for IS minutes. Evaporated milk is fortified with vitamin D.

Condensed Milk : Evaporated milk is incorporated with 40 per cent carbohydrate to form condensed milk. Sucrose or dextrose or a mixture of these two is added.

Dry Milk : Whole milk or defatted whole milk is used in its preparation. From milk most of the water is removed. There are two methods applied to form dry rajlk-spray process and roller process.

In the spray process fresh milk is sprayed under high pressure into a drying chamber where dry air is passed. The current of dry air evaporates water and the dried solids settle to the bottom of the chamber.

In roller process fresh milk is spread over the surface of hot revolving metal rolls. The water evaporates and milk solids remain in thin sheets. If the fat is removed from whole milk the keeping quality is more. Dried whole milk does not keep so long because the fats tend to become rancid.

Milk Products

Cheese, yoghurt or curd, buttermilk, low-fat milk, non-fat milk cream and butter are the common milk products used.

Cheese is mainly from casein of milk with fat and a little whey. To precipitate cheese lactic acid or rennin are used. For commercial cheese making rennin is used and for homemade cheese lactic acid is used.

Cheese is rich in protein and fat and a considerable amount of lactose is also present. In homemade cheese calcium is less but other minerals like phosphorus are there in both. Compared to milk vitamin A content is the same but thiamine and riboflavin are dissolved in the whey. Cheese is used to spread in items like butter as it contributes flavour. In vegetables, fish and meat preparations and in custards and souffles cheese can be used. Low temperature is suggested for cheese cookery. Buttermilk is prepared from curd; it is churned and the butter removed. Yoghurt or curd is a custard-like product prepared by fermenting milk. In commercial production streptococcus thermophilus is added to milk to ferment it. Low-fat milk is commercially prepared for people who prefer low calories from milk with a fat flavour. Non-fat milk is skimmed milk where the fat is removed entirely. Vitamin A and D are fortified in it. Creams with different levels of fats are commercially produced. The fat content ranges from 18 to 40 per cent. Cream combined with sugar, stabilisers and emulsifiers is often sold in cans. Commercially, cream is prepared by using a cream separator which enables the separation of fat at different levels. Whipped cream is an air-in-water foam in which fat globules are piled up in the film and it is used in salad dressings and in desserts.

Milk is used in many preparations for its solvent capacity and its nutritive contribution. Milk is used as a dispersing medium, especially in puddings and for scum formation with its coagulating property. Vegetables with tannin or acid cause milk to curdle. Scum formation and curdling occur when the protein in milk becomes insoluble in heat or in the pH change. Use of low temperature and short heating periods prevent scum formation. Acidity of the content produces curdling. Adding acid food towards the end and mixing it in low temperature reduce curdling.

Types of Milks in the Market : Homogenised milk, separated milk, toned milk, sterilised milk, separated milk (skimmed milk), recombined milk, filled milk and flavoured milk are the common fluid milks available in the market.

Homogenised Milk : In this, fat globules are broken up mechanically to less than one microne in diameter. It is easily digested but cream is not obtained on separation.

Separated Milk (Skimmed milk) : This is prepared by removing fat from milk in a cream separator. Very little fat is in it.

Toned Milk : Toned milk is prepared by mixing milk reconstituted from skimmed milk powder with buffalo milk which has 7.0 per cent fat.

Recombined Milk : It is a homogenised product prepared from milk fat, non-fat milk solids and water. It should be pasteurised.

Filled Milk : Filled milk is the homogenised product prepared from refined vegetable oil and non-fat milk solids and water.

Sterilised Milk : Cow or buffalo milk is sterilised in bottles by fixating to 15 minutes.

Flavoured Milk : Chocolate, coffee, cardamom or any other edible flavours, colour and to flavour milk.

6

The Beverages

Beverages are materials used as drinks for the purpose of relieving thirst and introducing fluid in the body, nourishing the body and stimulating or soothing the individual. Beverages are classified into four groups according to the reactions they produce.

(1) *Refreshing* to relieve thirst

   (a) Plain or carbonated water.

   (b) Ginger ale and other bottled beverages.

   (c) Fruit juices and fruit-ades.

   (d) Iced tea or coffee.

(2) *Nourishing*

   (a) Milk pasteurised, skimmed, evaporated, dried, malted, buttermilk, and chocolate and cocoa.

   (b) Eggnog made with rum, whisky, brandy, fruit juices, coffee and chocolate.

   (c) Fruit juices with egg-white or whole egg.

   (d) Glucose lemonade or orangeade.

   (e) Tube feeding feeds.

(3) *Stimulating*

(a) Eggnogs made with whisky, rum, brandy, tea or coffee, other alcoholic beverages.

(b) Tea or coffee.

(4) *Soothing*

(a) Warm milk.

(b) Hot tea.

Coffee and tea have no food value unless served with crdam and sugar. Cocoa in itself will increase the food value of the milk which in used its preparation. Coffee, tea and cocoa, all have a certain amount of stimulating effect. Iced coffee, tea and cocoa are used all the year round between meals as a refreshing drink. Milk, eggnog and other cold beverages made with milk or milk and egg are valuable for serving between meals when the protein content of the diet needs to be increased. Wines and other liquors which are used occasionally to provoke appetite, or to stimulate or flavour furnish additional calories. Cold beverages may be used with meals or between meals. They are more appetising when chilled. The choice of the beverage will depend largely upon whether it is taken for its food value or merely as a refreshment. In any case a cold drink served between meals adds interest to the monotonous hours in the daily routine of a person.

Tea

Tea is obtained from the leaves and flowers of teabush, which grows in sub-tropical countries, such as China, India and Japan. The top tiny leaves and buds or flowers at the end of the shoot furnish the choicest and the most expensive brands designated as flowery pekoe. Orange pekoe is the most commonly marked tea which consists of the first and second leaves just below the buds.

The kind of tea is determined by its treatment. Subsequent to picking, the green tea undergoes no fermentation process and is light in colour and rich in tannin. Black tea is fermented to a dark colour and loses some of its tannin.

A good cup of tea will contain a high proportion of volatile oils and a minimum amount of tannin. Tannin is a stimulant and has approximately the same psychological effect as caffeine. Tea, which is allowed to stand too long in the container with the leaves, contains a considerable amount of tannin which causes the finished beverage to be bitter and unpalatable.

Coffee

Coffee is the beans of coffee plants which grow in tropical countries. It is tasteless in its green form. Roasting develops the volatile oils, specially caffenol which give the characteristic aroma and flavour to coffee. The typical individual brands depend upon the various types of coffee used and the length of the time of roasting. Bitterness is increased when combined with substances like chicory.

Coffee alone has no food value, but it does furnish a certain amount of stimulation of caffeine which raises the blood pressure and stimulates renal activity and momentarily masks fatigue. As a rule a cup of coffee contains from 1.5 to 2.5 grains of caffeine depending upon the strength , of the brew. Tannin, another alcoholite present, interferes with digestion. The amount of tannin in coffee depends upon the method of preparation. Since it is soluble in hot water, the longer the coffee is brewed, the greater will be the tannin content.

Coffee beans are ground to increase the surface area for maximum extraction of the flavour. However, coffeole is rapidly lost, the loss being greater in freshly ground coffee. Ground coffee should be kept in a tightly covered tin in a cool place and should be used promptly after purchase.

Drip and vacuum coffee contains the least amount of tannin, caffeine and coffeole. Percolated coffee contains slightly more caffeine and less coffeole than drip and vaccum coffee. Boiled coffee should be strained. It contains more caffeine and tannin than either the drip or percolated coffee.

Instant coffee is made from pure coffee which has been pulverised. The extraction of caffeine and tannin is high, but the amount in the finished beverage will depend upon the amount used. Some instant coffees are mixed with malted cereals.

In decaffeinated coffee 75 per cent of caffeine is removed. It does not possess the stimulating property of coffee. Coffee substitutes are made from a combination of roasted grains and cereals.

Cocoa

Cocoa is the ground product of cocoa beans and often some of the fat has been removed. The cocoa beans are in fleshy pods. The well ripened pods are feathered and dried for 24 hours after which they are cut open and the beans are removed. The beans then undergo a series of processes like fermenting or sweating, removing the pulp, washing, drying and roasting. Chocolate and cocoa, unlike tea and coffee, have a high nutritive value. The roasted beans, likewise the bitter chocolate, contain approximately 50 per cent of fat, about 80 per cent protein, 10 per cent starch, and 7 per cent sugar. Like tea and coffee, they also contain a stimulant and tannin. The stimulant known as theobronine has approximately the same physiological effect, though not to such a marked degree. There is only a small amount of stimulant in one cup and the chief nutritive value of the drink lies not in the cocoa but in the amount of milk and sugar used in it.

Coconut Water

Tender coconut water is high in sugar and electrolytes like Na and Potassium.

Fruit Juices

Fruit juices in themselves are not only pleasantly refreshing but they are easily digested and especially useful in increasing the fluid intake. The citrus juices and tomatoes are excellent for the ascorbic acid content Fresh, frozen and canned juices may be served plain or flavoured. They may be reinforced with egg white or whole egg. The following fruits include commonly used juices: (1) citrus—orange, lemon, grape fruit, lime; (2) Berries—blackberries, loganberries, raspberries, strawberries; (3) tree fruits—apple, apricot, peach, pear, plum, prune; (4) other fruits—grapes, pineapple; (5) vegetables—tomato; (6) Sugarcane juice, sweet toddy and coconut milk are some of them.

Milk

Milk is most of all liquid food since it has more nutritive value than any other beverage. Milk beverages add protein, calories, calcium and other nutrients to the diet. The milk flavour of food permits its use in a great number of ways. It may be plain, malted, or acidulated. It may be reinforced with egg and sugar depending upon the type of nutrient needed and it may be flavoured with chocolate and other flavouring extracts.

Egg

Eggs are specially useful for increasing the protein content of beverages. Eggnog in which whisky or other alcoholic beverages are used is the most important stimulating beverage. Eggnog made with strong coffee is only mildly stimulating but lends variety to the general liquid diets. Eggnogs contain egg, cream, and possibly milk and sugar in addition to fruit juice.

Sugar

Sugarcane, fructose and lactose may be used to sweeten beverages and to increase the caloric value.

Bask Rules in the Preparation of Tea and Coffee : The best tea produces a pale coloured infusion. Freshly infused tea has good flavour and it is harmless for digestion. Reheating or overheating of tea results in continued infusion and it extracts more tannin which gives a bitter taste Id tea. Moreover, this is often harmful to the digestive system.

The quality of tea is an important factor in giving good taste to the product. Since tea leaves absorb water, they must be stored in airtight containers.

Water used for the preparation of tea also has the key role in producing good result. Water should be freshly drawn before it is used for tea. Previously drawn water and previously boiled water should not be used for preparing tea as it makes the tea insipid. To get the best flavour water should not be boiled before or after tea leaves are added. As soon as the tea leaves are added to boiling water it circulates from the bottom to the top and from the top to the bottom.

To prepare good tea, one teaspoonful of tea leaves for 1 cup is ideal. For one litre of water 15 gms of tea leaves are enough. It is always better to measure tea leaves to get good result. The vessel in which tea is being prepared must be warmed before making tea. Instead of adding tea leaves to water, pour boiling water into a wanned pot in which tea leaves are added. Do not disturb this for five minutes. Let the leaves infuse. If it is kept for more than five minutes, it gives a bitter taste and less then five minutes yields a weak tea.

Milk and sugar can be added according to individual taste. To get a good cup of tea, it must be served hot.

Coffee : There are different methods of preparing coffee. The pot method, the filter method, the urn brewing method and the vacuum brewing method are the common methods.

Coffee powder must be fresh to get the best flavour and taste. The quantity of coffee powder used must be just enough. For one cup of coffee 8.5 gms or 1/2 teaspoon of brewing coffee is required. If instant coffee is used one teaspoon is enough to make 1 cup of coffee.

Freshly drawn water must be used to prepare coffee.

Coffee is generally prepared by any one of the following :

1. The pot method.
2. Filter method.
3. The percolator method.
4. Instant coffee preparation.

The Pot Method: In this method an earthenware pot or jug is used to prepare coffee. Warm the earthenware pot or jug, put in 3 level spoons or 11/2 heaped teaspoons of fresh coffee powder (coarse grind) for an 8-ounce cup. Pour boiled water over the powder and stir. Cover the vessel and let it stand for 5 minutes. Strain the coffee through a fine meshed sieve or cloth. Add milk and sugar to taste.

Filter Method: Stainless steel or brass filters can be used to filter coffee in this method. If brass filter is used it should be tinned properly to get good results. Put 2 to 3 level teaspoons coffee powder for 8 ounces water. Press the plunger down lightly over the powder and pour boiling water over the plunger in a circular motion. Allow to stand for 5 to 7 minutes. Pour out coffee from the lower vessel and add milk and sugar to taste.

The Percolator Method: Keep coffee powder in the middle of the percolator and pour boiling water through it to the upper section. Coffee is collected from the bottom section. It takes thirty minutes to percolate down. Vacuum coffee makers are also available in some places.

Instant Coffee: Strong brew is made by percolation of roasted and ground coffee which is sprayed into fine atomised particles from the top of a chamber into a mass of very hot air. The flavour is retained through highly developed technology. It is quickly packed into airtight bottles. Varying degrees of chicory is mixed with coffee powder to give slight bitterness and colour to brew. Now mixtures of coffee soluble with solids and sugar in the form of thick pastes are also marketed. By adding hot water it can be used instantly.

Alcoholic Beverages

Beer, ale, toddy, wine whisky, gin, brandy, rum, vodka, arrack and cider are some of the alcoholic beverages. Beer is made by fermenting barley malt extract with yeast and ale is the same thing with more stimulation. Toddy is from fermenting sap of palm, coconut and palmyra or date palm.

Wine: Wine is prepared by the fermentation of grapes with yeast and cider is apple wine.

Whisky: It is an alcoholic distillate from a fermented mash of grains.

Gin: It is an artificial drink from diluted pure ethyl alcohol.

Brandy: This is made from fruit juices like grapes and apples.

7

Meat and Fish

Flesh of all edible animals including mammals and poultry is termed as meat. Meats of cattle origin, beef and veal, of hog origin, pork, sheep, lamb, mutton, flesh of partridge, duck and game birds are used commonly as food.

Meat

Muscle of meat is composed of 75 per cent water, 20 per cent protein and 5 per cent fat, carbohydrate and minerals. The composition of meat varies with its sources, season of the year and pH of the meat. Protein comprises about four-fifths of the solid material. True fats, fat-like substances called phospholipids, inorganic substances and vitamins are present in fats.

Veal is the meat from cattle slaughtered three to fourteen weeks after birth. Calf is from fourteen to fifty-two weeks old animal. Beef is from cattle over one year old. A heifer is a female cattle that has never borne a calf and a cow is a female cattle that has borne a calf. Meat has skeletal or muscular cuts and organ cuts. Skeletal or muscular cut has muscle, tissue, connective tissue, bone fats and nerve tissues. Muscle tissue and connective tissue supply the qualities of meat in cookery. Muscle structure is made up of bundles of muscle fibres called fasciculi. Connective tissues are scattered throughout the muscles where they play a connecting and supporting role. White fibrous and yellow elastic fibres with proteins like collagen and elastin are present in all muscles. Collagen contributes strength and elastin supplies elastic quality to meat Collagen reacts with water in hydrolysis to produce gelatin and brings about a tenderising effect to the meat. But no sack change is brought in elastin and if elastin is more in muscle fibres meat will be hard even after cooking. This can be remedied by chopping or grinding the elastin fibres.

Fats in meats are distributed between fasciculi. Fat is laid as a protective layer around the organs. The fat deposited with the muscle structure is known as marbling. Muscle haemoglobin is present in small amounts. Muscle cells of living animals contain glycogen which is otherwise known as animal starch. But immediately after the animal is dead glycogen is converted into lactic acid. The vitamin content of different meats varies. The B vitamins—thiamine, riboflavin and niacin—occur in significant amounts in all meats. The meat of hog origin is rich in thiamine liver is rich in vitamin A. Iron-containing substances and B vitamins are more in organ meats.

Rigor Mortis : After the animal is slaughtered certain changes occur in the meat and it stiffens. This process is known as rigor mortis. On slaughtering the animal adenosine triphosphate is synthesised from glycogen and lactic acid, ammonia and other products are formed from it. Myosin and actomyosin are found in contracted muscles and stiffening occurs to it. If the slaughtered animal has a high glycogen level, it produces a low pH and the quality of meat will be good with less rigor mortis. Fasting, exercise, nervous exhaustion or insulin injection cause low glycogen level and the carcass has a high pH. Undesirable dark colour and poor flavour and texture of meat result from this. Slow and long slaughter produce more adrenaline in the animals and this results in sticky, gummy and dark coloured meat. Rigor mortis reaches its peak within 24 hours and then the muscle softens.

Quality and Grading : Appearance, tenderness, juiciness and flavour of the meat decide the quality of meat. Good quality flesh must be firm, the fasciculi should be small and velvety in texture, with characteristic colour of the meat-bright cherry red for beef, pinkish-red for lamb, greyish pink for veal and pork; bone must be small, red and spongy in appearance and the fat must be white and firm. To grade the meat quality grades are given and they are prime, choice, good, commercial, utility and cutter and conner. For lambs and veal cull is the term used instead of utility. For pork grades of 1,2, and 3 are used. Meat of top quality is more tender than lower quality grade. Tenderness of meat increases palatability and it varies according to the cuts of meat. The most tender cuts are from backbone and ribs. Less tender cuts are from the shank, shoulder, neck and under parts of the body. Under part cuts of beef are flank, plate and brisket. The same cuts from lamb and veal are known as breast.

Meat Cookery : Palatability, appearance and flavour are enhanced by cooking meat. Cooking also makes meat more tender and digestible. There are two methods generally applied to meat cookery—dry heat and moist heat methods. In all methods of cooking the interior of the meat changes. Heat penetration, substances added to meat and composition of the meat affect these changes. Mainly the method adopted for cooking depends upon the cuts of meat. Less tender cuts contain more connective tissues than tender cuts and so moist heat methods convert collagen to gelatin and it has a tenderising effect upon tissues. Roasting and broiling are the methods used to cook by dry heat methods and these methods can be applied to tender cuts of meat. For less tender cuts moist heat methods are used. Moderate temperature is recommended for this method. Water, juices and fluids are used for most heat methods. The addition of sodium chloride or sodium and potassium phosphate has been reported to increase the tenderness of meat.

Dry Heat Methods : *Roasting:* Roasting is cooking by means of radiant heat. Meat is roasted either in the open or in a closed tin in an oven. In covered roasting evaporation is diminished, and it cooks faster too. The water runs out of the meat instead of evaporating from the surface. An enclosed roast loses more weight and water than the open roast. Cooking to rare stage only increases the juiciness and flavour. If the pieces of meat lack fat, a thin layer of fat on top will be very effective. Water should not be added to oven roast.

Broiling and Pan Broiling: Slices from tender cuts of meat can be cooked rapidly without adding moisture or fat by radiant heat from a broiler or in an uncovered pan. The time required depends on the stage of cooking required and the size and shape of the pieces. This method gives a uniform result. The temperature should be higher than in roasting. It can be regulated either by changing the heat input or by changing the distance of meat from the heat source.

This method is suitable for cooking tender cuts of beef and lamb or ground meat on the pattern of ham, bacon and sausages.

Frying : Meat is fried when it is partly or completely immersed in hot fat during cooking. When the meat is completely immersed the method is called deep-fat-frying. When a small amount of fat is added it is called pan-frying or sauteing. The fat should be kept below smoking point. A temperature of 350° F for pre-cooked meat and 300° F for raw meat is usually satisfactory. This method is suited to lean meats such as liver or cooked steak.

Cooking by Moist Heat: The cooking medium is steam of water.

Braising and Pressure Cooking: Meat may be cooked in steam with or without pressure. When the meat is first browned in its own fat or a small amount of added fat and allowed to cook in its own steam without pressure the method is known as braising. Meat braised as a large piece is called a pot roast and that as small pieces, a fricassee. Braising is preferable to boiling for fresh pork, because it produces less hardening and drying of fibres when meat is cooked to a safe stage of cooking.

Stewing or Simmering: Meat may be cooked by immersing it in water. To minimise the hardening effect of high temperature simmering is preferable to boiling. Vegetables are added to stews to enhance the flavour.

Generally shrinkage of meat cooked by moist heat method is greater than that in dry heat methods.

Tenderising Meat : Papain, an enzyme derived from papaya, tenderises beef. Solution of papain is often injected into the beef shortly before slaughter. Meat of old animals is tough and stringy and usually this meat requires tenderising. Tougher cuts of meat require a long, slow cooking period. Prolonged cooking changes collagen into gelatin in moist heating. Dry heat hardens it. Acid hastens this conversion of collagen into gelatin; tomato juice and vinegar tenderise meat. High temperature gives a tasteless dry-product-like quality to tough meat. Cutting the tough meat into smaller pieces enhances the flavour of meat. Mining, grinding or pounding of meat helps to break the muscle fibres.

Vitamin retention is more in dry heat methods. Roasting or broiling retains 65-70 per cent of thiamine whereas moist heat retains only 25 per cent of thiamine.

Gelatin : Gelatin is a protein obtained from collagen on hydrolysis. Connective tissues of cartilage, tendons, bones, skins and hoofs are used to prepare gelatin. After preliminary treatments a dispersion of collagen in water is obtained. And this dispersion is boiled for several hours. Water is evaporated and gelatin is obtained from collagen.

From the nutritional point of view gelatin is inferior because it is an incomplete protein. The property of gel formation is used in cookery. In desserts gelatin is used to provide a spongy appearance to the product.

Processing of Meat : Canned meat, cured meat, smoked meat, dehydrated meat and sausages are the common processed form of meat.

Canned-meat : The meat is cooked and filled in the can along with the gravy, leaving a little head space.

Cured-meat : Meat is pickled by the use of curing agents like sodium chloride, sodium nitrate and vinegar.

Smoked-meat : Smoking helps to preserve the meat and develop flavours in it The temperature and period of smoking vary with the type of meat

Dehydrated-meat : The meat is cut into pieces and cooked in steam for 30 minutes at 10 lb pressure. It is passed through a meat chopper and the chopped meat is dried in a drier.

Sausages : Cooked chopped meat is ground to a paste with seasonings and packed casings. It is available as fresh, cooked, smoked, cooked and smoked, dry, and semi-dry.

Fish

Fish is a very good source of calcium, protein, vitamins A, D and iodine. There are two main classes into which fish may be grouped as vertebrate fish or fin fish and shell fish. Cod, mackerel and salmon have backbones and fins, whereas a shellfish is more or less completely covered with a shell. Oysters, lobsters and crabs are examples of shellfish.

Protein is the main component in fish and its content ranges from 15 to 20 per cent by weight of flesh. The fat content in fish varies from 1 per cent to 20 per cent or more. Hilsa has a fat content of about 19.4 per cent. The fat content of shellfish is less than 5 per cent. Since the fats in fish are unsaturated fats even in therapeutic diets this can be recommended. The flesh of fish is devoid of carbohydrates except in the case of shellfish. The mineral content of fish is variable; 1 to 1.5 per cent of the flesh of fish is with mineral elements. Small fish which can be consumed with their bones are rich in calcium. Oysters and shrimps are rich in calcium. Most fishes are not very good sources of iron except oysters. Sea fish are high in iodine content. Oysters are also rich in copper. The flesh of most fish contains fair amounts of thiamine and riboflavin and some amounts of niacin. Fatty fish are rich in vitamins A and D. Fish livers are rich sources of these vitamins.

Freshness of Fish : Fish deteriorates quickly and so requires careful handling. The quality of freshness of fish is indicated by bright red gills, clear eyes, bright and slightly bulging but firm flesh when pressed with a finger, and elasticity. The fresh odour of fish suggests its freshness. Suggestion of putrefaction is easy to detect from its foul smell. The tail of a fresh fish is stiff.

Spoilage of Fish : Fish is highly perishable. The spoilage of fish is due to microbiological, physiological and biochemical changes.

Processing of Fish : Fish protein concentrates, fish meat, smoked and salted fish are the common forms of processed fish. Fish protein concentrate is used for supplementing diets.

Fish meals is not suitable for human consumption. It is used for poultry and for animal ration.

Preservation of Fish : Cold storage, freezing, canning and drying are the common methods of preservation of fish. Salting before drying is essential.

The Eggs

Eggs are used in many preparations due to the variety they provide to food. Eggs are important ingredients of recipes for breakfast, lunch or dinner for any person. Nutritive importance, flavour, pleasing colour, thickening power, foam-forming ability, emulsifying capacity, stabilising capacity and the binding property of eggs are utilised in menu planning.

An egg has two parts—outer egg-white and inner egg yolk. The egg is held in position in the shell by two membranes which lie close together. About 63 per cent by weight of an egg is made of egg-white. It has three layers—an outer portion of thin egg-white, a thick jelly-like layer and a thin portion which surrounds the yolk. The central part of the egg is the egg yolk, and the yolk is surrounded by a thin covering called vitellin membrane.

The egg-white is rich in protein. Albumin, globulin and mucin are the proteins in egg-white. It is easily coagulated by heat and it provides the jelly-like character of thick egg-white. Minute quantities of sodium, potassium, magnesium, calcium, clorides and phosphates and riboflavin are present in the egg-white.

The egg yolk contains ovpvitellin as the proteins and it is phosphoprotein, and it resembles -(he casein of milk. The egg yolk is rich in fat, that is, about one-third of the yolk is fat, especially of cholesterol and lecithin variety. The fats in egg are finely dispersed and are in emulsified forms. The mineral content is more in egg yolk compared to the white. Egg yolk is an important source of iron and it is also rich in sodium, potassium, calcium and magnesium. Lipovitellin and lipovitellinin are the lipo proteins in egg yolk which act as emulsifying agents. Egg yolk is rich in vitamins, especially thiamine riboflavin, vitamins A and D and their content in egg depends upon the hen's diet.

The whole egg is a rich source of all nutrients except vitamin C. The protein in egg is considered as a complete protein because of its high biological value and digestability. It is often mentioned as reference protein. But a raw egg-white contains a protein known as avidin which renders the vitamins biotin unavailable to the body. Duck egg-white contains another substance known as trypsin-inhibitor which inhibits the action of trypsin on protein. Heating destroys avidin and trypsin-inhibitors.

Role of Egg : Egg protein coagulates on heating. The coagulum formed is a smooth and jelly-like mass. It supplies a thickening effect to the products. If the temperature is high, the coagulum is toughened and it shrinks and water comes out of the product. This is called syneresis and it occurs in custards and scrambled egg preparations.

Egg is cooked in its shell, scrambled, poached, fried or used in the preparations of deserts, and cakes and other confectioneries. Egg is used in food mixtures to contribute flavour and colour to the products. Cakes and puddings owe their colour and flavour to eggs. In the preparation of mayonnaise the protein of egg stabilises the oil in water dispersion. The protein of egg stabilises the air in liquid dispersion. Egg is also used for thickening various desserts and creams and patties. Since egg is highly perishable the quality has to be judged by internal appearance flavour and external shell appearance.

Candling : Candling is used as a method to detect the interior quality of egg. In a closed box with a small opening a lighted candle has to be placed. The depth of air cell can be observed by placing the egg in front of the opening. In a fresh egg the white is stationary and slight movements of yolk are permitted by the white. Yolk is in the centre and it does not move nearer to the shell. In a poor quality egg the yolk moves. The shell of good quality egg is firm and not stained or cracked.

Formation of Dark Greenish Discolouration : When egg is boiled for more than 15 minutes a dark greenish colour may be formed on the surface of the egg yolk. This green colour is due to the formation of FeS on the surface of the yolk. Yolk contains more iron and the white has more sulphur in its amino acids. On boiling it yields H_2S. The H_2S combines with the iron of the yolk and it gives the dark green colour in the middle. In high temperature this change takes place immediately.

8

Vegetables and Fruits

A fruit is the edible fleshy seed-bearing part of a plant or tree. Fruits are favoured by all due to their attractive appearance, appealing flavour and pleasing odour. Apart from their nutritive value they are also appreciated for their appetising effect. Fruits contain starch. On ripening the starch is converted into fructose and glucose and these sugars contribute to the sweetness of fruits. Water is the chief constituent of fruits. About 75 to 95 per cent of the total weight is from water. Fruits as a whole are rich sources of vitamin C and vitamin A and minerals like sodium, potassium, magnesium, fibre and cellulose. Characteristic flavours of fruits are due to the presence of organic acids in them. Citric acid and malic acid are the acids present in fruits. Grapes contain tartaric acid. Acid content of various fruits varies. Citrus fruits are the best sources of vitamin C. Dried fruits like raisins, dates and figs are rich in iron. Calcium is present in citrus fruits. Banana contributes a very high amount of carbohydrates and about 23 per cent of it is sugar. Jackfruits, figs and cherries are also very good sources of carbohydrates. For vitamin C guava is a cheap source. About 300 mg vitamin C is present in 100 gms of guava. Cashew fruit is also an excellent source of vitamin C. Lime, mango, papaya and orange are other sources of vitamin C. Fruits have alkaline reaction in the body.

Fruits

Fruits with different colours have carotenoids, anthocyanins and flavanols in them. Fruits also contain enzymes like ascorbic oxidase and phenol oxidase. The vitamin A value of fruits is in the carotene in yellow fruits. Organic acids and salts are present in fruits. Dried fruits like dates and raisins are sources of iron. Potassium is the most important salt in fruits. Potassium is present in citrus fruits as citrates and in grapes as tartarate.

Pleasing colours of the fruits are due to their pigment content. In ripe fruits carotenokl pigments are present in different forms. Red, purple and blue shades of fruits are due to the presence of anthocyanine pigments in them.

Tannins and Browning Reaction : Tannins are present in almost all fruits. They contribute bitter taste and astringent quality and discolouration of cut surfaces of fruits on exposure to air. When certain fruits are pared a brown colour appears on their surface which becomes darker as the fruit is allowed to stand. Apple, banana, plum, guava and peach show this change. Tannins and flavone pigments give brown coloured pigments on oxidation. Oxidising agents present in fruit cells catalyse the reaction. If the fruits are cut with an iron knife, tannins and flavones react with iron salts forming a greenish purple compound which discolours the fruits. Acidic juices of fruits like lime, grape or pineapple are sprinkled on the fruits to prevent browning action.

Enzymatic browning occurs only when the fruits or vegetables are cut into pieces. The natural phenolic compounds present in tissues of plants do not come into contact with the phenoloxidase which is also present in some tissues. Practical methods for preventing enzymatic browning are to change the pH and the use of anti-oxidants which retard oxidation. By using the acidic juices of fruits the pH can be brought down.

Non-enzymatic browning occurs due to the formation of brown condensation products from reducing sugars and amino acids on heating. To prevent non-enzymatic browning storing the food materials at a low temperature, between 0° and 5°C, and addition of sulphur dioxide in the case of dehydrated vegetables are recommended.

Browning reaction affects the nutritive value of food, especially the lysine content of it. Loss of nutrients of fruits on cooking is negligible.

Prevention of Browning : Thermal inactivation of oxidase enzymes, elimination of O_2 from the systems, change of pH to prevent enzyme action, and use of antioxidants are the measures to prevent enzymatic browning.

For non-enzymatic browning low temperature (0°-5°C), addition of SO_2 and keeping the moisture content low are some measures.

Vegetables

Edible parts of plants like bulbs, tubes, roots, seeds, fruits, stems, leaves or flowers are considered as vegetables. Vegetables as a whole are important sources of minerals and vitamins and certain vegetables have starch, cellulose and water. In meal planning and for dietary calculation vegetables occupy a prominent place. To check their sufficiency and also for convenience, vegetables are grouped into green leafy vegetables, and roots and tubers and other vegetables. Legumes like double beans, french beans, cluster beans and peas are also included into other vegetables. Vegetables are the best sources for calcium, iron, sodium, chlorine, cobalt, copper, magnesium, manganese, phosphorus and potassium. Carotenes, the precursors of vitamin A and ascorbic acid are present in abundant quantities in leafy and yellow vegetables. Vegetables are poor sources of calories except in the case of roots and tubers. Raw legumes and beans are good sources of proteins. Vegetables as a group supply a high amount of water and the mineral elements present in them contribute the alkaline substances in the body. This enables to maintain the acid-base balance of the body.

Cabbage, cauliflower, coriander leaves, lettuce, spinach, amaranth and drumstick leaves are the common leafy vegetables. Of all the vegetables green leafy vegetables are the richest in carotene and the carotene content depends upon the green colour. If the colour is deep the carotene is more in it. Green leafy vegetables are fair sources of proteins and good sources of folk; acid, ascorbic acid and calcium. Some leafy vegetables contain oxalic acid which interferes with the absorption of calcium in the diet. Leafy vegetables act as a buffer and maintain the Proper alkalinity of the blood by balancing the acidity of acid-producing foods like meat.

Roots and Tubers : Roots and tubers are good sources of starch and fair sources of protein, vitamin B and vitamin C. Roots and tubers include tapioca, potato, sweet potato, carrots and beetroots. Carrots are rich sources of Vitamin A. The starch in roots is easily digestible. Potato, a root vegetable, is a fair source of proteins, ascorbic acid, and B vitamins. Sweet potato is a poor source of proteins but it is a fair source of B vitamins and ascorbic acid. The yellow variety of sweet potato is a rich source of pro-vitamin A or carotene.

Other Vegetables : Other vegetables are good sources of minerals, cellulose and water. They are also good sources of vitamins. Common vegetables like ladyfingers, gourds, brinjals, pumpkins, green peas and all other vegetables fall in group Tomato is also considered as another vegetable. Green peas and yellow pumpkin are fair sources of carotene.

Cooking of Vegetables : Cooking affects the colour, texture and nutritive value of vegetables. Cooking methods can destroy the good qualities if not handled properly. Common methods of cooking are boling, steaming, baking, frying and pressure cooking. Before applying these methods the chemical structure of vegetables must be considered. Vegetables contain different chemical compounds like pigments, tannin, enzymes and flavouring substances. Common pigments in vegetables are chlorophyll, carotinoid, flavones and anthocyanins. On heating these pigments undergo certain chemical changes which affect the colour of the finished product. Vegetables also lose organic acids and nutrients in cooking water. Acid or alkali content, and the cooking media also affect the appearance and quality of the product

Chlorophyll : Chlorophyll is the green pigment in leaves which is not readily soluble in water. In an acid medium chlorophyll turns an olive green and in alkali it becomes bright green. But the use of alkali to retain the green colour in salad vegetables destroys vitamin C and thiamine. Cooking in copper vessels turns chlorophyll into deep bluish green colour. Addition of sodium carbonate affects the texture and the cellulose becomes mushy. Cooking with a lid and in minimum time retains the colour . The Carotenoid is yellow orange in colour and insoluble in water. It is soluble in ether and acetone. Carotenoids are present in different forms like carotenes, Ly copence, xanthophylls, cryptoxanthin and citroxanthin. Carotenoids are present in yellow vegetables like carrots, sweet potatoes, tomatoes, and xanthophyll in all vegetables. Cryptoxanthin is present in papaya, orange peel and yellow maize. Carotenoids are unchanged in acid or alkali and so vegetables with carotenoid pigments may be cooked with a cover. Such vegetables can be cooked by any method as the carotenoids remain unaffected.

Flavanoids : Flavones and flavonols are together known as flavanoids. Anthocyanins, anthoxanthins, leucoanthoxanthins and catechins are together known as flavonoids.

Anthocyanins are reddish water soluble pigments occurring in fruits and vegetables. They are destroyed by heat and oxidation. Passion fruit, apple, cherry, plum, grapes, blackberries are some of the fruits with anthocyanin in them. Beetroot has anthocyanin as its pigment Anthocyanin is water soluble and in acid medium it turns blue. In some vegetables flavones and anthocyanin are combined together which gives a green colour on cooking in alkaline medium.

Another chemical substance present in plants is tannin. Tannin and ferric salts form a greenish purple compound. On oxidation this turns brown which imparts a brown colour. Certain vegetables show discolouration if an iron knife is used to cut it. If vegetables rich in tannin are cooked in milk, the milk curdies.

Effect of Cooking : At a high temperature the cellulose is disintegrated and protopectin, a non-dispensible substance in water, undergoes hydrolysis and makes it dispersible. This changes the structural framework of the vegetables and makes it soft. If it is overcooked the vegetable becomes mushy and unpalatable. Some vegetables lose their natural flavour on cooking. Strong flavoured vegetables like onion, cabbage and cauliflower must be cooked by suitable method either to retain or to lose the flavour. If a mild flavour is desired the vegetables can be cut into small pieces and cooked in a small amount of boiling water in a covered pan. If vegetables are in large pieces, large quantity of water can be used in an open pan. The plant acids are dissolved in water and the open lid allows the acids to escape.

Green vegetables turn brown because the plant acids and cholorophyll are disintegrated on cooking. Cooking time must be reduced so as to prevent their contact. Use of pressure cooker, large quantity of boiling water to cook the vegetables and use of an uncovered part are some of the measures that can be used to reduce the colour change.

Retention of Nutrients : Vegetables are rich in mineral salts and vitamins and most of them are water soluble. Some of them which are not soluble in water are destroyed by heat or by oxidation. If cooking water is thrown away considerable loss of mineral salts, especially of sodium, potassium and chlorine due to leaching occurs. To avoid this either use minimum , amount of water or the excess water after cooking can be used for other preparations. Cut the vegetables as big as possible to reduce the cooking lose bring the cooking water to boiling before inserting vegetables. Cook just enough. Washing vegetables after cutting enables considerable loss of minerals in the water. About 40 per cent minerals and thiamine are lost if the rice is washed twice or thrice -before cooking. Washing once and utilising the rice water are better methods for retaining nutrients. Absorption method of cooking can be adopted for rice cooking.

Vitamin A or carotene is not affected when food is cooked in water. But shallow frying or roasting brings considerable loss of these vitamins. If sodium bicarbonate is added especially for dals or pulses most of the thiamine is destroyed. Roots like potato, sweet potato and other roots which can be cooked with their skin must be cooked so as to prevent leaching out of nutrients. They can be boiled first and later cut into pieces. Vitamin C is the most unstable vitamin among water soluble vitamins. Washing, cutting with an iron knife, cutting into smaller pieces and prolonged cooking lead to oxidation of vitamin C content of food items. If tamarind or lime, which is high in acidity, is added to cooking water it has a preservative effect on this vitamin. Since vitamin

C is leached the least possible amount of water must be used and the cooking done in a covered vessel.

Cooking just before consumption is another method to preserve vitamin C content in the foods. Use of pressure cooker minimises the loss of nutrients. Consumption of raw vegetables provides all nutrients without any loss. If boiling is the method used to cook vegetables put them only into boiling water and bring the cooking water again to boiling temperature. For strong flavoured vegetables use little boiling water and let them remain uncovered. For other vegetables smaller amounts of water and tightly covered untensils are recommended. Avoid the use of baking soda.

The Losses of Vitamins in Some Vegetables during Cooking

Name of vegetable	*Method of cooking*	*Loss of* *Tiamine*	*Riboflavin*	*Vitmin C*
Beans	Boiling	29%	20%	42%
Cabbage	"	32%	18%	45%
Carrots	"	20%	10%	30%
Potato	"	25%	9%	23%
Peas	"	28%	17%	40%
Spinach	"	15%	10%	28%
Sweet potato	"	20%	13%	24%
Beans	Steaming	22%	14%	32%
Cabbage	"	18%	9%	38%
Carrots	"	15%	10%	25%
Peas	"	24%	11%	30%
Potato	"	20%	12%	25%
Spinach	"	16%	12%	25%

9

Pulses and Cereals

Dried edible seeds of certain plants are called pulses. Split pulses are known as dals. Pulses are good sources of proteins and B vitamins. Even though vitamin C is not present in pulses, sprouting of pulses makes it a very good source of vitamin C. Pulses are the best supplements for cereals. Generally pulses contain about 20 to 28 per cent of protein but they are inferior in quality compared to animal foods. Methionine, the limiting amino acid in cereal proteins, is present in good amount in pulses. Thus cereal pulse combination improves the nutritional quality of the diet. This combination is the most economic way to provide protein. Pulses contain fair amounts of minerals like calcium and iron. About 8 to 10 mg of iron and about 100-200 mg of calcium are present in 100 gms of pulses. Pulses contain fair amounts of thiamine, riboflavin and niacin. Sprouting of pulses improves the carotene content and vitamin C.

Pulses

Green gram, red gram, bengal gram, black gram, cow gram, horse gram, lentils, peas and soyabeans are the common pulses used in our cookery. All these pulses are good in protein, carbohydrate and vitamin B and fair in mineral content. But vitamin C is not present in them. On sprouting this deficiency can be rectified. To sprout, cleaned pulse is soaked for 24 hours and then it is spread on a muslin cloth, sprinkling water occasionally on it. After 24 hours tiny sprouts appear on the surface. These sprouts are very rich in vitamin C and carotene. Sprouted Pulses are very tender and cooking is not necessary. They can be used as raw salad by sprinkling lime juice and salt or as raitas by adding curd and salt to it. As cooking destroys vitamin C it is better to use them as a raw item.

Vitamin 'C' Content of Sprouted Legumes

Period of Germination	*Vitamin C*
Dry legumes	Trace
After 24 hrs germination	7-12 mgms/100 gm
After 48 hrs germination	11-18 mgm/l00 gm
After 72 hrs germination	13-20 mgm

If pulses are not cooked properly they contain much undigestible materials. But if pulses are used as flours they are completely digested and absorbed. In raw pulses trypsin inhibiter is present and it is anti-digestive in nature. Among the various pulses green gram is digested best. If more than 100 gms of pulse is consumed in a day it is not properly digested and gas is produced. If diarrhoea or other stomach complaints are present gas is produced by consuming small quantities of pulses. Bengal gram has the efficiency to reduce serum cholesterol level by increasing faecal excretion of total bile acids.

Lathyrism : Lathyrism is a paralytic disease seen in India and Spain where a type of pulse called khesari dal. -— Lathyrus sativus — is used liberally in food. In India Khesari is extensively cultivated in Madhya Pradesh, Bihar and Orissa. In Uttar Pradesh, West Bengal, Maharashtra, Gujarat, Karnataka and Andhra Pradesh it is cultivated in limited extent. Kesari dal is very cheap and highly nutritious. The flour prepared from this dal is used as a staple food in these areas. In these areas most of the agricultural labourers get khesari dal as their wage and about 250-500 gm of it is consumed by a person in a day. As a result about 4 per cent of population in these areas are crippled. Surveys in two districts of

% Madhya Pradesh (Rewa and Satna) showed 32,000 paralytic victims with lathyrism. The cause of lathyrism is the production of a toxin known as BOAA-B oxalyl amino L alamine which affects the nervous system. The toxin is produced more by the busk and its consumption causes high incidence of disease. The Indian Council of Medical Research has developed certain methods to destroy the toxin.

The disease occurs in different stages. The onset is sudden. But in the first stage itself it can be detected. On exposure to cold the calf muscles contract and backache occurs. An awkward manner of walking with short jerky steps and crossed gait are common. Muscular stiffness necessitates the use of a stick to maintain balance. Later spastic paralysis of the lower limbs, and exaggerated knee and ankle jerks occur. And at this stage the patient needs crutches for support. Finally correct posture becomes impossible as the knee joints cannot support the weight of the body. The patient crawls by throwing his weight on his hands. The serum globulin and urinary glutamic acid are increased in this condition.

The National Institute of Nutrition identified the toxic factor and a simple feasible solution was suggested by them to destroy the toxin. The household procedure is to steep the dehusked grain is boiling water. The seeds are drained of water and sundried and thus the toxin is eliminated. About 90-95 per cent of the toxin is leached out by this method.

Soyabean : Soyabean is otherwise known as the miracle bean because it is a storehouse of proteins, fat, vitamin A, thiamine, riboflavin and nicotinic acid. The protein content in soyabean is about 43 per cent and fat content is 19.5 per cent. The proteins of soyabean consists of all the essential amino acids in proper proportion except methinine and cystine. It has a good amount of lysine and it can be used for supplementing rice which is deficient in lysine. Apart from the nutritive importance it is also made use of in the production of penicillin and other antibiotics for industrial purposes. Enamel paints, varnishes and printing inks are made out of soyabean.

Even though it is rich in nutrients, unless it is processed

properly, proteins and carbohydrates of soyabeans are poorly digested. Soyabean contains a factor which inhibits the protein digesting enzyme, trypsin. Heat destroys it and after cooking if it is used the protein is properly digested. Soaking, steaming, germinating and fermenting are other processes which improve the nutritive value. Soyamilk is very wantious, but soyabean seems to produce goitre at times because it interferes with iodine metabolism. Soyamilk has about 4.2 gm of 3.4 gm of fat and 0.75 gm of salts. Soyabean can be parched and used like parched bengal gram. It can be used as a substitute for Dal gram flour in recipes.

The Cooking : Soaking, use of sodium bicarbonate, acids, alkali or calcium salts quantity of cooked pulses. Dals absorb water before they can be cooked. If dal is soaked in cold water overnight or in warm water 1 4 to 5 hours it can be cooked within 30 to 40 minutes. But if hard water is used for cooking it takes more time. (a) If sodium bicarbonate is added to cook pulses cooking time is reduced. (b) vitamin like thiamine destroyed. Addition of acid helps the cooking of pulses whereas as retard the cooking. Addition of formic acid of the pulse also influences the cooking quality. The exact causes for these variations are not found out.

Cereals

Cereals and millets form the staple food of human race. Wheat, rice, maize (corn), barley, oats, jowar, ragi and bajra are the common cereals and millets used. Cereals as a whole are rich sources of starch or carbohydrate, and good sources of protein in an Indian diet. Millets like ragi are a rich source of mineral calcium. Cereals are also good sources of B group vitamins.

Wheat : Wheat is one of our important food crops. Next to rice wheat is the most extensively grown foodgrain.

Wheat varies in its nutritive value, especially its protein content depending upon the seeds, variety in soil and water supply. The protein content varies from 7 to 20 per cent. In the 'green revolution' new varieties of superior quality wheat were produced. Sonalika, safed lerma, kalyan sona, sharbati sonora are some of these new varieties of wheat.

The wheat grain consists of three main parts. The germ or embryo is rich in protein and oil, the endosperm is rich in starch and fair sources of proteins and the outer coverings such as pericarp.

Various wheat products available are broken wheat, semolina, wholewheat flour (atta) and white flour. These flours are made from different kinds of wheat. Hard wheat, durum wheat and soft wheat are the different types of wheat from which flours are made. Flours of different percentages of extraction are obtained. Depending on the extent of milling, flour made from hard wheat has an extraction of 83 per cent and it is used for chapati making. With 75 per cent extraction white flour or soft wheat products are obtained. Biscuits, cakes, crackers and breakfast items are prepared from this flour. Semolina is prepared from hard wheat and durum wheat. It is granular wheat and it is used for the manufacture of macaroni products like noodles.

Composition of Wheat Flour : Wheat is a very good source of carbohydrates and proteins; 100 gms. of wheat provides 350 calories. Wheat contains carbohydrate in the form of starch; 66 to 68 per cent of flour is carbohydrate. Wheat flour has cellulose and hemicellulose and its quantity enhances the quality of dough prepared from it

Protein : Wheat flour contains glutenin and glidin as proteins which are commonly known as gluten. Wheat flour contains about 12 to 12.5 gms of protein per 100 gm. The strength of wheat flour is based on die quality of gluten in it Giuten is comparatively poor in amino acid lysine. Pulses, soya bean and groundnut are good in lysine content and they can be supplemented with wheat.

Lipids : Wheat flour contains 1.88 per cent fat; its kernel has about 3 per cent and bran contains about 4.8 per cent fat. Fat is present in the gluten of the flour.

Minerals : Wheat flour is poor in calcium content. The iron content of whole wheat flour is high but refined flour has less iron. Phosphorus content in whole flour is high compared to refined flour. But this high phosphorus content interferes with the absorption of calcium and iron.

Vitamins : Wheat is comparatively a good source of thiamine, riboflavin and nicotinic acid. But cooking results in loss of these

B complex vitamins. During chapati making about 33 per cent of thiamine is destroyed and during fermentation much more thiamine is lost. In the preparation of puri 40 to 90 per cent of thiamine is lost. During bread making about 10 to 30 per cent lysine is destroyed. If the bread is toasted again this vitamin is lost Bulgar wheat is much more nutritious. Wheat is parboiled in the preparation of bulgar wheat. During this process vitamins and other nutrients in the outer layers of the seeds move to the centre. Thus bulgar wheat is better in its auditive value.

Processed Wheat Products : Bread, biscuits, toast and breakfast cereals like wheat flakes and shredded wheat are processed wheat products. The autrivtive value of bread, especially fresh bread, is similar to two-thirds of wheat flour in its quantity. In the preparation of biscuits thiamine and protein is lost. On baking, apart of thiamine is lost and lysine is also lost. The average protein content in biscuit is 5 to 8 per cent as against 12 per cent in wheat flour. In the preparation of toast heat is applied on bread slices and part of thiamine is again destroyed. Lysine is also lost considerably. Readymade breakfast items are also inferior in nutritive value.

Rice : Rice is the staple food not only in India but also in many of the South-East Asian countries like Ceylon, Burma, China and Japan. Rice with its husk is called paddy. Husk is the loose covering of rice. It has no nutritive value. About 20 per cent by weight of paddy is busk. Removing the husk is called milling. Bran, aleurone layer, embryo or the germ and endosperm are the other layers. Bran is die outermost layer which contains nutrients and cellulose. Aleurone layer which lies beneath the bran is rich in vitamins, minerals, proteins and fats. Embryo or the germ is at the base of the grain which is rich in all nutrients and vitamin E. Endosperm forms 75 per cent of the grain and it comprises mainly starch. On milling vitamin B lost considerably. Parboiling improves the vitamin content of rice.

Parboiling of Paddy : In the conventional method of parboiling paddy is steeped in cold water for two to three days. Then it is steamed for about 5 to 10 minutes and men dried under the sun. During soaking paddy is fermented and at times gives out a foul smell. B complex vitamins in the pericarp and germ penetrate into the endosperm and get fixed into the grain. If parboiling is not done with care certain moulds present in the grain grow and produce toxins like aflatoxin. This can be rectified by using modern techniques in food processing. Through hot soaking process developed by CFTRI (Central Food Technological Research Institute) in Mysore is can be rectified. In this method paddy is soaked in water at 65° to 70°C for about 3 to 4 hours. Then water is drained and the soaked paddy is steamed in the same vessel for 5 to 10 minutes. This paddy is dried in sun or in mechanical driers.

The nutritive value, especially of thiamine and niacin, is improved by parboiling.

Nutritive Value of Rice : Rice contains carbohydrate in the form of starch. Though it is a source of protein, it is a poor source of fat, iron and calcium, provides about 335 calories per 100 gm of dry weight. The protein content is about 7 per cent which is not an appreciable amount since it is consumed in large quantities, rice supplies a good amount to proteins. In poor Indian diets cereals or millets are the main source of proteins. Though the protein content is low in rice compared to wheat the quality of rice proteins is superior to wheat proteins. Rice protein deficient in lysine which can be improved by supplementing the with pulses. High yielding varieties of rice are cultivated in our country. The protein count of hybrid variety rice like IR-20 is 9 per cent against 7 per cent of ordinary rice.

Rice is poor in its mineral content especially calcium. Iron content is also negligible, therefore rice-eaters must depend on other sources of minerals. Green leafy vegetables can be used for miner supplementation.

Rice is a poor source of carotene or vitamin A, and vitamin B present in good amount. Since it is present mainly in the outer lay polishing reduces the vitamin content. As polishing is increased content of B vitamins decreases. In 20 per cent polishing thiamine (B. content is only 40 mg/gms whereas in parboiled rice at 20 per a polishing thiamine is 200 mg/100 gm. Niacin content is 0.6 mg/100 gm, in 20 per cent polished rice whereas in parboiled rice it is 2.8 mg/1 gm. Among people who depend entirely on polished rice, a burn sensation in the feet, tingling and numbness occur. This is due to deficiency of thiamine in the diet. Even though rice is a good source of B complex vitamins. Its high solubility in water makes it non-available after various processes of cooking. About 60 per cent of it is lost through washing, if rice is washed twice or thrice. In the cooking water , remaining B complex gets dissolved. If the "conjee" water is thrown i this amount is also lost. Cooking rice in just enough water and utilise the excess water, if any, are the useful tips to make use of this vitamin.

Gelatinisation : Starch is present in cereals. This is insoluble in water, but in hot water it dissolves and forms colloidal dispersions. It swells and forms a sticky substance. In cereals about 65 to 80 per cent starch is present. Next to cereals roots like tapioca, potato and sweet potato are with starch granules. These roots and tubers contain about 22 to 39 per cent of starch. In starch there are two components—amylose and amylopectin. Amylose is about 15 to 30 per cent and amylopectin about 70 to 85 per cent. In different sources of starch it is present in different proportions. Amylose is soluble in water and amylopectin is insoluble in water. On 5 per cent concentration starch forms gels. Among the various sources of starch corn starch is the finest. This property of starch to gel is exploited in cookery. In many preparations starch is used as a thickening agent. To get a good gel, gelatinisation temperature of different starches must be known. Usually it starts to gel at 65° to 72°C. The enzymes in the starch are the other factors which affect the gelatinisation. Commercial products like sago, custard powder, corn syrup, glucose or dextrose and dextrimaltose are prepared from tapicao starch or from corn starch.

Millets : The common millets used in our country are jowar, ragi and bajra. Though millets are comparatively cheap they are superior in nutritive value. Usually people prefer rice or wheat as their cereals. If these millets are used more many deficiency symptoms among various groups can be minimised as they are excellent sources of vitamins and minerals.

Ragi : In India ragi is used mainly in Karnataka and Tamil Nadu as a staple food. In Kerala, ragi is used as a weaning food. Actually ragi is considered as a 'poor man's milk' by nutritionists because of its high calcium content. It can be easily substituted as a breakfast item. If we consider its cost it is cheap and at the same time rich in minerals that are lacking in rice or wheat. Like wheat or rice, ragi also contains 7 to 10 per cent protein and is equivalent to the above cereals in its carbohydrate content. It is considered superior because of high calcium content. Ragi is also a good source of iodine. No other cereal contains iodine. Another special feature of ragi is that the carbohydrate in ragi is digested slowly and so it can be given to diabetic patients because the sudden rise of blood sugar can be avoided by this. In certain hybrid varieties of ragi protein and iron contents are more and thus it is superior in its qualities. Ragi can be incorporated with pulse instead of rice in all breakfast items.

Bajra : Bajra is used in certain parts of the country as a staple food. It consists about 12 per cent of protein and calcium and iron content is compared to rice. Bajra is a better source of vitamin B, especially thiamine and riboflavin.

Jowar : Jowar is used in Maharashtra, Andhra Pradesh, Karnataka, Tamil Nadu and Madhya Pradesh. The protein content in jowar is 10 per cent and the amino acid proportion is better. It has more B complex vitamins compared to rice. But certain varieties of jowar inhibit the availability; of niacin and cause the deficiency symptoms of pellagra.

Effect of varying concentration of sugar syrup and crystallisation

Boiling temperature° C	***Sugar percentage (Syrup)***	***Cold water test***
110-112° C	78-80	Does not form a ball
113-115° C	83	Soft ball, eg., fondant
118-120° C	86	Firmer ball, eg., caramels.
128-130° C	90	Hard ball, but still plastic, eg., candy
138-140° C	94	Very hard ball, eg, toffees
149-154° C	98	Very hard and brittle, eg., brittles.

To prepare the syrup add a small amount of sugar in water and heat, when the whole sugar is dissolved examine the consistency.

10

Mineral Substances

In an adult 4 per cent of the body weight is from minerals. There are 24 minerals in the body. Mineral substances are important in maintaining good health because they are part of our tissues, skeletal structure and the body fluids. Hormones, enzymes and vitamins contain minerals. Vital organs like heart and nervous tissues cannot function normally in its absence. Water balance, regulation of fluids, and the acid base balance in the body depend to a great extent on certain mineral balance in the body. Neuromuscular activities, coagulation of blood and fluid movement in the tissues are under the control of certain inorganic substances in the body.

Minerals are inorganic substance that are present in the body tissues like phosphomin, iron, sulphur, zinc or copper or soluble minerals in the body fluids like sodium chloride or mineral components in the bones and teeth and hormonal secretions. Water balance in the body is regulated by minerals like sodium and potassium. Our body has calcium, phosphorus, sodium, potassium, chlorine, iron, magnesium, copper, iodine, cobalt, zinc, aluminium, arsenic, bromine, fluorine, nickel, chromium, cadmium, selenium, silicon, vanadium, and molybdenum as minerals. Each one of this has some role in maintaining the structure and function of vital organs of our body. Few milligrams of a mineral element decide the skeletal growth or the mental ability of a person.

Optimum intake of minerals and favourable environment for its utilisation are equally important to maintain mineral balance. During the growth period and also during pregnancy, lactation and old age, certain minerals are very essential to maintain health and growth.

Calcium

The most abundant mineral element in the body is calcium. Nearly 99 per cent of the calcium in the body is present in the skeleton. The remaining part of calcium performs a variety of functions in the body. About 850-950 gms of calcium is present in a healthy man.

Functions : The most important function of calcium is the ossification of bones. If calcium content in the blood is below normal, bone calcium is depleted and serum level is brought to normal. Parathyroid hormone regulates calcium level in blood and its metabolism in the bone. The bones constitute a mineral reserve which may be drawn upon in need. Calcium is also essential for the formation of teeth and for the contraction of the heart and muscle. The regulation of muscle irritability is dependent upon the serum calcium level. Calcium is also required for normal nerve impulse transmission and excitability of nerve fibres and nerve centres. Calcium forms part of acelytcholine which is necessary for the transmission of nerve impulses. Calcium helps in the coagulation of blood and is needed for permeability of capillary walls. Calcium exerts its influence in the absorption of vitamin B12 from the intestine. The mineral has been shown to activate certain enzymes, especially lipase and certain proteolytic enzymes, adenosine triphosphate and rennin.

Factors Affecting Calcium Absorption : The most important factor promoting calcium absorption is vitamin D content in the body. Utilisation of calcium is more when it is needed by the body. For example, during growth the absorption of calcium is more than in adults. Calcium content in the diet increases the absorption of calcium. Phosphates and calcium ratio in the diet also influences the absorption. If the ratio is 1:1, absorption rate is high. If phosphate ratio is high, calcium absorption is low. Phytic acid which is present in cereals combines with phosphorus and forms insoluble calcium salts and calcium phytate and inhibits calcium absorption. Abnormal intake of fats and fatty acids reduces the calcium level in blood and its metabolism in the bone. The bones constitute a mineral reserve which may be drawn upon in need. Minerals like iron which forms insoluble phosphates interfere with phosphorus absorption which indirectly affects calcium absorption. Absorption of calcium is facilitated by low intestinal pH. Normal acidity of stomach due to the presence of hydrochloric acid is essential to stimulate the absorption of calcium. Parathyroid hormone also aids in the absorption of calcium from the stomach. If cereals with husk are used in the diet, calcium absorption is reduced. Whole flours, oatmeal, and preparation with these, if not fermented, increases the inhibition of calcium utilisation. On fermentation of the enzyme phytase splits up calcium phytate and then increases its absorption.

Like phytic acid, oxalic acid is present abundantly in leafy vegetables and certain fruits. This forms an insoluble compound calcium oxalate and thus diminishes the absorption of calcium. Magnesium, if taken in abundant quantities, often interferes with calcium absorption and results in poor calcium absorption. During sprue, steatorrohea or in biliary fistula fat is absorbed poorly and thus accumulated fats help the formation of calcium soaps. Good quality protein content in the diet helps calcium absorption. During rickets citric acid increases calcium absorption.

There are many factors which hinder the absorption of calcium. Phytic acid, oxalic acid, magnesium and fat reduce the absorption of calcium. Cereals contain phytic acid which combines with calcium to form insoluble calcium phytates. Phytic acid is present in the outer parts of certain proteolytic enzymes.

Deficiency : Calcium deficiency symptoms are usually accompanied by a lack of phosphorus and vitamin D. Any of the factors which affect calcium absorption can lead to its ineffective utilisation. Vitamin D deficiency alone can produce calcium deficiency symptoms. Pure calcium deficiency is osteoperosis. In osteoperosis serum calcium level is normal but body store is reduced. During old age the capacity of the body to absorb and utilise calcium is diminished. In women hormonal imbalance during menopause or removal of ovaries at earlier stages results in osteoperosis. Corticosteroid injection for any purpose may result in osteoperosis. In osteoperosis pores are seen in the bones due to decalcification. Back pain, loss of weight, fracture of bones for no reason and loss of teeth are some of the clinical symptoms. Calcium and vitamin D deficiency results in rickets and osteomalacia among children and adults, respectively.

Tetany is another symptom of calcium deficiency. In tetany there is hyperexcitability of the nervous system. A series of spasmodic movements take place in the muscles. Depending upon the level of serum calcium symptoms of tetany vary. Irritation, confusion, pain, involuntary muscle spasm, bronchial spasm, spasm in the gastro-intestinal tract, pain and convulsions are again symptoms of tetany. If parathyroid gland is in hypoactivity, calcium is not properly assimilated and results in chronic tetany. In this condition nervous changes occur and impairment of memory takes place. Depression, irritation, loss of hair and nails, roughness of the skin, pitting of teeth and epileptic fits are other symptoms of this condition. Cataract is common in such patients.

Hypercalcemia : Hyper-vitaminosis-D or Hyperparathyroidism produces hypercalcemia. Normally infants of 5-8 months get this problem. Loss of appetite, vomiting, constipation and flabby muscles are the symptoms.

Sources of Calcium : Milk and milk products are the best sources of calcium. Calcium is absorbed maximum from milk and milk products because of the presence of lysine and lactose in it. Ragi is the cheapest source of calcium; therefore it is known as 'poor man's milk'. Sesame seeds and leafy vegetables are also good sources of calcium. But the availability from these sources is not insured as in sesame seeds calcium is present in its husk and on dehusking it is removed. In leafy vegetables the presence of oxalates inhibits its absorption. Molasses and vegetables like peas, beans, potatoes, pulses, and small fish and dried fruits are rich in calcium. Unabsorbed calcium is excreted through faecal matters.

Requirements : The FAO/WHO expert group (1984) has recommended 400-500 mg of calcium per day for an adult man and woman. During pregnancy and lactation the requirement increases to 1,000 mg/day. For infants 500-600 mg and for children of between 1 and 9 years 400 to 500 mgs and for age group 10 to 15 years 600-700 mg are recommended.

Phosphorus

No other mineral takes part in a number of functions as phosphorus in different parts of our body. An adult human body contains about 400-700 mgs of phosphorus. About eighty per cent of the body's phosphorus is deposited in the bones and teeth as calcium phosphate. The remaining portion is located as sodium phosphate and potassium phosphate in the body fluids and in cells. Phosphorus is also present in many enzymes and co-enzymes and co-carboxylase. Thus it takes part in metabolism.

Functions : Phosphorus takes an important role in the calcification of bones. It also plays an essential part in carbohydrate metabolism in phosphorylation of glycogen. Phosphorus is an essential constituent of co-enzyme and co-carboxylase enzyme system in the oxidation of carbohydrate, proteins and fat metabolism. Phosphorus compounds control the storage and release of energy through ATP, ADP system. Phosphorus is present in the nucleoproteins which are an integral part of the cells. Phospholipids, which are present in various cells of our body, contain phosphorus. Inorganic phosphates in the body fluids act as a buffer system in the body.

About seventy per cent of ingested phosphorus is absorbed. If iron, aluminium and magnesium are present in excess amount in the diet they interfere with the absorption of phosphorus. The proportion of calcium and phosphorus is an important factor which regulates its absorption.

Requirements : The daily requirement is about 1 gm. Phosphorus requirement is similar to calcium for various groups.

Sources : Animal foods like meat, fish, poultry and eggs are excellent sources of phosphorus. Milk and milk products are good source of phosphorus. Though cereals and legumes are good sources, phosphorus is less available from them as it is present in the form of phytic acid which is not utilised well.

Iron

Iron is a component of the complex protein haemoglobin, a haemoprotein component of the red blood cells. It is the pigment of red blood cells. Of the 3 to 5 gms of iron present in our body more than half of it is present in haemoglobin. The rest of it is stored in the liver, spleen, bone marrow and muscles. In the muscle it is present in the form of myoglobin. Iron is also present in some enzymes like cytochromes, catalase and peroxidase.

Functions : The most important function of iron is to form a constituent of haemoglobin which takes part in the transportation of oxygen from the lungs to the tissues. This oxygen is used by the body for the processes of metabolism. In its deficiency the oxygen-carrying capacity of the blood is reduced. In red muscles myoglobin, a pigment substance, is formed with the participation of iron. This also has the same oxygen-transporting function like the haemoglobin. During strenuous activity this muscle is involved. In our body there are certain cytochrome protein pigments with iron as an essential component. They act as respiratory enzymes. Certain flavin enzymes also contain iron in them.

Absorption of Iron : Dietary iron is absorbed primarily from the duodenum by the mucosa. Depending upon the need of iron in the body its absorption varies. If the haemoglobin concentration of the blood is satisfactory, little or no iron is absorbed and it is excreted in the faeces. Ferrous iron is absorbed more efficiently than ferric iron. In the intestine most of the iron is reduced to ferrous state. Absorption of iron is enhanced in an acid medium. Vitamin C also enhances the absorption of iron, in the intestinal tract, because of its strong reducing capacity. Protein content in the diet helps the absorption of iron to some extent Sulphur containing amino acids reduce iron which helps its absorption. Excess of calcium phosphate, phytates and oxalic acid interfere with iron absorption by forming insoluble iron salts or oxalates. Minute traces of copper help the formation of haemoglobin. The rate of absorption of iron is affected by the magnitude of iron store in the body and the rate of erythropoiesis. Malabsorption syndromes also reduce iron absorption.

Deficiency of Iron : Anaemia is a conditon where the haemoglobin level is lowered in the blood. Shortage of iron in the diet is the most common cause of anaemia. Iron, folic acid, proteins, B12, copper, cobalt, vitamin C and pyridoxine are the nutrients required for erythropoiesis. Folic acid deficiency produces megaloblastic anaemia.

Symptoms of Anaemia : When the haemoglobin level falls below 12.5 gm/100 ml of blood, it results in the diminished oxygen-carrying power of the blood. Iron deficiency anaemia is otherwise known as hypochromic anaemia. In this condition the number of RBC is not much reduced but the quantity of haemoglobin is less. The pallor of skin and tissues, fatigue, weakness, giddiness, blurred vision, anorexia, heartburn, palpitation, dysproea, oedema over feet, headache and a feeling of tiredness are some of the symptoms. The nails show koilonychia and tongue and papillary atrophy are seen. Among women menstrual disturbances like menorrhagia or phases of amenorrhoea take place.

Nutritional megaloblastic anaemia occurs due to dietary deficiency during old age, increased requirement during pregnancy, lactation, prematurity, rapid growth and in endocrine disturbances. Haemolytic anaemias occur during malignancy, rheumatoid arthritis, pulmonary tuberculosis and in skin diseases. Alcoholism and side effects of drugs also result in anaemia. In diseases like idiopathic steatorrhoea, tropical sprue and in partial gastractomy anaemia takes place. During pregnancy this type of anaemia usually occurs in the second trimester. Weakness, palpitation and fatigue are followed by fainting bouts.

Anaemia must be treated with ferrous sulphate tablets and iron-rich diet.

Hemosiderosis : It is a disorder of iron metabolism. In this condition large deposits of iron are made in the liver and in the recticulo endothelial system. Iron with carrier protein which is known as transferrin of circulation becomes saturated with iron. When transferrin is unable to bind all the absorbed iron this metabolic disorder of iron takes place. If iron vessels are used for almost all cooking; much more iron is ingested in the body, which may result in hemosiderosis. The Bantu population of Africa is an example of this condition.

In hemolytic anaemia a large number of red blood cells are destroyed and this can lead to hemosiderosis. Prolonged prophylaxis with massive doses of iron can often produce hemosiderosis.

Sources of Iron : Liver, kidney, heart, lean meat, egg yolk and shellfish are the best sources of iron. Dried beans, legumes, dried fruits, nuts, green leafy vegetables, whole cereals, enriched grains and molasses are good sources. Milk is a poor source of iron.

Requirement of Iron : Recommended intake of iron is calculated, based on its loss and absorption. For an adult man weighing 55 kg the iron requirement is 24 mg/day and for a woman of 45 kg weight 32 mg is the requirement. During pregnancy 40 mg and during lactation 32 mg of iron is required.

For infants till one year 1 mg/kg body weight and for children of 1 to 9 years of age 20 to 25 mg of iron are required. For boys and girls between 10 and 12 years of age 30 to 25 mg and for boys 13 to 15 years of age 25 mg and for girls from 13 to 15 years 35 mg of iron are required. During later adolescents for boys (16-18 years) 25 mg and for girls 35 mg is the recommended allowance.

Magnesium

The human body contains minute quantities of magnesium in the bones along with calcium and phosphorus and the rest in soft tissue. An adult human body has about 25 gm of magnesium.

Functions : Magnesium plays an important role in normal calcium and phosphorus metabolism in man. It is also present in certain enzymes like co-carboxylase which decarboxylase pyruvic acid. Magnesium also takes part in the phosphorylation in metabolic pathways.

Magnesium is easily absorbed from the intestine. Excess magnesium is excreted in the faeces. Normal human serum contains about 2 to 3 mg/100 ml.

***Deficiency*:** Prolonged starvation brings magnesium deficiency, vomiting and intestinal fistula. Diabetic acidosis is associated with raised serum level of magnesium. Neuromuscular irritability, tremors, cramps, twitching and convulsions occur in its deficiency. Confusion, disorientation and visual hallucinations are some other complaints with magnesium deficiency.

Sodium

Sodium is essential for the normal functioning of the body. Sodium plays an important role in the regulation of acid-base balance and water metabolism in the body. Sodium in the form of sodium chloride is ingested directly through food and many food materials contain this mineral. Almost all the body sodium is found in the extracellular fluids of our body. Through sodium the osmotic pressure is also regulated by the body. A healthy kidney regulates the excretion of sodium according to the requirement. During kidney diseases sodium is retained in the body and oedema takes place. Under normal conditions 90 per cent of ingested sodium is excreted in the urine in the form of sodium chloride and sodium phosphate. Sweat contains some amount of sodium. Sodium helps to contribute alkalibility of the gastrointestinal tract.

Sodium plays a special role in controlling the heartbeat by helping in its origin and maintenance. Contraction of muscles and regulating the permeability of cell membranes are the other functions of sodium.

The adult human body contains about 100 gm of sodium. On an average diet about 3 to 5 gins of sodium is excreted in the urine. In profuse perspiration sodium chloride is depleted from the body. In sodium depletion muscular cramps, weakness, headache and vascular collapse take place. In decreased water intake the serum level of sodium is increased and this condition is known as hypernatraemia. Blood urea is increased and cellular water is depleted during dehydration and in hyponatraemia. In infants shrinkage of brain causes rupture of cerebral veins and further cerebral thrombosis. Retention of sodium produces oedema and ascites. In renal diseases, cardiac failure, hepatic cirrhosis with ascites, toxaemia of pregnancy, prolonged administration of corticosteroids sodium must be restricted through diet. Normally 4-5 gm sodium is injested through food and it is restricted to 250 mg, 500 mg, or 750 mg depending upon the condition.

Requirements : For an adult 1-2 gms of sodium and for children 350 mg-1 gm is recommended. For infants 150-400 mg is enough.

Sources : Animal foods like meat, meat extractives, fish, poultry, milk, milk products, nuts, bakery items, salted biscuits, egg-white, dried fruits, roots like beetroot, carrot, radish and leafy vegetables like spinach are excellent sources of sodium.

Abnormalities of Sodium Level in Blood : Hyponatraemia and hypernatremia are two conditions which manifest the sodium level in the blood.

Hyponatraemia : When serum sodium level falls below normal the condition is known as hyponatraemia; various conditions lead to this symptom. Severe dehydration, decrease in blood volume, decrease in blood pressure, circulatory failure are some of them, prolonged vomiting, diarrhoea, chronic renal diseases with acidosis, Addison's disease due to adrenocestical deficiency are the common causes for hyponatraemia.

Hypernatraemia : In this condition the plasma sodium iron content is higher than normal. Hyperactivity of adrenal cortex as in cushing's syndrome, prolonged treatment with cortisone, ACTH and sex hormones are some of the causes of hypernatraemia. Deficient water intake due to vomiting, nausea, difficulty in swallowing with sore throat, oesophegal disease, cerebral impairment, increased water loss through respiration diarrhoea, excessive urination in diabetes mellitus, chronic renal diseases, increased sodium intake and deficient sodium excretion brings hypernatraemia.

Potassium

In contrast to sodium, potassium is present in the intracellular fluids of the body. It is an important constituent of cells. In adult body 250 gm of potassium is present. Most of it, that is, 97 per cent is found in the intracellular fluids and the remaining 3 per cent in extracellular fluids. The other important function of potassium is that it influences the cellular excitability of smooth, skeletal, cardiac and nervous tissues. If potassium content is high muscular irritability is increased. Diminished calcium and magnesium also result in such a condition. Involuntary muscular function is controlled by potassium content in the body. When the potassium level in the myocardium is changed it is reflected in the electrocardiogram. Potassium is also necessary for the acid base balance in the body.

Potassium deficiency shows weakness in muscular functions, mental disorientation and cardiac irregularities. Hyperpotassemia results from kidney diseases when renal excretion is less. In dehydration these symptoms occur. Deficiency of potassium is not common unless as a complication of diseases. Deficiency condition shows hypokalaemia during operation when food cannot be ingested. In diabetic acidosis this condition occurs. Deficient intake due to starvation and excessive loss from gastro-intestinal tract leads to deficiency.

Hypokalaemia occurs due to deficient intake of it while a surgical patient is exposed to prolonged parent therapy, excessive loss from gastro-intestional tract through persistent vomiting, fistulae or diarrhoea before and after surgery, the patient may be on a poor diet which may enhance the excretion of potassium. Surgical cases due to low intake produces hypokalaemia.

In diabetic acidosis potassium is lost through dehydration on insulin therapy, sugar is converted to glycogen and starch. This glycogen formation depletes some potassium from extracellular fluids and serum potassium level falls and it is dangerous.

While diuretics are administered considerable loss of potassium through urine leads to hypokalaemia. Nephritis and other kidney diseases, tumours of large intestine, treatment with steroids, deficiency disease like kwashiorkor, megaloblastic anaemia, heatstroke and violent exercise are some of the other causes which produce hypokalaemia. Prolonged hypokalaemia causes injury to myocardium and kidney.

Hyperkalaemia : When the potassium level in the serum is high the condition is known as hyperkalaemia. This occurs due to diminished urine volume, intake of potassium salts of penicillin, iodide, acidosis, endogenous tissue breakdown, Addison's disease, kidney failure due to kidney disease or chronic renal failure and rapid intravenous administration of potassium are some of the reasons for hyperkalaemia. Cardiac and the central nervous system are affected by this condition. Anorexia, weakness of the muscles, diminished tendon jerks, low heart sounds followed by peripheral vascular collapse and general apathy are some of the symptoms.

Trace Elements

Elements which occur in our body in small quantities are known as trace elements. Copper, iodine, zinc, manganese, molybdenum, fluorine, selenium, nickel, chromium, cadmium, silicon, vanadium, strontium, cobalt and chlorine are some of the trace elements. Though these trace elements are present in traces they are very essential to maintain the structure of various organs in good condition. RBC formation, thyroxine production, formation of healthy bones, haemoglobin formation and proper enzymatic actions are some of the important functions of trace elements in our body.

Iodine

The significance of iodine as an essential trace element lies in its role in thyroxine production. Thyroxine is the hormone produced by the thyroid gland. Iodine is the/main component of thyroxine and a normal -adult has only about 10 mg of iodine in the thyroid. The absorbed iodine combines with a thyroxine binding protein. Thyroxine plays an important role in carbohydrate metabolism. Deficiency of thyroxine interferes with the normal physical and skeletal development of the child during the growing period. Iodine also helps the tissues in its oxygen consumption. Conversion of glycogen to glucose occurs only in its presence. Thyroxine increases the heart rate and depletes bone calcium.

Deficiency : If the iodine intake is deficient the thyroid gland enlarges and this condition is known as simple or endemic goitre. The normal thyroid gland weighs about 25 gin. In goitre it enlarges from 200 to 500 gm. Iodine deficiency goitre results in hypofunctioning of the gland. If diagnosis is done early and iodine supplements are given the thyroid may become normal. If timely treatment is not given the enlargement will become permanent. During the growing period if severe deficiency occurs it results in retarded growth and the condition is known as cretinism. People living mainly in submountain zones are affected. In India residents of sub-Himalayan regions and Kashmir and parts of adjoining plains are the victims. The most affected are among girls between the ages of 12 and 18 years and boys from 9 to 13 years. The areas where goitre is common is known as the goitre belt. According to the latest data available, about 40 million people are affected with goitre. In these areas air, water, soil and foods grown are poor in iodine which results in its high incidence. Goitre can be prevented by the regular use of iodized salt.

Goitrogenic Substances in Foods : Certain foods of brassica genus like cabbage, cauliflower, turnip, soyaflour and oil contain substances which react with iodine present in the food and make it unavailable. These substances are known as goitrogenic substances. Consumption of these foods in large quantities makes the iodine non-available to the body.

Requirement of Iodine : An adult requires about 0.15 to 0.2 mg of iodine and for infants and children 0.05 to 0.1 mg daily is enough.

Sources : Iodine is present in minute quantities in all foods. The iodine content of foods depends on the iodine content of the soil. In certain parts the soil contains less iodine and the foods grown in such places are iodine poor. The soil of mountainous regions is usually iodine-poor soil. The best sources of iodine are seafoods, common salt and vegetables; cereals and legumes are poor sources of iodine. Most of the iodine in the vegetables is in their leaves and flowers.

Copper

Copper is found in all tissues of our body. It is an essential element in the formation of haemoglobin. About 100 to 150 mg of copper is present in the human body. The brain, liver, heart and kidney contain the highest amounts. In the blood, copper is present in the plasma and RBC. Copper is absorbed rapidly into the blood.

Copper is a component to certain enzymes in our body. It is part of the enzyme cytochrome oxidase which is essential for glucose metabolism. It is also part of tyrosinase, an enzyme necessary for converting the amino acid tyrosine to melanin, the dark pigment of the skin. Copper stimulates the absorption of iron. Copper is a constituent of the elastic connective tissue protein elastin.

Deficiency : Copper deficiency is not common in human beings. In malnourished persons iron is not absorbed properly and anaemia occurs. Premature babies fed on cow's milk develop anaemia due to copper and iron deficiency.

Requirement : An adult person requires about 2 mg/day. During pregnancy and lactation the copper requirement is increased to 3 mg/day. Infants from 0 to 12 months require 0.5 mg to 1 gm and children 2 gm and adolescents 3 mg of copper.

Sources : Millets like cholam, bajra, raw rice, whole grams, cow gram, dried peas, red gram dal, vegetables like drumstick are good sources of copper. Liver is the richest source of copper. Organ meat like kidney is also a good source of copper.

Zinc

In an adult 2 to 3 gms of zinc is present. It is present in certain parts of the eye, certain glands and in their secretion. Liver, muscles, bone and hair contain some of the remaining zinc. Blood, especially the RBC, has some zinc. In small amounts it is present in tissues and bone cells.

The important function of zinc is its role in enzymatic action. At least in 25 enzymes of digestive and metabolic action zinc takes part. It is also present in an enzyme carbonic anhydrase which is essential for the transport and elimination of carbon dioxide. Co-carboxypeptidase and alkaline phosphatase contain zinc in them. Zinc plays an essential role in the formation of DNA and RNA. It is a constituent of insulin and it is necessary for the formation of connective tissues. Zinc also aids in the healing of burns and wounds.

Deficiency : Zinc deficiency in animals results in growth failure, loss of appetite, degeneration of the male sex glands, testes, skin lesions, alopaecia or loss of hair, changes in the epithelial and cutaneous tissues, dermatitis and scaling and cracking of the paws. Abnormal bone formation, thickening and metabolism also occur during zinc deficiency. In human beings zinc deficiency produces anaemia, growth retardation and delayed genital maturation.

Requirement : Infants require 3 to 5 mg. For children 10 to 15 mg and during pregnancy and lactation 20 to 25 mg of zinc are required.

Sources : Seafoods, meat poultry and eggs are good sources of zinc. Cereals, legumes, and nuts contain considerable amounts. Fruits and vegetables are poor sources. On milling eighty per cent of zinc is lost

Cobalt

In an adult body about 1.1 mg of cobalt is present. It is found in the blood, liver, kidney, spleen and pancreas. It is also present in minute quantities in all tissues.

The most important function of cobalt is that it forms an integral part of vitamin B_{12}. In animals, if the feed is deficient in cobalt severe anaemia occurs. Addition of cobalt to feed or injection of vitamin B_{12} cures this condition. In human beings vitamin B_{12} synthesis occurs in the colon. At that phase absorption is not taking place. So preformed B_{12} must be ingested for the cobalt requirement. But deficiency of cobalt is not common among human beings. Cobalt also takes part in the enzymatic action and also in thyroid function.

Requirement : Since deficiency is not known, the requirement is not studied. In pernicious anaemia B_{12} deficiency is present and on administration of B_{12} it disappears.

Sources : Cereals, beans, peas and organ meats like kidney and liver are good sources.

Manganese

About 20 mg of manganese is present in an adult. Bones, liver, pancreas, kidney and pituitary gland contain manganese.

The manganese functions in many enzyme systems. It activates certain enzymes which take part in the digestion and metabolism of carbohydrates, proteins and lipids. In the synthesis of ATP cholesterol, fatty acids and RNA, the manganese plays an important role.

Manganese deficiency in human beings is not known. Experimental deficiency symptoms are slowing of growth, abnormalities in bone, depressed reproductive functions, neurological weakness of the limbs, ataxia and lack of balance. Among birds a condition known as perosis occurs on manganese deficiency. It is characterised by bone deformities and dislocation of bones.

Sources : Cereals, bran, dried beans, peas, green vegetables and nuts are good sources. Tea and coffee are high in manganese content. But animal foods are relatively poor in it.

Chromium

An adult body contains about 6 mg of chromium. Chromium is present in skin, hair, muscles, brain, tissues, adrenal glands and in body fats.

An important function of chromium is its role in glucose tolerance factor formation in the body. Chromium niacin complex factor helps the action of insulin on glucose. Chromium is also essential for the activating enzymes involved in carbohydrate, protein and fat digestion and metabolism. The serum cholesterol level was raised on experimental animals in chromium deficiency.

Requirement : The exact requirements is not known. A normal balanced diet supplies enough chromium.

Sources : Organ meats like liver, meats, wholegrain cereals and brewer's yeast are good sources of chromium.

Fluorine

Fluorine is present in the body as fluoride. It is present in teeth, bones, thyroid gland, skin and liver.

Fluoride is necessary for the formation of caries-resistant teeth. Small amounts of fluorine prevent dental cariesin children. It promotes calcification.

Fluoride and Tooth Structure : For healthy tooth formation some amount of fluorine is essential. Though the exact mechanism by which it prevents dental caries is not fully understood, it is assumed that fluorine is incorporated into the tooth enamel. This reduces the solubility of enamel in the acids produced by bacteria. In its absence the enamel is soluble in the acid and tooth decays. In experimental animals fluoride seems to help normal growth during the growing period, but in human beings its role is not yet proved.

Toxicity or Fluorosis : If water contains more fluorine, dental fluorosis occurs. The enamel of the teeth loses its natural lustre and becomes rough. Opaque white Patches with yellow or brown stains are found irregularly on the teeth. Ridges or diffused spots and pitting occur over the teeth. The surface of the teeth gives a corroded appearance.

Skeletal Fluorosis : The bones of vertebral column, especially the cervical part of the spine, Pelvis and bones of the lower extremities, are affected in this condition. Hypercalcification of the bones and the collagen take place due to excessive fluorine deposition. Stiffness and pain are the earlier tymptoms. Weakness and spasticity of lower limbs, wasting of the distal muscles of hands, deep jerks on lower extremities and numbness or girdle pain are the other symptoms of skeletal fluorosis. Gradually the person becomes crippled as he cannot bend or squat for become stiff.

Requirements : Water containing 1 to 2 ppm prevents caries and so it is considered as the requirement.

Molybdenum

The adult human body contains only 9 mg of molybdenum. It is present in the liver, kidney, adrenal glands, bones, skin and tissues. Molybdenum is essential for the normal enzymatic system and its, function. Enzymes like xanthine oxidase and aldehyde oxidase make use of molybdenum. These enzymes are essential for the proper nucleic acid and iron metabolism. Molybdenum seems to aid in preventing teeth decay. The requirement of molybdenum is not known. Whole-grain cereals, dried beans, peas and dark green leafy vegetables are good sources. Meat is also an excellent source.

11

Oils and Fats

Fats and oils are used in cookery to enhance flavour and taste of food. Moreover, fat gives much more energy to body; from every gramme of fat 9 calories are given. In India most of our fats and oils are extracted from vegetable oilseeds. A solid fat-like ghee has a decisive place in Indian cuisine and it is often substituted by vanaspati. Liquid vegetable oils are hydrogenated and converted to solid fats. Hydrogenated fat is incorporated with vitamins A and D.

The Oils

Common oils used in cooking are groundnut oil, sesame oil, mustard oil, safflower oil, coconut oil, soyabean oil, cottonseed oil, palm oil and sunflower oil.

Choice of fat for cooking depends on various factors. Flavour, taste, saturation of fatty acids and presence of double bonds, melting points, smoking temperature and nutritive value are some of them.

Liquid oils rich in linoleic acid like safflower, cottonseed, and groundnut are good in suppressing the formation of scales on the walls of blood vessel. The characterisitc flavours of various oils are due to the presence of very small quantities of certain susbtances which have no nutritional qualities. Cooking in fat gives fuliness of satiety along with taste to the dishes. Now most of the oils are refined to make them colourless and odourless.

Composition of Fats : Fats and oils are classified into two groups based on their origin as vegetable and animal sources.

Vegetable fats and oils are again broadly divided into three groups based on their composition. They are: Group I which has fatty acids with 16 or 18 carbon atoms; Group II has erucic acid; and Group III has short chain fatty acids. Groundnut oil, cottonseed oil, sesame oil, sanflower seed oil, sunflower seed oil and soyabean oil are with 16 or 18 carbon atoms and they are termed as Group I. Mustard oil and rapeseed oil have erucic acid in it and they are in Group II. Coconut oil and palm oil are with short chain fatty acids in Group III.

Oil	*Linoleic acid content (%)*
Coconut oil	2
Mustard oil	15
Ghee	2
Palmoil	10
Palmolein	12
Safflower oil	70
Cottonseed oil	50
Sunflower oil	50
Groundnut oil	40

Animal fats are also classified into three groups based on their composition. Group I is milk fat which contains butyric acid and short chain fatty acids. Group II comprises animal body fat with fatty acids of 16 to 18 carbon atoms. Ox fat, pig fat and egg fat are examples of this group. In Group III fish oils with large amounts of polyunsaturated fatty acids are present herring, cod, halibut and sardine oils fall in this group.

Melting points of fats depend on the saturation of fatty acids. Solid fats like lard with high saturated fatty acids have high melting point and liquid oils have a low melting point.

Smoking Temperature of Fats : When fats and oils are heated to a high temperature decomposition of fats occurs and at a point visible fumes are given off. This temperature is called the smoking temperature of fats. The smoking temperature of fat is low when amounts of free fatty acids are high.

Smoking Temperature of Some Fats and Oils

Oil or fat		***Smoking temperature***
Butter	...	208° C.
Cottonseed oil	...	322° C.
Coconut oil	...	138° C.
Lard	...	194° C.
Olive oil refined	...	234° C.
Groundnut oil	...	162° C.

Fat absorption is high if foods are cooked in fat with low smoking points.

Smoking temperature of fat varies depending on the surface area of oil exposed while heating. The penetrability of the produce to be fried also determines the fat absorption while frying.

Shortenings : Fat is used as a shortening agent due to its ability to give a soft crisp texture to fried or baked products. In bakery items fat plays a significant role.

In wheat products fat brings shortening of gluten strands and in other dishes shortening is through gelatinisation of starch. To bring shortening effect one fat or a mixture of fats is used. In biscuit-making plastic shortenings are used which contribute a white, opaque and creamy appearance to the shortening. Butter and ghee are generally preferred as shortening agents. In certain preparations fat blends homogeneously with non-fat substance to form an emulsion and examples of such preparations are sweets like halwa, mysore pak or cake.

Changes in Fats : Changes in fat during heating are related to its smoking points. Free fatty acid is increased on heating. This is due to the change brought about by the moisture content of food materials. It hydrolyses the cooking fat and increases the free fatty acid content. Therefore, deep frying foods in fats with high smoking temperature is preferred. Ghee has a lower smoking point and therefore other oils or hydrogenated fat are suggested for deep frying.

When fat is used for deep frying, temperature control is essential. If high temperature is used to heat fat, it affects its nutritive value and in experimental cases it produced a toxic effect. Certain changes take place in fat during heating and this brings down the unsaturated fatty acid content. Repeated heating of fats brings chemical changes and results in poor absorption. It also increased blood cholesterol level in experimental animals. If the fat is heated too high, the food will char outside and remain raw inside. If the temperature of fat is too low the food will get soaked in it. In correct temperature food will be crisp and uniformly cooked. To get the best result fat must be heated to its smoking point. When fumes appear insert food material, fry and lower.

Fats and Oils : The heat after frying the first batch allow the fat to heat to required temperature and add the second batch to fry. Absorption of fats by food materials varies depending on its smoking temperature. If fat with low smoking point is used to fry, more fat is absorbed and fat with high smoking point is required less to fry items. Food items with milk, sugar, or eggs require low temperature to fry them. If food items are fried at high temperature more fat is absorbed. If the food item has greater surface area fat absorbed is more and if surface area is small less fat is absorbed. The moisture content of food item also affects fat absorption. With more moisture content fat absorption also increases. Hard materials also take more fats.

Characteristics of Some Common Fats

Fat source fatty acid	*Poly unsaturated fatty acid*	*Saturated fatty acid*	*Mono-unsaturated*
Groundnut	31	19	50
Sesame	43	14	43
Safflower seed	67	7	26
Cottonseed	47	23	30

Contd...

Fat source fatty acid	*Poly unsaturated fatty acid*	*Saturated fatty acid*	*Mono-unsaturated*
Coconut oil	2	92	6
Mustard oil	63	5	32
Butter	3	70	27
Soyabean	64	18	18
Lard	10	96	40

Salad Dressing : Fat is used as an emulsion in salad dressing. Along with fat, egg yolk, vinegar, salt, spices and starch are used in salad dressing. Mayonnaise, french salads and cooked salad dressing are the common methods used. Winterised oils are used in all these methods.

Mayonnaise : It is a semi-solid emulsion of vegetable oils, egg yolk or whole egg, vinegar and seasoning. Egg yolk functions as a seasoning agent.

French Dressing : Vinegar, oils (soyabeans, cornoil, sesame oil or sunflower oil) and salt and spices are added. It is mixed vigorously. Mustard powder is also added.

Lipids

Lipids are heterogeneous groups of compounds with same properties. Fat is the common household name given to lipids.

Like carbohydrates, fats are organic compounds composed of carbon, hydrogen and oxygen. But they differ from carbohydrates in that they contain much less oxygen and much greater proportions of carbon and hydrogen. Seventy-six per cent carbon, twelve per cent oxygen and twelve per cent hydrogen are present in fats. True fat has one molecule of organic ester of glycerol and three molecules of fatty acids. Fats vary in their structure due to the presence of different fatty acids with glycerol molecules. If the glycerol in a fatty acid is combined with three fatty acid units such fats are called triglycerides. Fats are insoluble in water and soluble in organic solvents like ether, benzene or chloroform.

Their cooking properties depend upon the type of fatty acids present in them.

Classification : Lipids are classified into (a) simple lipids, (b) compound lipids, (c) waxes, and (d) derived lipids.

Simple Lipids : These are esters of fatty acids with various alcohols. Fats are esters of fatty acids with glycerol Oils are liquid at 20°C while fats are solid at 20°C. Combination of fatty acids and sterol alcohol form cholesterol. Food fats contain neutral fats. Fat is stored in the body as adipose tissue. This is also neutral fat.

Compound Lipids : When fatty acids and glycerol combine with other organic compounds they are known as compound lipids. Examples of compound lipids are phospholipids and glycolipids.

(i) *Phospholipids* (phosphatides) are substituted fats in which phosphoric acid and a nitrogenous base take the place of one fatty acid radical. These are present in vegetable oils and in the fats of animal tissues in the blood and in egg yolk. Phospholipids help the absorption of fat in the body, and also help the transfer of fat from one place to another. They have prominent role in the functions of the nervous tissue. Lecithin and cephalin are the two important forms of phospholipids in our body.

Lecithin is composed of glycerol, two fatty acids in ester linkage, phosphoric acid and choline. This is the most common form of phospholipid.

Cephalin contains nitrogen compounds other than choline, phosphoric acid, glycerol and fatty acid esters. Phosphatidyl ethanolamine and phosphatidylserine are the common forms of cephalin.

(ii) *Glycolipids* are compounds which contain a carbohydrate molecule with fatty acids and a nitrogenous base. Cerebrosides are the best example of glycolipids. Cerebrosides occur largely in the brain. Galactolipids, glycolipids, and sulfolipids are examples of cerebrosides.

Waxes : These are esters of higher fatty acids and alcohol other than glycerol. Cholesterol esters, sterol esters, esters of fat-soluble vitamins are the common waxes.

Derived Lipids : Substances derived from one fat on hydrolysis or by enzyme activities are called derived fats. Examples of derived lipids are fatty acids like stearic acid, linoleic acid, butyric acid, alcohols like glycerol, cetanol and lanol. Sterol alcohols like cholesterol, ergosterol and sistosterol are also derived fats.

Sterols are nutritionally important. They are classified as animal sterols, plant sterols and mycosterols.

Cholesterol is the precursor of steriod hormones and bile salts. Hormones like progesterone, oestrogen and adrenal corticoids contain cholesterol. Sodium glycocholate and sodium taurocholate and the bile salts have cholesterol in them. Cholesterol esters produced by the liver regulate the cholesterol content in the blood. It is present in all animal and human tissues. The white matter of the brain has 1-14 per cent cholesterol while the grey matter has 60per cent on dry basis. Sebum secreted by sebaceous gland also has large amounts of cholesterol.

Properties of Fat : The properties of fat are based on its composition and structure. Melting point of fat is the most important physical quality. Fats which have a high proportion of saturated fatty acids are solid at room temperature and unsaturated fatty acids are liquid at room temperature; If the melting point of a fat is above 50°C or 122° F it is poorly utilised by the body.

The most important chemical properties of fats are hydrogenation, saponification, emulsification and rancidity. All these reactions are related to their chemical structure. The double bond linkages in the unsaturated fats make them readily available for all chemical reactions.

Iodine Value (number) : Iodine value gives the number of grammes of iodine which combine with 100 grammes of fat Iodine value depends on the number of bonds it makes at the site of double bonds. This gives the degree of unsaturation of fats.

Saturated and Unsaturated Fatty Acids : Saturated fatty acids have all the carbon atoms in the chain saturated with hydrogen atoms. When a double bond is present between two carbon atoms the fatty acid is termed as unsaturated fatty acid. The degree of unsaturation varies according to the presence of double bonds. If one or more double bonds are present in a fat it is called unsaturated fatty acid. If there are more double bonds in a fat it is called polyunsaturated fatty acid.

Palmitic, stearic and butyric acids are saturated fatty acids. Oleic, linoleic, linolenic and arachidonic acids are unsaturated oils. They have one, two, three and four double bonds, respectively. The unsaturated fatty acids are often described as essential fatty acids. Vegetable oils are rich in linoleic acid and fish oils and animal fats are rich in arachidonic acid.

Saturated and Unsaturated Fatty Acids

Common name (saturated fatty acids)	*Molecular formula*	*Unsaturated fatty acids*	*Molecular formula*	*Double bonds (Nos)*
N-Butyric	C4 H8 02	Palmitoleic	C16 H30 02	1
Caproic	C6 H12 02	Oleic	C18 H34 02	1
Caprylic	C8 H16 02	Elaidic	C18 H34 02	1
Capric	CIOH20 02	Vaccenic	C18H34 02	1
Laurie	C12H24 02	Linoleic	C18H32 02	2
Myristk	C14H28 02	Linolenic	C18H30 02	3
Palmitic	C16H32 02	Ylinolenic	C18H30 02	3
Stearic	C18H36 02	Eleostearic	C18H30 02	3
Arachidic	C20H40 02	Arachidonic	C20H32 02	4

Percentage Composition of Some Fatty Acids in Common Oil and Fats

Fat	*Oleic acid (C18.1)*	*Linoleic acid (C18.2)*	*Polyunsaturated fatty acids*
Butter fat	3.3	4	2
Coconut oil	8	1.6	-
Cottonseed oil	30	56	2
Egg fact	52	9	3
Groundnut oil	65	17	Trace
Olive oil	65	15	Trace

Contd...

Fat	*Oleic acid (C18.1)*	*Linoleic acid (C18.2)*	*Polyunsaturated fatty acids*
Pig fat	47	4	3
Rapeseed oil	28	15	59
Red palmoil	52	5	Trace
Safflower seed oil	31	51	-
Sesame oil	45	41	Trace
Soyabean oil	29	51	6
Sunflower seed oil	14	73	1
Sardine oil	17.3	2.5	30.1

Some of the chemical reactions of fats are related to its unsaturated status.

Hydrogenation : Unsaturated fats have a tendency to combine with hydrogen at the site of its double bonds. Hydrogenation is such a process. Vegetable fats are exposed to hydrogen at high temperature in the presence of an alkali. Nickel or cobalt is used as a catalyst. For example, groundnut, coconut, cottonseed and other oils are usually used for hydrogenation. On hydrogenation liquid fats turn into solids. Vanaspati is an example of hydrogenated fat from groundnut oil.

Rancidity : On keeping, fats develop an unpleasant odour and this is known as rancidity.

There are two types of rancidity: hydrolytic and oxidative rancidity. One of the common reactions which cause the spoilage of fat is hydrolytic rancidity.In hydrolytic rancidity lipase hydrolyses fats into free fatty acids. Fatty acids with low molecular weight give an unpleasant odour to fats which is easily spoiled.

In oxidative variety unsaturated fatty acids are oxidised. The addition of oxygen may give rise to peroxides which on decomposition yield aldehydes and ketones. They have a foul odour. These peroxides are oxidising agents. They destroy important vitamins present in fats. This oxidation is prevented by the presence of antioxidants in the fat. Vegetables oils have tocopherols or vitamin E as a natural antioxidant in them. There are α β γ (alpha, beta, gama) tocopherols. Among these alpha tocopherol is the most powerful antioxidant.

There are some other substances which can be used as antioxidants which give a prolonged shelf life to fats. Ascorbic acid or vitamin C, gum guaiac acid and propylgilate are some of them.

Saponification of Fats : The ester linkage of fatty acids with glycerol brings this process of saponification. Neutral fats are rich in ester linkage. When neutral fats are heated with sodium or potassium hydroxide the fat molecules may readily rupture at the ester linkage. It is thus hydrolysed into glycerol and sodium or potassium salts of fatty acids and this is known as soap. Thus, the formation of soap from the hydrolysis of fat with heat and alkali is termed as saponification. Saponification of fats, when performed under standard conditions, is used as an index to find out the molecular weight of fatty acids in a fat.

Emulsification : Fats and oils are insoluble in water because they are lighter than water.

When fat forms a homogeneous mixture with water it is called emulsification. This property is made use of in cookery.

Essential Fatty Acids : Nutritionally important fatty acids are known as essential fatty acids (EFA). Since the body cannot synthesise them the EFA must be supplied to the body through foods. Linoleic acids, linolenic acids and arachidonic acids are the essential fatty acids.

Functions of Fat in the Diet : The most important function of fat is due to its highest calorie density in foodstuffs. One gramme of fat supplies 9.3 kcal while 1 gramme of carbohydrate and protein supply 4.3 and 5.6 kcal respectively. Fats are the carriers of fat-soluble vitamins A, D, E and K. Fats also exert a thiamine-sparing action. Fats have a protein-sparing action also. Essential fatty acids help to maintain tissues, their normal structure and efficiency. Fats, especially phosphatides, form an essential part of nervous tissues. Adipose tissue, where fat is stored, serves as an insulation material in injury. Vital organs of the body are thus protected. In growth period it promotes growth. It also enables proper sexual maturity, especially in female experimental animals. Fat provides flavour and palatability to food.

Digestion and Metabolism of Fats : There is no digestive change for fat in the mouth. In the stomach a little fat is mechanically separated from other foodstuffs. Since carbohydrate and proteins are partially digested in the stomach fat is set free. In the stomach fat forms a coarse emulsion and is eventually released into the duodenum as small portions. Even though there is not much digestive action in the stomach for fat, digestion is delayed in the stomach by fat. The emulsified fat, known as chyme, takes three to four hours to transfer it from the stomach into the intestine. Pancreatic lipase is the most powerful enzyme in fat digestion. This lipase breaks the alpha position of triglycerides and makes them free fatty acids and 1,2, diglycerides. 1,2 diglycerides are hydrolysed to 2, diglycerides and fatty acids. They combine with bile salts and emulsify fats. They reduce the surface tension of fat and make it available for enzyme activity. Bile salts promote emulsification and help the fat into partial solution of fats and fatty acids. The water-soluble glycerol is quickly absorbed. Short chain fatty acids which are also soluble in water are trasported into the portal blood. The products of fat digestion which reach the intestinal wall are resynthesised into new 'triglycerides by the intestine. In the jejunum triglycerides are absorbed. These triglycerides pass into the lymphatics in the form of small scattering particles called chylomicrons. Thorasic duct carries them into the systemic circulation.

Fats which are quickly absorbed are readily available for the tissues. If the fats remain in the gastrointestinal tract for a longer period for utilisation they add satiety value to the meal.

Absorption of fat is based on its digestibility, melting point, the size of the fat molecules and its quantity. The age of a person is another factor which affects the absorption. Old people utilise fat slowly.

Lipids in Blood : In the post-absorptive state plasma contains about 500 mgs of total lipids/100 ml of blood. In this 180 mg/100 ml is cholesterol. Neutral fat occurs in blood plasma largely in the form of chylomicrons. In the liver triglycerides are resynthesised and they combine mainly with cholesterol and form beta lipoproteins and are passed into the circulation. Cholesterol esters are converted to bile acids in the liver, re-excreted into the small intestine and reabsorbed. The concentration of bile may affect cholesterol synthesis in the intestine. Only forty percent of cholesterol in the circulating blood is from food. The remaining portion is synthesised in the liver and intestine. Cholesterol synthesis in the body is controlled by various factors like calorie intake, cholesterol intake, fat intake and essential fatty acid content of the fat. Liver, skin, intestine, adrenal cortex, kidney and lungs are the organs which synthesise cholesterol. Acetate is the precursor of cholesterol and so all substances which yield acetate become precursor of cholesterol. Apart from falty acids, carbohydrates, ethyl alcohol, alanine, and certain amino acids yield acetate or pyruvic acid or glycogen or glucose which can in turn form acetate. Thus every foodstuff forms acetate and cholesterol while calorie restriction and essential fatty acid deficiency inhibit hepatic cholesterol synthesis.

The adult human body synthesises about 2,000 mgs of cholesterol daily. The dietary intake of cholesterol ranges from 500 mg to 1,200 mg depending upon the consumption of milk, butter, eggs, meat and fish. Apart from these foods coconut oil and hydrogenated fat increase the blood cholesterol level. Unsaturated fats or essential fatty acids reduce the blood cholesterol level. Safflower oil, sesame oil, sunflower oil and soyabean oil are the unsaturated fats.

Metabolism : The metabolism of fatty acids and glycerol are separate. Glycerol enters the glycolytic pathway through the formation of 2 glycerophosphate. Fatty acids are oxidised to acetyl-CoA and it is again oxidized to Co_2 and water through Kreb's cycle of carbohydrate metabolism. Energy is the end product in the metabolism of fat also. In the stepwise reactions two carbon atoms are added or subtracted using co-enzyme A. For this reaction ATP or Adenosine triphosphate is required. The enzyme which catalyzes the reaction is thiokinase. In the next step dehydrogenation of fatty acids takes place. Fatty acids are dehydrogenated into alpha beta unsaturated fatty acid. Alpha beta unsaturated fatty acids are converted or hydrated to hydroxyacyl-CoA. Hydroxy acid is oxidised to keto acid. The last step in the fatty acid oxidation is the thiolytic cleavage of Keteacyl-CoA ester to yield a fatty acid-CoA derivative. This is repeated and Acetyl-CoA molecules are released. Different fatty acids yield varying numbers of acetyl-ÇoA. For example, one molecule of steraic acid gives 9 molecules of acetyl-CoA and 1 molecule of palmitic acid gives 8 molecules of acetyl-CoA and 1 molecule of lauric acid gives 6 molecules of acetyl-CoA.

Acetyl-CoA thus formed is oxidised through Kreb's cycle and glycerol is converted into glyceraldehyde - 3 - phosphate and it enters into the glycolysis pathway.

Summary of Fat Metabolism

(1) Fatty acid - α β unsaturated fatty acid

+ H2O

↓

(Beta) hydroxy fatty acid

Keto acid

↓

Acetyl CoA

(2) Glycerol———————→Glycerophosphate

↓

Glyceraldehyde - 3- phosphate.

Ketone Bodies : Normally acetyl-CoA combines with oxaloacetic acid which is oxidised in the TCA cycle. When large amounts of fat are metabolised in the absence of carbohydrate more acetyl-CoA is formed. When there is not enough oxaloacetic acid from carbohydrate, two molecules of acetyl-CoA condense to form acetoacetyl-CoA. This is hydrolysed into acetic acid and later converted into P-hydroxybutric acid and acetone. These metabolic end products of fats are known as ketone bodies. When the concentration of ketone bodies in the blood increases the condition is known as ketosis. When it is excreted through urine it is called ketonuria. In normal condition acetoacetate is either oxidised directly to carbon dioxide and water or converted to two molecules of acetate which are later oxidised. In diabetes and during fasting ketone bodies are formed in excessive amounts. Apart from fats, amino acids like leuine also produce ketone bodies.

Lipo Proteins and Lipotropic Factors : Lipo proteins are lipids containing protein in blood and other parts of the body. The insoluble lipids when attached to proteins form soluble lipo proteins. Cholesterol, neutral fats and phospholipids bind them with protein to form lipo proteins.

Lipo protein in the liver along with choline enhances the oxidation of fatty acids. On diets which are high in fats with saturated fatty acids and which are low in protein or in choline there is an increased deposition of fat in liver. A high cholesterol diet also leads to the production of fatty liver. An increased choline intake will bring down the fat in the liver. This effect of choline on fat release is known as lipotropic action. Methionine likewise exerts a marked lipotropic action owing to its ability to promote the synthesis of choline by the transfer of methyl groups to suitable precursors. Inositol (B-complex vitamin), tryptophan and glutamic acid exert this lipotropic action. Essential fatty acid deficiency can also produce fatty liver, pantothenic acid or riboflavin deficiency may also lead to it

Requirements : The requirement of fats is about 50 to 60 gms for a normal person. The dietary fat is used for energy and it is also a source of essential fatty acids. At least fifty per cent of the fat must be from vegetable oils which are rich in essential fatty acids.

An average Indian adult derives about 20 per cent of energy from dietary fat.

Sources : Vegetable oils, like groundnut oil, coconut oil, mustard oil or gingelly oil, hydrogenated oils, butter and ghee and lard are the common sources of fats.

Deficiency of Fats : During infancy a deficiency of essential fatty acids results in perianal irritation and skin changes like dry skin and itching. In adults phrynoderma is easily cured by essential fatty acids with vitamins B2 and A.

12

Other Foodstuffs

Spices and Condiments

Spices and condiments are accessory foodstuffs mainly used for flavouring food. Some condiments are good to fair sources of carotene. Red chillies and coriander are examples of this. Green chillies are rich in vitamin C. Turmeric and tamarind are fair sources of iron. Garlic promotes intestinal synthesis of certain vitamins. Among the spices and condiments fenugreek seeds have the highest amount of protein, that is, 26.2 per cent. Cuminseeds, dry chillies and coriander seeds have a protein content of 18.7,15.9 and 14.1 per cent, respectively. Caraway has a protein content of 17.1 and fat content of 21.8 per cent in it. Coriander, cuminseeds and dry pepper have a fat content of 16.1,15.0 and 6.8 per cent, respectively. Cuminseed is also rich in calcium. About 1.080 gm of calcium is present in 1·00 gms. Asafoetida and coriander are also good sources of calcium. Asafoetida, cuminseeds, cloves, fenugreek seeds, turmeric, tamarind and omum are good sources of iron. Cardamon, green chillies, coriander, cuminseeds, caraway and green pepper are rich sources of carotene. Dry chillies, nutmeg, cuminseeds, coriander and fenugreek seeds are goqd sources of thiamioe. Even though these condiments and spices are rich in vitamins and minerals they are used only in minute quantities. Black pepper, coriander, cumin, ferugreek, mustard seeds, chillies, nutmeg, cloves, cinnamon and cardamom are used as spices in our country.

The popularity of spices is due to their stimulating effect on appetite next to its flavour. Certain spices and condiments have medicinal effect. Turmeric is believed to be anti-inflammatory. In anti-allergic drugs turmeric is used. Turmeric also reduces blood cholesterol level. Consumption of spices and condiments in large quantities produces irritation in the mucous membranes of the alimentary canal. Peptic ulcer is common among people who use condiments and spices liberally.

Herbs like curry leaves, coriander leaves, mint leaves and leaves of citrus fruits are used for preparing chutney powders and they are also used in many curries.

Gelatin Dishes : When the protein in connective tissues and bones—the collagen—is heated with water gelatin is obtained. To bring this change in collagen it has to be heated to 95°C. Some peptide bonds of collagen are hydrolysed to gel.

Commercially it is available in granulated, flake or sheet forms. It easily dissolves in hot water and on cooling it sets to a gel. Concentration of gelatin, cooling temperature, the presence of minerals like calcium and concentration of sugar affects the quality of the product. With a 2 to 3 per cent of gel concentration and at temperatures below 15°C a good gel is formed. Sugar in 6 to 10 per cent concentration enhances the speed of gel formation. If the gel is heated it melts and on cooling it sets again. Now vegetable gels are also available in the market.

Gelatin is used for decorating various types of desserts. For ice, souffles, baked ice-cream, or other creams, jelly trifles, caramel puddings, flans, milk puddings, sponge puddings, with stewed fruits, milk jellies, purees of fruits, junkets and with baked fruits gelatin can be used to make them more attractive. Gelatin acts as a stabiliser and reduces crystal formation of ice-creams.

Emulsifying and Stabilising Agents : Substances which, when added to food, gives a uniform dispersion of oils and fats in aqueous media are called emulsifiers or stabilising agents.

In food industry it is often liberally used and they are harmful. Agar, edible-gums, brominated vegetable oils, gelatin, albumin, lecithin, starches, hydrolysed proteins and stearyl lastaric are some of the examples.

Sweetening Agents

Sugar, sugarcane juice, jaggery, molasses, honey, saccharin and cyclamates are the common agents used as sweetening.

Sugar (sucrose) : Sugar is almost pure sucrose made up of glucose (dextrose) and fructose (laevulose). One gramme of sugar supplies 4 kcals and a teaspoonful (5 gms) provides 20 kcals. Brown sugar and white sugar are similar in nutritive value. Sweets made from sugar produce a sense of satiety and so they must be taken after a meal. If sweets are consumed at the beginning of a meal they reduce appetite. Sugar syrup is used as a preservative for fruits, jams and preserves.

Excessive consumption of sugar and sweets affects the teeth adversely.

Sugarcane Juice : The juice obtained by crushing sugarcane is a refreshing drink. Addition of sugar and lemon juice improves its taste and flavour. Sugarcane is good for hepatitis patients. Jaggery is produced by heating sugarcane juice. It has been reputed to be good for the heart. Molasses is a byproduct in the manufacture of sugar. Fermentation of molasses produce alcohol.

Honey : Honey is a golden-coloured syrup made by bees from the nectar of flowers. It is a mixture of glucose and fructose. The quality of honey depends on the flower blossom. Honey is a soothing drink for patients with pharyngitis and tracheitis. Honey contains 17.7 per cent water; 36.5 per cent dextrose, 403 per cent laevulose and 1.9 per cent sucrose.

Gur or Jaggery is obtained from sugarcane juice after processing. Palmyra palm, dale palm, or coconut palm is also used for it.

Saccharine and Cyclamates : Saccharine and cyclamates are used as sweetening agents for calorie-conscious people. Saccharine was discovered in 1879. It is 350 times sweeter than sugar and has a bitter taste in concentrated form.

Cyclamate was discovered in 1937. It is 30 times sweeter than sucrose. Cyclamates are harmful to the body if consumed in large quantities. Malignant bladder tumours were produced in experimental animals. Excessive use of soft drinks with cyclamates produce renal tubular acidosis.

Leavening Agents : Leavening agents are substances which produce carbon dioxide in the product. This carbon dioxide contributes a porous appearance to the product as in bread, cake or in biscuits. Baker's yeast and baking powders are usually used as leavening agents. Baker's yeast when added to the dough, ferments sugar and produces carbon dioxide and alcohol. This gives the characteristic structure to the product. Baking powders are leavening agents produced by mixing an acid-reacting material and sodium bicarbonate with or without starch. There are three types of baking powders—fast acting, slow acting and double acting based on their speed of action.

Flavour Components in Foods : Food industry is a fast growing field. Much progress has been made in understanding the chemical nature of flavours in foods. Many artificial flavouring materials are now available in the market. The compounds which give the flavour characteristic in one food is given below:

Food		*Main flavour compound*
Banana	—	Isopentyl acetate
Lemon	—	Citral
Grape-Concord	—	Methyl anthranitrate
Potato	—	2-Isopropyo-3 methosy pyrazne
Beetroot	—	Geosmin
Apple	—	Ethyl -2- methyl butyrate

13

The Proteins

In 1838, a Dutch scientist, Mulder, suggested the name 'Protein' is complex nitrogen-bearing substance in tissues. The word 'protein' is derived from the Greek work 'proieot' meaning to take the first place. Proteins play a significant role in all activities of living organisms. Due to the multiplicity of their functions the name given to them is very apt. In our skin, muscles, skeleton and in body fluids protein is present. They are essential during growth because it is the main constituent in out tissues. In a person's life the first eighteen months are very important as eighty pet cent of the brain development is over by this time. If protein is not sufficient during this time menial retardation takes place. Acute shortage of protein in early months of life brings intellectual dwarfism and it remains so for the rest of the life. That is why this nutrient is called 'protein'.

Protein Chemistry

Proteins are complex organic compounds which contain carbon, hydrogen, oxygen, nitrogen, sulphur and phosphorus. Certain proteins contain iron, iodine, copper and other inorganic elements. The presence of nitrogen distinguishes protein from carbohydrate and fat. Protein contains an average of 16 per cent nitrogen.

On hydrolysis large molecules of protein give out smaller units called amino acids. Amino acids contain a drboxyl group (CuoH) and an amino group (NH.) They are otherwise known as die acidic and basic group, respectively. There are twenty-three amino acids in proteins. Amino acids are grouped into five groups based on the number of amino groups and carbosyl group in them. They are amino-monocarboxylic acids, monoamino-dicatboylic acids, dianino-moftocarboxylic acids, sulphur containing amino acids and aromatic and hetezocyclic amino acids.

1. Monoamino-monocarboxylic acid.

   eg., Glycine, Alanine, Viliae, Leuciw,

2. Monoamino-dicarboxylic acids

   eg., As paartic acid, Glutamitic acid

3. Diamino-monocarboxylic acid

   eg., Arginine, Lysine

4. Sulphur containing amino acids

   eg., Cystine, Cysteine, Methionine

5. Aromatic and heterocyclic amino acids

   eg., Phenylalanine, Histidine, Tyrosine, Tryptophan, Proline, Hydroxyproline.

A protein contains a number of amino acids. Two amino acids are linked together through peptide linkage. In this, the basic group of one amino acids combines wiih the carboiyl or acidic group of another group of amino acids. Two amino acids are linked together to form a dipeptide and it combines with another amino acid. When a number of adds are linked together it is known as polypeptide.

Classification : Proteins are biochemically classified into three groups based on their chemical composition. They are simple proteins, conjugated proteins and derived proteins.

Simple Proteins : On hydrolysis simple proteins give nothing but only amino acids. Such proteins are called simple proteins. Examples of simple proteins ate albumins, globulins and glutelins, prolamins, fibrousprotein, histamins and protamins.

Albumins and globulins are found within all body cells and in the blood serum. Potatoes and peanut also have globulin in them. Globulin is in haemoglobin and myoglobin.

Glutelin is present in wheat glutens, zein in corn and gliadin in wheat has prolamine.

Collagen, keratin and elastin have fibrous proteins and histones are present in thymus and in globin. Protamins is present in salmine from salmonsperm.

Conjugated Proteins : On hydrolysis conjugated proteins give a protein fraction and a non-protein fraction. In our body conjugated proteins are present in various parts. Examples of conjugated proteins are haemoglobin, nucleoprotein, glycoprotein and lecithoprotein. Nucleoproteins are present in nucleus, glycoproteins in mucin and lecithoprotein in lecithin.

Derived Proteins : Derived proteins are derived from a mother protein either through hydrolysis or digestion. On digestion derived proteins are formed. Examples of derived proteins are metaproteins, coagulated proteins, peptides and peptones.

Metaproteins are products resulting from the action of acids or alkalies.

Coagulated Proteins : By the action of heat on protein or by the action of alcohol on proteins it coagulates. Cooked egg is an example.

Peptides, Proteoses and Peptones : Peptide linkage forms peptides from amino acids. Two amino acids linkage forms dipeptide and three amino acid tripeptides. Digested proteins form peptides.

Proteoses are derived from proteins by the action of digestive enzyme pepsin and precipitated by saturating their solution with ammonium sulphate.

Proteins are again classified based on their nutritive value. From the nutritional standpoint amino acids are of two types, essential and non-essential. Essential amino acids are termed essential because the body requires them through our daily food. Non-essential amino acids are also essential to the body but the body can manufacture them from the available chemicals. Thus, the nutritional quality of a protein is based on the presence of essential amino acids in them. There are eight essential amino acids in the case of adults and ten essential amino acids in the case of children and others are non-essential amino acids.

The essential amino acids are isoleucine, leucine, lysine, methionine, phenylalanine, threonine, tryptophan and valine. For infants and children arginine and histidine are also essential.

Distance : The non-essential acids are alanine, aspartic acid, cystine, cysteine, glutamic acid, glycere, hydroxyproline, proline, serine, tyrosine, citrulline and norleucine.

Proteins *at* classified into complete proteins, partially incomplete proteins and incomplete proteins based on the presence of essential amino acids in them. The presence of amino acids helps a protein to perform all the functions of proteins in our body. A complete protein food contains all essential amino acids in proper proportion. It helps the protein to promote growth, maintenance and repair: Examples of complete protein foods are milk, egg and fish.

Partially incomplete proteins lack some essential amino acids and so they will help to maintain our body but growth is not promoted. Vegetable proteins are examples of this group.

Incomplete proteins neither help maintenance nor growth. Gelatin is an example of this type of protein.

There are certain methods tc evaluate the quality of a protein. Biological value of protein and growth gain or protein efficiency ratio (PER) are the important methods used for this.

Biological Value : Biological value of protein foods depends to a great extent on their amino acid composition and they are grouped as protein of low or high biological value. The biological value of protein also depends on the digestibility of a protein food. Cooking enables the digestibility whereas overheating and frying reduces the biological value of a protein because it destroys the essential amino acid, lysine. Egg, milk, fish, poultry and meat are of high biological value whereas proteins from cereals like wheat, rice, pulses and nuts are of low biological value. Vegetable proteins are low in essential amino acid pattern, especially of lysine, tryptophan, methionine and threonine. When a mixture of proteins from different sources is consumed mutual supplementation takes place. For example, rice or wheat protein is low in lysine whereas pulse is rich in lysine and low in methionine. Rice is rich in methionine and a mutual supplementation results when rice and pulse are consumed together.

Protein Efficiency Ration (Growth Method): The ability of a protein to bring increase in weight or growth gain is considered on the efficiency of a protein or its nutritive quality. Nutritive value is expressed as growth per gram of protein consumed.

$$PER = \frac{\text{Gain in body weight (gm)}}{\text{Proteinintake (gm)}}$$

The Functions

Building Block : The most important function of protein is to supply amino acids to cells for the continuous replacement of cells throughout life. It is the most abundant organic compound in our body. It is the building block of our body. From conception to growth at various levels, foetus, infant, child, and during pregnancy and lactation period protein is very essential.

In our body, tissues are not in a static position. Degradation and resynthesis of protein and other nitrogenous compounds always take place in our cells. Tissue proteins are broken down and new substances are continually synthesised in our body. Muscles and other tissues, bones, cartilage and trabecula contain fairly high percentage of proteins. Our hair, nails and skin also contain protein. All body fluids except bile and urine have protein in them.

Regulatory Functions : Different proteins perform highly specialised regulatory function in the body. Haemoglobin, which is the chief constituent of the red blood cells, carries oxygen to the tissues. Oxyhemoglobin and their alkaline salts permit oxygen to enter into the tissues and receive carbonic acid from the cells. This process helps the removal of 92-97 per cent of all carbon dioxide from the tissues. Plasma proteins, especially albumin and globulin, play an important role in regulating osmotic pressure and water balance within the body. When plasma proteins are decreased the water balance is upset and accumulation of fluids in the body takes place. Proteins enable the blood to maintain their slight alkalinity. Proteins function as buffers, thus helping various sites of chemical reaction to maintain its pH. Nucleo proteins contain the blueprint for the synthesis, of all body proteins.

Formation of Enzymes, Hormones and other Secretions : Proteins supply raw materials for the body to synthesise enzymes like trypsin and pepsin. Hormones like insulin and thyroxine are protein in nature. Digestive juices contain protein in them. Antibodies which give resistance power to the body are protein in nature. They are known as immune proteins.

Sources of Energy : Protein is generally considered as the building material of our body. But when the diet contains insufficient carbohydrate and fat for fuel, proteins are used as a fuel by the body. Each gramme of protein yields four calories. But it is not a wise contribution because the nitrogen excretion increases along with the cost of the food.

Proteins form Part of Vital Compounds in Body : Nitrogenous compounds are present in certain substances of immunological and antigenic reactions. Examples of these are globulin of blood serum and chromatin in nucleus. Methionine, an amino acid, provides methyl groups for the formation of creatinine and choline. Tryptophan, an amino acid, is converted to niacin, a B-complex vitamin, in emergencies, Glutathione, a tripeptide of crystine, is present in small amounts in all our active tissues and this helps oxidation- reduction reactions. Histidine, an amino acid, is decarboxylised to histamine and it is a stimulator of gastric glands to secrete gastric juice. It is also used as a dialator of capillaries. Threonine, another amino acid, acts as a lipotropic agent which prevents deposition of excess fat in the liver. Proteins also help the transport of drugs by binding them into protein molecules. Contractile proteins (myocin, actin) regulates muscle contration.

Acts as a Binding Factor : Retional binding protein, lipo proteins, transferin and serum proteins are very essential to our body for transporting many chemical substances.

The Digestion

There is no protein-splitting enzyme in saliva and so in salivary digestion protein is not affected. Hydrolysis of protein begins in the stomach. Pepsin, the enzyme secreted by gastric glands in the stomach, breaks down proteins into proteoses and peptones. Milk proteins are first converted to casein by a special enzyme called rennin. Casein combines with calcium to form calcium casemate. Pepsin converts this into peptones. All the peptide linkages are not broken by gastric digestion. Stronger enzymes of pancreatic and intestinal juices complete the digestion of protein into peptones. Pancreatic juice contains trypsin and chymotrypsins. They hydrolyse peptones and proteoses into polypeptides. The final breakdown of all protein fractions to amino acids is brought about by erepsin secreted by intestinal mucosa.

Amino acids are absorbed by the small intestine and they are carried to the tissues or to the liver. A small amount, that is, about eleven per cent of the total protein is absorbed by the stomach, sixty per cent by the intestine and twenty-eight per cent by the colon. The absorbed amino acids reach the tissues where then metabolism takes place.

The Metabolism

In the tissues amino acids undergo breakdown and synthesis. They are also transaminised, doaminised or decarboxylated. The anabolic activities consist of a formation of new cells or the repair and maintenance of existing ones and secretion of various substances. If there is excess of amino acids after these anabolic activities, they will be either used for energy purpose or are converted to fat

Deamination : One of the catabolic action is deamination. In deamination, the amino group is removed from an amino acid. This reduced amino group is used for the synthesis of new substances like simple amino acids or eliminated as urea. The carbon, hydrogen and oxygen may be used for carbohydrate or fat metabolism.

Transamination : The deaminated amino group may be utilised for transamination process. Transamination is a process whereby a new amino acid is formed using the deaminated amino group and another precursor. If all the amino group is not used for transamination, nitrogen freed from the amino group forms urea.

Formation of Urea : Amino acid is oxidised to ketoacids and ammonia. The ammonia, if it accumulates in the tissues, is toxic. So in the liver ammonia is converted into urea and excreted as urine. Therefore, the end product of protein metabolism is urea.

Ammonia is converted into carbamyl phosphate. This reacts with ornithine to form citrulline. Citrulline combines with aspartic acid and forms arginosuccinic acid which is converted to arginine. Arginine is hydrolysed into ornithine and urea. This ornithine again combines with carbamyl phosphate and this cycle is repeated.

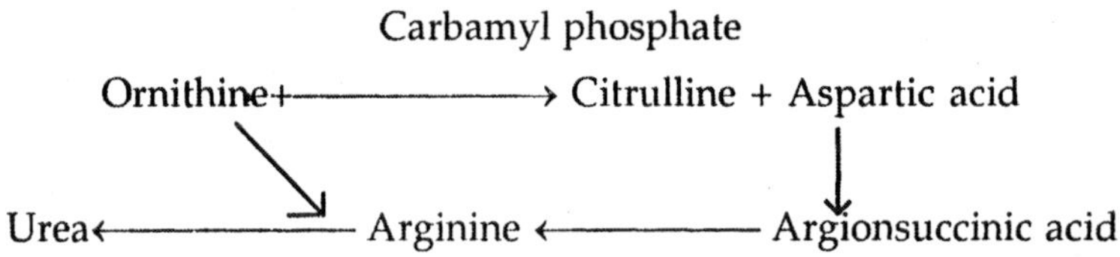

Fig. : Urea Cycle

Another catabolic change which takes place in an amino acid is decarboxylation. In this carbon dioxide splits off from carboxyl group and amines.

Nitrogen Balance : Tissue proteins are broken down at various times. When tissue proteins are utilised by the body for different functions nitrogen is excreted in urine and faeces. When protein food is taken new tissue proteins are formed. Nitrogen balance is the difference between the amount of nitrogen taken in through the diet and the sum of nitrogen and nitrogenous substances excreted from the body. When the nitrogen intake is more and the nitrogen loss is less the body is said to be in a positive nitrogen balance. When the nitrogen intake is less and nitrogen loss is more the body is said to be in a negative nitrogen balance. When the nitrogen intake equals the nitrogen lost from the body, the body is said to be in nitrogen equilibrium. During deficiency state the body shows a negative nitrogen balance.

Factors Affecting Protein Utilisation : If sufficient calories are present in the diet, protein will be effectively used for its function. Certain minerals and vitamins are essential for the proper utilisation of protein. Potassium, phosphorus, vitamin A, riboflavin, niacin and vitamin B6 are the factors which enable the utilisation of proteins. Heat treatment or processing improves the utilisation.

Heat destroys inhibitors of certain enzymes and that is how it helps digestion. But if direct heat is applied to meat it retards protein utilisation. Examples of this action is roasting meat in direct heat or frying.

Requirement of Proteins : Determination of protein requirement is based upon the nitrogen expenditure pattern and the estimation of nitrogen intake. Assessing nitrogen intake or protein quality of a mixed or complex diet is difficult. Efficiency of a protein source is based on its amino acid patterns and milk and egg proteins have 100 per cent efficiency. This is used to compare the quality of other proteins.

Covering a 30 per cent addition to the variability of different proteins 1 gm of protein/kg body weight is recommended by FAO/ WHO expert group. Protein allowance during infancy and childhood is more as there is rapid growth. Studies have shown that well nourished pre-school children can attain heights and weights close to Western children. During pregnancy and lactation there is an increase in protein content. Protein requirement during various phases of life is given in table .

Protein Allowance

Man		1 gm/kg/day...55 gm (55 kg)		
Woman		1 gm/kg/day...45 gm (45 kg)		
Pregnancy		2nd, 3rd trimester 14 gm/day		
Lactation		0-6 months + 25 gm/day		
		0-3 months - 2.3 gm/kg (for milk protein)		
		3-6 months -1.8 gm/kg		
		6-9 months -1.8 gm/kg		
		0-12 months -1.5 gm/kg		
Children	1-3 years -		1.83 gm/kg day	- 22 gm (12.03 kg)
	4-6 years -		1.56 gm/kg	29 gm (18.87 kg)
	7-9 years -		1.35 gm/kg	36 gm (26.37 kg)
Boys	10-12 years -		1.24 gm/kg	43 gm (34.30 kg)
	13-15 years -		1.10 gm/kg	52 gm (47.03 kg)
	16-18 years -		0.94 gm/kg	53 gm (56.50 kg)

Girls	10-12 years -	1.17 gm/kg	43 gm (36.47 kg)
	13-15 years -	0.95 gm/kg	43 gm (45.53 kg)
	16-18 years -	0.88 gm/kg	44 gm (50.00 kg)

Protein Deficiency : Deficiency of protein-predominant foods during growth period is known as kwashiorkor and protein-calorie malnutrition is known as 'marasmus'. Protein calorie malnutrition is one of the largest nutritional problems in India.

Kwashiorkor : Kwashiorkor is a term used by Ga tribe in Ghana and the meaning of this word is 'the sickness the old child gets when the next baby is born'. Dr. Cicely Williams in 1933 reported the occurrence of kwashiorkor (red baby) in poorer sections of population in Africa. Internal and external changes occur in the structure and functional efficiency of various organs in the body. Starting from hair and skin to tissues, blood, water content in the cells, blood cells, organs like liver, alimentary canal and nervous tissues are affected by protein deficiency. The child will maintain an apathetic, anaemic, anorexic, ocdematous and diarrhoeic condition.

The early symptoms of kwashiorkor include loss of weight, fatigue, reduced resistance to infection, delay in recovery from illness and retarded growth. As the protein deficiency prolongs, changes in the body take place. Discolouration of the skin, oedema with ascites, peeling of skin, cracks in the skin, angular stomatitis, cheilosis and atrophy of the tongue and- anaemia occur. Degeneration of the epithelial tissues and reduced secretions of digestive juices result in diarrhoea. Absorption of nutrients is delayed and result in general weakness and mental retardation. Liver may be palpable and soft. Fatty liver, especially in the periphery, is seen. Functions of liver are affected adversely, especially the synthesis of lipoprotein is reduced, hence fat accumulates in the liver. Deficiency of all vitamins occurs. Deficiency symptoms of B-complex vitamins lead to stomatitis and ulcers in the mouth and stomach, vitamin A deficiency leads tonight blindness and xerophthalmia, severe protein deficiency results in permanent retardation of mental development, psychomotor development and behaviour.

Marasmus : Marasmus is the result of starvation in children. Marasmus occurs when the diet is low in both calories and in protein. The term Marasmus is derived from the Greek word meaning 'to waste. Tissues are wasted and unlike kwashiorkor the body looks shrunken. The child has an emaciated look. Irritation, apathy, loss of appetite, dehydration, wasting of muscles and weight loss and incessant diarrhoea are the other symptoms of marasmus. All these result in retardation of growth and severe wasting of subcutaneous fat Deficiency of vitamins A and B complex is common in these patients.

Marasmic Kwashiorkor : This occurs when a combination of marasmus and kwashiorkor of varying degrees is present in a child. The child shows signs of both kwashiorkor and marasmus. Gastrointestinal and respiratory infection may accompany the other symptoms.

The Sources

The best sources of protein are animal sources like meat, fish, poultry, eggs, milk and milk products. They are otherwise known as first class proteins because they are rich in essential amino acids.

Cereal grains and legumes are the important sources of plant proteins. They are also known as second class proteins. Pulses are commonly knows as "poorman's meat" because they are comparatively cheap and fair in the nutritive content. The biological value of vegetable proteins can be improved when cereals and pulses are taken together.

Mutual Supplementation of Amino Acids : Amino acid content of the diet can be improved by supplementing one protein source with the other. In Indian diet this method has a significant role to play. For example, cereals and pulses. Lysine, the limiting amino acid in cereals, is supplemented by pulses, which is rich in it Methionine, the limiting amino acid in pulse, is supplemented by cereals. Soyabeans is also rich in lysine but deficient in methionine. A small quantity of animal foods can also do the same to vegetable proteins.

14

The Carbohydrates

Carbohydrates are compounds which contain carbon, hydrogen and oxygen. Oxygen and hydrogen are present in carbohydrates in the same proportion as in water. They are the main sources of energy for our body and the only source of energy for our nervous tissues.

Carbohydrates are mainly distributed among plant foods with few exceptions, like glycogen, lactose and ribose which are present in muscles or in liver, human milk and animal cells, respectively.

In various parts of the plants like leaves, stems, fruits, seeds and roots, carbohydrate is stored as starches and sugars. And the plant sources supply about 45 to 80 per cent of the energy requirement of the people throughout the world. Plants through the process of photosynthesis manufacture carbohydrate and after its use store them as polysaccharides.

The Chemistry

Carbohydrates are chemically known as saccharides, as carbon, hydrogen and oxygen form saccharide groups. Carbohydrates are classified, based on the number of saccharide groups in them:

a. Monosaccharides

b. Disaccharides, and

c. Polysaccharides.

Monosaccharides : On hydrolysis or by digestion, monosaccharides cannot be simplified as they are the simple form of carbohydrates. This is the absorbable form of carbohydrate in our body. Based on the number of carbon atoms present in them, monosaccharides are grouped into triose (3 carbons), tetrose (4 carbons), pentose (5 carbons), and hexose (6 carbons).

Monosaccharides

Biose, ($C_2H_4O_2$) (e.g. Glycolical-dehyde)	Triose ($C_3H_6O_3$) (e.g. Glyceral dehyde)	Tetrose ($C_4H_8O_4$) (e.g. Erythrose)	Pentose ($C_5H_{10}O_5$) (e.g. Xylose, ribose)	Hexose ($C_6H_{12}O_6$) (e.g. Glucose, fructose)

Biosetriose and tetrose are not nutritionally significant

Pentoses like ribose, xylose and arabinose are of little dietary significance, xylose and arabinose are widely distributed in many root vegetables and in many vegetables. Ribose is part of riboflavin and DNA and RNA Body can synthesise it and so it is not a dietary essential.

Xylose and Arabinose are not present in free state in nature. Both of them are present in gums of various origins like woodgum, cherry gum or gum arabic.

In human nutrition only hexoses are of importance. Commonly found hexoses are aldose and ketose. The three monosaccharides of hexose group are glucose or dextrose, fructose or levulose and galactose and mannose.

Glucose: Also known as grape sugar, glucose is an aldose sugar. It is a white crystalline substance easily soluble in water with a sweet taste. Glucose is readily absorbed from die stomach. Glucose is used to detect the disorders in carbohydrate metabolism because it is in the form of carbohydrate circulating in blood. A normal healthy person has 80-100 mgs of glucose/100 ml of blood. Glucose is present in fruits and honey.

Fructose known as fruit sugar is a ketosugar. It is sweeter than glucose. Fructose enters the cells without the aid of insulin and so it is recommended for diabetics. Honey is the richest source of fructose (30-40%). Fructose on reduction yields a mixture of sorbitol and mannitol.

Galactose is not present in nature. It is present only in milk sugar lactose; on hydrolysis lactose yields galactose and glucose. Galactose is present in cerebrosides in brain and nervous tissues.

Disaccharides : They are complex sugars with two saccharide groups and are formed by the condensation of two monosaccharides. On hydrolysis or toy digestion they are split into monosaccharides. Disaccharides have a structure of $C_{12}H_{22}O_{U}$. They are water soluble, diffusible and crystallisable. They vary in their sweetness.

Commonly found disaccharides are sucrose, maltose and lactose.

Sucrose, otherwise known as invert sugar, is the commonest form of sugar in the diet. Either by hydrolysis or by enzyme action sucrose is simplified into glucose and fructose. It is present in sugarcane, beetroot and in many fruits and vegetables, and honey. Sucrose is formed by the condensation of one molecule of glucose and one molecule of fructose.

Maltose is a disaccharide which contains two molecules of glucose. It is otherwise known as malt sugar. Starch is converted to maltose before breaking down into glucose. It is present in sprouted grains. Cereal grains are rich in maltose.

Lactose is also known as milk sugar and contains glucose and galactose. Cow's milk contains about five per cent of lactose and human milk has about 6.8 per cent It is easily digestible but it is not as sweet as cane sugar. Lactose is synthesised in the mammary gland. It is less in sweetness compared to other sugars.

Trisaccharides : They are with three saccharide groups but are not of any nutritional significance. The fibrous parts of certain beans and root vegetables which are not digestible contain trisaccharides, e.g., Raffinose is present in beetroots and cottonseed meal.

Tetrasaccharides **:** They contain four monosaccharides and like trisaccharides they are of no nutritional significance, e.g., stachyose and scorodose.

Polysaccharides **:** They are large complex molecules of monosaccharides linked together. Several hundreds of glucose units are linked together in polysaccharides. They are insoluble in water. Starch, dextrins, glycogen, pectin and cellulose are the common forms of polysaccharides.

Starch: Plants store carbohydrates in the form of starch and it is the main source of nourishment for human race. Cereal grains, seeds, roots like potato, tapioca, yam and plantain contain considerable amount of starch. On cooking, starch absorbs water and it swells and ruptures. This thickening quality of starch is used in cookery to produce a variety of dishes. Different sources of starch behave differently. Maize starch and cornflour are better 'thickening agents' than rice or wheat starch. All starches are broken down into glucose in the digestive system.

The characteristics of starch molecule depend upon the way in which 2,000 or more glucose units are linked. The number of glucose molecules in starch vary from 2,000 to 15,000. The glucose chain in starch are of two types. They are amylose and amylopectin.

(i) *Amylose:* Amylose is a long chain of glucose and starch has about 10-20 per cent of amylose molecule in it One molecule of amylose has 500-5,000 glucose molecules. Amylose is easily soluble in water and it is mainly responsible for the stiffening of cooked rice on keeping.

Amylose chain - G-G-G-G-G-G-G-G-G

(ii) *Amylopection:* Starch contains about 80-90 per cent of amylopectin in them. Amylopectin molecule has more glucose units. Usually one amylopectin molecule has 50,000-5,00,000 glucose molecule. Glucose molecules are arranged in branches.

```
              G
              G
              |
        G-G-G-G-G-G G
              G
                -G-G-G-G
                -G-G-G-G
       -G-G-G-G-G-G G
     G
     G
     G
     G
     G
```

Amylopectin molecule

Small branches of amylopectin has 20-30 glucose units and it has a colloidal property and this is responsible for thickening of starch water on heating.

Dextrins: When starch is partially broken into fragments either by digestion or by acids they are called dextrins. They are like starch itself. Dextrin is broken down into maltose. Starch is broken down into-amylodoxtrin, erythrodextrin, achrodextrin and maltose.

Pectin is a polysaccharide with no nutritional significance. But it is useful in the preparation of jam and jelly and thus it contributes to the palatability of foods.

Glycogen can be described as the animal starch as it is in this form animals store carbohydrate in the body. Glycogen is stored in the liver and muscles. This is the form of immediate energy for the body. About 350 gms carbohydrate is stored as glycogen in the body.

Cellulose is an insoluble; indigestible polysaccharide. More than 3,000 glucose units are there in cellulose but it is not of human utilisation. Cattle can digest cellulose. Even though it is not of much food value it provides bulk to the diet and thus helps movements in the large intestine. It prevents constipation and to an extent cancer of the bowel. It reduces the cholesterol level in blood as well as body weight. A high fibre diet can help in the treatment of obesity as it delays digestion and contributes satiety to obese people.

Sugar Alcohols : There are three sugar alcohols available commercially. They are sorbitol, mannitol and dulcitol. Sorbitol is used by diabetic patients. D-sorbitol or D-glucitol is an alcohol made from glucose by hydrogenation. By this process the aldehyde (CHO) group of glucose is reduced to alcoholic groups (CH,OH). Mannitol is produced from mannose, a monosaccharide which does not occur free in nature.

The Functions

Energy Supply : The most important function of carbohydrate is to supply energy or fuel for the body. Energy from carbohydrate is utilised by the body for immediate use. Nervous tissues and erythrocytes use glucose for their energy whereas other tissues can make use of fatty acids for energy.

Special Functions of Carbohydrates in the Liver : Carbohydrates have protective and detoxifying action in liver, liver is protected against bacterial toxins by producing glycuronic acid or acetify groups from carbohydrates. They are excreted through kidneys. Since the defences of the liver against toxic agents are of great importance to the body, this function of the liver is very valuable. For this purpose glycogen is made use of and thus carbohydrates play a significant role in removing poisonous substances from the liver. Carbohydrates, especially lactose, help the growth of desirable bacteria in the small intestine and help the synthesis of some B-complex vitamins in the intestinal tract. Lactose also helps in the absorption of calcium.

Carbohydrates and the Heart : Carbohydrate is used as the most efficient fuel for muscular exercise. Glycogen stored in the heart muscle is used for this purpose, especially in an emergency.

Carbohydrates and the Central Nervous System : Continuous supply of carbohydrate is essential for nervous tissues to function normally. Glucose as such has a specific influence in maintaining the functional integrity of nervous tissues.

Carbohydrates and Protein-sparing Effect : If carbohydrate supply is enough in the diet, protein is spared for important functions of it. This effect exerted by carbohydrate is called protein-sparing effect Insufficient intake of carbohydrate forces amino acids to be de-aminated and converted to fatty acids for energy purpose.

Carbohydrates and Non-essential Amino Acids : Carbon skeleton of carbohydrate is used by the body for the synthesis of non-essential amino acids.

Carbohydrates and Fats : Carbohydrate is essential for the oxidation of fat. In extreme restriction of carbohydrate, fat may be metabolised rapidly.

Carbohydrates in Diet : Carbohydrate provides flavour and variety to the diet Cellulose and other indigestible polysaccharides add bulk to the diet Carbohydrate also retains water content of colon which helps intestinal motility.

The Sources

The best sources of carbohydrates are cereals like rice, wheat, millets like ragi, maize, roots and tubers like potato, tapioca, sweet potato, yam and chalcosis, pulses, sugar and jaggery; honey, fruit and vegetables are the other sources of carbohydrates. Banana, apple, plantain and dried fruit are also good sources of carbohydrates.

In an average Indian diet carbohydrates are the cheap sources of calories. One gram of carbohydrate gives 4 kcals and an average Indian diet contains about 300 gms of carbohydrate.

Absorption and Digestion

In the mouth salivary digestion takes place. Carbohydrates are broken down by mastication and the saliva prepares the broken pieces of food for swallowing. Ptyalin, the enzyme present in saliva, acts on starch splitting it into dextrin and maltose. This action continues in the stomach till the acidity of the gastric glands interferes. There is not any digestive action on carbohydrates in the stomach as there is no enzyme in the stomach to act on carbohydrates.

The main digestive action on carbohydrate takes place in the small intestine. Amylase of the pancreatic juice hydrolyses the remaining starch into maltose. The intestinal/ mucosa cells contain sucrose or invertose, maltose and lactose. These enzymes convert sucrose into glucose and fructose, maltose into glucose and glucose and lactose into glucose and galactode, respectively.

The end products of carbohydrate digestion—glucose, fructose and galactose—are absorbed in the intestine. Bulk of these sugars are absorbed in the duodenum and jejunum and small amounts are absorbed in the ileum. Glucose and galactose are readily absorbed while fructose is absorbed slowly. These monosaccharides are absorbed by the capillaries in the walls of the intestines and they are carried by the bloodstream to the liver and the tissues.

The absorption of carbohydrate from the intestine is controlled by certain factors like condition of the intestinal tract and muscle tone, endocrine glands like anterior pituitary, thyroid and adrenal cortex and their functions and vitamin B-complex content in the diet.

Carbohydrate Metabolism : Through portal circulation the absorbed monosaccharides are carried to the liver. There galactose and fructose are converted into glucose. The liver cells release some glucose into the bloodstream and blood carries it to the tissues. In the tissues this glucose is metabolised to release energy. Excess of glucose is polymerised in the liver into glycogen and stored in the liver and muscles. This is again converted into glucose when energy is required. This process helps the bloodstream to maintain the normal blood sugar level.

Glycogen is formed from glucose, fructose and galactose and it is shown below:

Glycogen Galactose

↑ ← ↓

Glucose-1 – Phosphate→Galactose-l-Phosphate

Glucose→Glucose-6-Phosphate

Fructose—Fructose-6-Phosphate.

Formation of Glucose from other Monosaccharides

The carbohydrate metabolism in the body involves a series of chemical reactions. Glycogen formation, breakdown of glycogen and glucose to supply energy, formation of glucose from amino acid and glycerols of fat and formation of fat from carbohydrate are the main changes related to carbohydrate metabolism.

Breakdown of glycogen and glucose to yield energy involves a series of gradual reactions. Some of these chemical reactions take place in the absence of oxygen and it is known as the analrobic phase of carbohydrate metabolism. In the aerobic phase of metabolism oxygen is used. In total 38 ATP are yielded by each molecule of glucose: 8 ATP through glucolysis, 6 from pyruvic acid to Acetyl-CoA, 24 from TCA cycle.

Anaerobic Phase of Carbohydrate Metabolism (Glycolysis)

In glycolysis glycogen is degraded to glucose and it is converted to lactic acid.

The energy release in the process of glycolysis is calculated from the breakdown of ATP (Adenosine Triphosphate) and regeneration of ATP. Two molecules of ATP are used and four molecules of ATP are formed. Glycolysis takes place in muscular tissues.

Glycolysis is also known as Embden-Meyerh of-Parnas scheme of carbohydrate metabolism. The basic steps involved in glycolysis are given in Fig.

Aerobic phase of carbohydrate metabolism involves oxidation of lactic acid into pyruvic acid and pyruvic acid to Acetyl-CoA. Acetyl-CoA undergoes a series of a chemical reactions whereby carbon dioxide, water and CoA are formed.

Glycogen

Glucose-1-Phosphate

Glucose Glucose-6—Phosphate

Fructose—-6—Phosphate

Fructose-l-6-diphosphate

Glyceraldehyde-3-phosphate ——————>Dihydroxy acetone phosphate

1,3 Diphosphoglyceric acid

3 Phosphoglyceric acid

2 Phosphoglyceric acid

Pyruvic acid

Kreb cycle —— Lactic acid

The conversion of Acetyl-CoA is otherwise known as Kreb's Tricarboxylic acid or T.CA. cycle.

Oxalacetic acid reacts with Acetyl-CoA and this is repeated.

Most of the absorbed glucose is oxidized releasing energy. During this process carbon dioxide and water are also formed. Some portion of glucose is converted to fatty acids and some amount combines with free amino group to form amino acids.

B Vitamins like thiamine, niacin, riboflavin, pantothenic acid and B6 and minerals like magnesium and phosphorus are very essential for the proper metabolism of carbohydrates.

Certain hormones like epinephrine, thyroxine and glupxine and glucagon and islands of Langerhans are essential to maintain normal blood sugar level in a person. A healthy person's blood shows the fasting sugar level as 80-100 mgV 100 ml. of blood. After a meal it rises to 130-150 mgV 100 ml. of blood. If the blood sugar level is above this the person is termed as a diabetic patient

If the carbohydrate intake in the diet is inadequate, it leads to malnutrition and other metabolic disorders. Tissue proteins and fats will be utilised for energy purpose. If excess carbohydrate is taken, it will contribute towards obesity and diabetes.

Regulation of Blood Sugar : Blood circulates glucose continuously to each and every cell of the body as a source of energy and for the synthesis of a number of substances. From the blood if glucose is taken, the liver immediately replaces the blood glucose.

In the fasting state blood sugar level is 60-85 mgm per 100 ml of blood. Soon after a meal it rises to 140-150 mgm per 100 ml.

If the body is in a good carbohydrate metabolic pathway this concentration will fall down to a normal level of 60-85 mgm. The liver very efficiently maintains the normal blood sugar level. If the blood sugar concentration is high this condition is known as Hyperglycaemia. Hypoglycaemia occurs when the blood sugar level is below normal level. This occurs in certain abnormalities of liver function or when insulin is produced in excessive amounts by the pancreas.

The liver is the only organ in our body which can either supply glucose to the circulation or to remove the excess sugar from blood. Glucose is made available to blood from absorption, from intestine glucogenesis, i.e., glycogen is produced from glucose and reconverts it; glucoseogenis, i.e., de-aminated amino acids are used for the synthesis of glucose, glycogenolysis is reconversion of stored glycogen to glucose and lipogenesis, i.e., excess glucose is converted to fatty acids and the use of glycerol to form glucose.

15

The Vitamins (At a Glance)

Vitamins are the discovery of 20th century scientists. Carbohydrates, proteins and fats were considered as the nutrients essential for health. In the history of nutrition, the findings of a Polish chemist, Casimir Funk; in 1911 was a turning point. He found out for the first time that a disease syndrome could be produced in an experimental animal if inadequate supply of certain factors in food other than carbohydrates, proteins and fats takes place. He gave the name for the water soluble substance as "vitalamine" which was later known as vitamin. Scientists from various parts of the world were engaged simultaneously in finding out these specific factors which promote growth and maintenance of various organs in our body. Takaki, a physician in the Japanese Navy, first, demonstrated that fatal diseases like beriberi could be treated with rice bran, vegetables, fish and meat. Thus the study of vitamins brought to light about 17 different vitamins. Each one of them has its own history, chemistry, structure, function, sources, requirements and disorder symptoms.

Vitamins are defined as organic compounds which are necessary for good health and vitality. Vitamins are required in minute quantities and their deficiency results in structural and functional disorders of various organs in the body. Thus vitamins are essential for life.

Classification of Vitamins : Vitamins are classified broadly into two groups: (a) fat-soluble vitamins, and (b) water-soluble vitamins.

Fat-soluble Vitamins

Fat-soluble vitamins are soluble in fats and fat solvents. They are insoluble in water. So these vitamins are utilised only if there is enough fat in the body.

Vitamin A and carotene, vitamin D, vitamin D2, vitamin D3, vitamin E and vitamin K are fat-soluble vitamins.

Water-soluble vitamins are soluble in water and so they cannot be stored in the body. Therefore, a day-to-day supply of these vitamins is essential.

Water-soluble Vitamins

B-complex Vitamins: 1. Vitamin B (Thiamine), 2. Riboflavin, 3. Nicotinic acid and Nicotinamide (NIacin), 4. Pyridoxin (B6), 5. Pantothenic acid, 6. Folic acid, 7. Biotin, 8. Choline, 9. P-amino benzoic acid, 10. Inositol, and 11. Vitamin B12.

II. Vitamin C or Ascorbic acid, and

III. Vitamin P. (Bioflavonoid)

Vitamin A : Vitamin A was the first fat soluble vitamin to be discovered. In 1913 Osborne and Mendel found that certain natural fats and oils stimulated growth in rats. Again McCollum and Simmonds (1917) demonstrated that Xerophthalmia was due to the lack of a fat-soluble vitamin. Steenbock (1919) discovered the vitamin A activity of carotenoids. In 1920 Roseheim and Drummond demonstrated that the carotene of plants had similar biological action to that of vitamin A. In 1930 Moore discovered that when carotene was fed to vitamin A-deficient experimental animals vitamin A was found in their liver. In 1931 Karser obtained the structure of vitamin A. In 1937, Kuhn and Morris synthesised vitamin A. In the same year Holmer and Corbet obtained vitamin A in a crystalline form.

Chemistry : Vitamin A in its pure form is a pale yellow substance soluble in fat. It is unsaturated alcohol which is stored in the body as esters. Vitamin A as such occurs only in animal foods. Retinol is the generally accepted chemical name for vitamin A. It is stable to heat Vitamin A is present in the form of retinol (vitamin A), Retinal (retinene) or vitamin A aldehyde, retonic acid or vitamin A acid, dehydro retinol (vitamin A2) and carotene (provitamin A). Carotene has alpha, beta and gamma isomers. Vitamin A, is poor in its biological activity. Upon hydrolysis each molecule of beta-carotene yields 2 molecules of vitamin A. But the biological value of beta carotene is only half of vitamin A.

Functions : Vitamin A is essential for the building and growth of all cells, especially skeletal cells. Vitamin A is also needed for proper tooth structure. It also helps in the integrity of the epithelium, especially the mucous membranes, which line the eyes, mouth, alimentary canal, salivary glands, respiratory and genito-urinary tracts. In the deficiency of vitamin A the epithelial tissues of these parts are keratonised.

Vitamin A maintains normal reproductive function in males. Vitamin A is again essential for the maintenance of normal vision in dim light. This vitamin combines with protein to form rhodopsin or visual purple which is present in the retina. Visual purple absorbs light and is decomposed to retinene and protein. The protein and retinene are partially recombined to form rhodopsin in the dark. Vitamin A is essential for this reaction and so continuous supply of this vitamin is required.

Rhodopsin Cycle

(Rods of the retina)

Reteinal dehyde + opsin (Dark) ————↓

Rhodopsin (Visual purple)

↓

Retinene + protein←————Lights←————

Vitamin A plays an important role in maintaining the structure of myelin sheath. Vitamin A also helps the synthesis of mucopoly-saccharides.

High serum retinol levels reduce the risk of cancer. This is a new finding in vitamin A studies and further research is going on.

Vitamin A is also known as anti-infective vitamin as it

provides resistance power to the body by maintaining normal structure of the organs.

Deficiency : Vitamin A deficiency in man may be due to low intake of vitamin A sources or due to interference with absorption or storage of vitamin A in the body. In certain diseases proper assimilation of vitamin A is affected. Loss of vitamin A occurs in the blood during some infectious diseases.

Clinical manifestations of vitamin A deficiency are retardation of growth and lowered resistance to infection. Deficiency of vitamin A shows in delayed adaptation to darkness on coming from the lighted area because it produces changes in the eyes. In a healthy person the eyes adapt quickly to dim light and bright light. In vitamin A deficiency state rhodopsin is not formed effectively, therefore vision in dim light is not easy. This symptom is known as night blindness. As the deficiency advances a condition known as xerophthalmia develops. The conjunctiva becomes dry and lustreless. The transparent appearance of the eye and its elasticity is lost. The eye becomes grey and opaque. If this condition persists the eye becomes infected and ulcerated. This is a serious conditon of the eye which results in blindness. Softening of the cornea brings blindness. Advanced neglected xerophthalmia leads to degeneration of cornea and blindness. This condition is known as keratomalacia.

Bitot's Spots : Grey coloured triangular spots are located in the conjunctiva in severe cases.

Phrynoderma : In vitamin A deficiency, the skin becomes rough because the epithelial tissues are affected. This condition is known as follicular keratosis or toad skin or phrynoderma.

Mucous membraneous structures of respiratory tract, alimentary tract and genito-urinary tract are affected and infectious diseases of these areas result due to a deficiency of vitamin A. Thickening and dryness of the alimentary canal reduce the secretion of digestive juices which leads to diarrhoea.

Experimental animals showed that vitamin A exerts a protective effect against artificially induced cancer and urinary calculus.

Hypervitaminosis: If excess amount of vitamin A is consumed it shows ill-effect in the body. The symptoms are drowsiness, headache, vomiting, itching, skin lesions and loss of appetite.

Sources of Vitamin A : Vitamin A activity is in the form of retinol and carotenoids. The best sources of retinol are colshark and halibut liver oils, and liver of animals like sheep, goat and cow. Butter, ghee, egg, especially egg yolk, and all milk powders are good sources of vitamin A.

Carotenoids are present in green leafy vegetables, yellow vegetables like carrot, pumpkin and yellow fruits like papaya and mangoes. Vegetable oils are available in processed forms (vanaspati) and they must contain specific amounts of retinol as per government rules.

Loss of Vitamin A during Cooking and Storage : Under Indian conditions SO per cent of available vitamin A in food is lost during cooking. However, cooking loss of carotene is negligible. Since most of our dietary vitamin A is from the conversion of carotene, cooking loss is not considerable.

Requirements of Vitamin A : In our diet vitamin A is present as preformed vitamin A or retinol or as carotene which is converted into vitamin A. Retinol or preformed vitamin A is present only in animal foods. In an average Indian diet vitamin A is present in the form of beta carotene. It is converted into retinol. Due to various physiological reasons only one-third of beta carotene is absorbed. But from leafy vegetables and from carrots and papaya the conversion rate is about 50 to 99 per cent. Considering these factors the ICMR Expert Group recommended a daily intake of 750 mg of retinol/day for an adult. If beta carotene is consumed the recommendation is 3000 mg of beta carotene for an adult/day.

Prophylatic Treatment : In India a majority of the people consume vitamin 'A' deficient diet. As per 1978 studies half of the world's 15 million blind persons are in developing countries. Out of this two out of every three cases are preventable nutritional or infective conditions. The age group most affected in our country is 1-5 years and males are affected more than females.

Low availability of vitamin A in the diet and its poor utilisation due to protein deficiency are the main causes for deficiency. Now the Government of India has introduced prophylatic measures to combat vitamin A deficiency. Intake of one teaspoonful of codliver oil daily provide 3000 I.U. Four ml of red palm oil, supplying oral dose of 1,00,000 microgrammes of vitamin A palmitate through 5 ml of arachis oil are some of the measures.

In severe cases of deficiency large doses of 50,000 mg are administered for a few weeks.

Vitamin D : Vitamin D is otherwise known as 'sunshine vitamin' as it can be synthesised from sunlight by our body. Palm (1890) first demonstrated that a bone deformity, ricket, can be prevented by exposure to sunlight. Mellanby (1919) discovered that codliver oil can cure rickets produced in experimental animals. McCollum (1922) showed that codliver oil contained two vitamins, vitamin A and vitamin D. Recrink and Van Wijk (1931) isolated crystalline vitamin D from activated ergosterol. Schenck (1937) obtained a crystalline vitamin D from activated 7 dehydrocholesterol.

Chemistry : Vitamin D is fat soluble. It is a group of different sterol compounds with antiarchitic properties. They are vitamin D, vitamin D2 and vitamin D3. Among these vitamin D2 or calciferol and vitamin D3 or cholecalciferol are biologically important. Ergosterol and 7 dehydrocholesterol act as provitamin of vitamin D. When ergosterol of vegetable origin is exposed to ultraviolet light calciferol is formed. Vitamin D3 is produced by the action of ultraviolet rays of sunlight on 7 dehydrocholesterol which is present in our skin. If the body is kept in 'purdah' cholecalciferol is not produced by the body.

Functions : Vitamin D itself is in an inactive form. Vitamin D is concentrated in the liver and in minute quantities in skin, lungs, brain, spleen and kidneys. In the liver vitamin D is hydroxylated to 25-hydroxy vitamin D, (25-OH-Dj). It is also known as 25-hydroxyl-cholecalciferol. From the circulating blood vitamin D, reaches the kidney and here it is further hydroxylated to form 1,25-Dihydroxy cholecalciferol (1,25, (OH)D,). This is the active form of vitamin D.

Vitamin D promotes the absorption of calcium and

phosphorus from the small intestine. In the DNA vitamin D helps to form active calcium binding protein. It also maintains the concentration of calcium and phosphorus in the blood. Vitamin D helps the deposition of calcium Phosphate in the bone. In rachitic children there may be delay in dentition and malformation of the teeth occurs.

Sources : Fish liver oils, egg yolk, milk, butter and ghee are the best sources of vitamin D.

Some brands of butter are fortified with vitamin D. Sunlight is the best natural source of vitamin D.

Vitamin D is absorbed from the intestine with dietary fat from where it is taken to the liver.

Deficiency of Vitamin D : Deficiency of vitamin D results in poor absorption of calcium and phosphorous from the intestine and excessive loss of them in the urine and faeces. Concentration of calcium in the plasma is diminished and it produces tetany which is charaterised by hyperexcitability of the nervous system. This stimulates the parathyroid glands to decalcify bone calcium to raise the plasma calcium level. Deficiency of vitamin D leads to rickets in children and osteomalacia in adults.

Rickets: The name ricket is given to vitamin D deficiency after a famous English bone-setter by Ell. In mild deficiency of vitamin D, the symptoms shown are restlessness and irritability. Ricket is a disease of infancy and childhood. Calcium and phosphorus are not deposited enough to form strong bones and as a result the leg bones become bent and deformed. Small round unossified areas are developed near skull bones and projections are formed on the osteochondal junction of the ribs. It resembles a rosary and it is known as 'rickety rosary'. The abdomen is distended and chest deformities like depression of ribs attached to diaphragm take place. This is known as Harrison's sukus. This gives an appearance of 'pigeon breast'. In a later stage pelvic deformities, bow legs and knock knees develop. Nutritional rickets occur in infants and children of poor families. Premature infants are more susceptible to rickets.

Osteomalacia occurs among adults. It is often known as 'adult rickets'. Osteomalacia is most common among women who live on a poor diet lacking in calcium and vitamin D. Pregnant women are more prone to this. It is mainly because of the increase in weight during pregnancy accompanied by the deficiency of vitamin D. It is again common among women who observe 'purdah'. Repeated pregnancies and prolonged lactation further deplete calcium stores which results in osteomalacia. Pain, softening and tenderness in the bones and fractures are the common complaints of the patients of osteomalacia.

Prophylatic treatment: For the treatment of rickets and osteomalacia 1,000 to 5,0001.U. of vitamin D should be administered orally for about a month. If there is marked improvement the dose can be reduced to 8001.U./day for at least six months. A massive dosage of 1,000,0001.U. can also be administered but small dosage is safer.

Storage : Cholecalciferol is converted to 25-hydroxy cholecalciferol in the liver. This is the active form of vitamin D. Vitamin D is stored in body tissues in the inactive form. Adipose tissues and skeletal muscles are the major storage sites. Brain, lungs, skin and spleen are other places where it is stored.

Vitamin E : Evans and Bishop (1923) discovered that for normal reproduction of experimental animals a fat soluble vitamin is needed. It was later named vitamin E or tocopherol. Female rats with vitamin E deficiency are unable to carry the young ones throughout the gestation period. The word tocopherol is derived from the Greek word tocos (childbirth or offspring).

Chemistry : There are eight naturally occurring tocopherols with vitamin E activity. Among these Alpha tocopherol is the most active. Tocopherols inhibit oxidation; therefore it is known as antioxidant vitamin. Vitamin E is soluble in fats and stable to acids and heat.

Sources : Cereal gem oils are the richest source. Vegetable oils except coconut oil and nuts are fair sources.

Functions : Vitamin E has a prominent role in maintaining the stability and integrity of cell membranes. One latest discovery about its function is its role in erythrocytes survival among premature infants. Experimental animals showed that deficiency of vitamin E results in anaemia in monkeys. Vitamin E deficiency produced muscular dystrophy in rabbits and reproductive failures in rats. One important function of vitamin E is its antioxidant properties. Through this activity it reduces the oxidation of vitamins A and C and sulphur containing amino acids. Through this activity it stabilizes the lipid parts of cell membrane. The presence of vitamin E as a natural component of vegetable fats protects against rancidity. Recent studies on rats have shown that vitamin E may protect lung tissues from damage by air pollution through nitrogen dioxide and osazone. In experimental animals vitamin E has been shown to promote the effectiveness of linoleic acid in preventing essential fatty acid deficiency. Studies have suggested the possibility of polyunsaturated fatty acids having a role in brain development during infancy. The antioxidant property of vitamin E helps fatty acids from oxidation and thus it indirectly helps the brain development.

Vitamin E Deficiency : Vitamin E deficiency is not common among human beings as it is widely distributed in foods. Experimental deficiency in female rats results in foetal death and in males testicular changes in sterility. Muscular dystrophy and paralysis are the other symptoms of vitamin E deficiency. Premature babies with vitamin E deficiency showed impaired fat absorption. If vitamin E deficiency is present in the pregnant modier transfer of placental blood to foetus is poor and so haemolytic anaemia may occur in the baby after delivery. Vitamin E deficiency in adults results in poor absorption of fat and increased haemolysis and increased excretion of urinary creatinine. Brown muscle pigment deposition catted ceroid occurs with tocopherol deficiency. Liver necrosis, erythrocyte haemolysis and anaemia are the other symptoms.

Requirements : For males 30 I.U. and for females 25 I.U. are recommended. Human milk supplies enough tocopherol for the infants.

Vitamin K : In the 1930's a Dutch scientist Dam and his associates found that "Coagulation vitamin" is necessary for clotting or coagulation of blood. It was named vitamin K, Karrer and coworkers (1939) isolated pure vitamin.

Chemistry : There are two naturally occurring forms of vitamin

K, vitamin Kl and vitamin K2. Vitamin Kl is vegetable origin and vitamin K2 is produced by bacterial sythesis.

Sources : Plant oils are the best sources. Rice bran oil and wheat germ oil, soyabean and cottonseed oils are the best sources. Wholegrain cereals, fleshy foods and organ meats are fair sources. Vegetables and fruits are poor sources. It is a fat soluble vitamin and is easily oxidised by acids, alkalies and light.

Functions of Vitamin K : Vitamin K is essential for the formation of prothrombin which helps normal clotting of blood. Vitamin K is not directly involved in the coagulation but it is essential for the coagulation of blood. It takes part in the carboxylation of glutamic residue.

Requirements : The exact requirement is not known. If lowered prothrombin level is there 1-2 mg of vitamin K/day is recommended.

Deficiency : Dietary deficiency is not common. Deficiency of vitamin K is indicated by a tendency to bleed from skin and mucous membranes. Deficiency occurs due to deficient production of it by the gut, with prolonged antibiotic therapy, due to poor absorption of vitamin K, in malabsorption of fats and due to low prothrombin synthesis in liver disease. Pyloric obstruction and diarrhoea or dysentery can result in poor absorption.

Deficiency of vitamin K among newboms, especially among premature babies, delays clotting of blood. If the mother's intake during pregnancy is poor it can lead to this condition.

Excess of vitamin K due to excessive administration produces hypoprothrombinemia and disorders in liver.

Vitamin K is not stored in large quantities in the body. Only in the liver vitamin K is stored.

Water-soluble Vitamins : There are 13 vitamins in this group. Vitamins C and B complex family, comprising twelve vitamins, are soluble in water.

In 1882 Takaki, the Director General of the Naval Medical Service in Japan, demonstrated that 'beriberi', a fatal disease in the navy, could be cured by including a mixed diet of whole cereals, vegetables and milk. In 1897 Eijkman found that fowls on a poor diet manifested symptoms like beriberi. He suggested that instead of rice, rice Polishings must be included in the dietary pattern to avoid beriberi. In 1911 Casimer Funk described a group of water-soluble substances present in food which can act as anti-scorbutic, anti-beriberi and anti-pellagra factor. In 1926 McCollum and Kennedy suggested different substances in food with different properties which were later identified as anti-beriberi and anti-pellagra factors.

Thiamine (B1) : The occurrence of beriberi led to the discovery of thiamine. Followed by the findings of Takaki, Eijkman established that rice polishings prevented a disease similar to beriberi in fowls. Dr. Grijins, an associate of Eijkman, suggested beriberi was a deficiency disease and rice polishings contain the factor. Funk, Suzuki and Edic attempted to isolate the 'anti-beriberi' factor. In 1926 Jausen and Donal's isolated crystals of thiamine. In 1936 Dr. R.R. William synthesised thiamine.

Chemistry : It is a white powder which is readily soluble in water. Alkali destroys thiamine. Thiamine is commercially produced as thiamine hydrochloride. It has a yeast-like odour and it is stable in its dry form.

Physiological Functions : Thiamine combines with pyrophosphate and forms thiamine pyrophosphate. (TPP) Thiamine pyrophosphate participates in intermediary metabolism of carbohydrates. Transketolase and enzyme has thiamine pyrophosphate as a co-factor. It acts as a co-enzyme in the metabolic pathway of carbohydrates at different places. Animal experiments showed that thiamine pyrophosphate is essential for protein metabolism. Thiamine is also required for the synthesis of glycine in the body. Thiamine pyrophosphate is present in the peripheral nerve cells. It has a specific effect on nerve tissues. In thiamine deficiency the peripheral nerves degenerate. Thiamine is also required for the maintenance of normal gastro-intestinal tone and motility. Thiamine seems to help the heart in its noitnal action. Thiamine is otherwise known as "morale vitamin" because it helps the transmission of nervous impulses and maintains the stability of nerves.

Deficiency of Thiamine : It is difficult to identify the deficiency earlier. In mild deficiency loss of appetite, fatigue, irritability, depression and constipation occur. Gastrointestinal and cardiovascular disorders with peripheral nervous problems initiate the deficiency state of thiamine.

In mild deficiency loss of appetite, loss of weight and strength fatigue, emotional upsets, irritability, depression, anger and fear are manifested. Constipation is common among such people. Headaches, insomnia, and techycarida with slight exertion occurs during thiamine deficiency. Severe deficiency of thiamine causes beriberi. Two forms of beriberi occur in human beings. They are dry beriberi and wet beriberi.

Dry Beriberi: In dry beriberi, the peripheral nerves of the legs and arms are affected first. Degeneration of the myelin sheath of the nerves is followed by the degeneration of the axis cylinder. Calf muscles become lender. Numbness in the toes and ankles takes place. The knee and ankle jerk and later on the sensation is lost. Tingling and numbness of the legs and hands are followed by wasting of muscles and difficulty in walking.

Wet Beriberi: All the symptoms of dry beriberi are seen in wet beriberi also. In addition to these oedoma in legs and in between cardiac muscle fibres is observed. These changes in the heart muscles lead to enlargement of the heart. Later palpitation, oliguria, and dyspnoea may develop. Oedema on the legs will be prominent. The patient may be bed-ridden and cardiac failure may take place. If untreated it leads to death within hours.

Infantile Beriberi: It affects infants within six months. The early symptoms are restlessness, sleeplessness and loss of appetite. The affected child is pale, oedematous and ill-tempered. Vomiting and green-coloured diarrhoea are common. If treatment is delayed it leads to death.

Requirements : Thiamine requirement is related with energy intake. The Nutrition Expert Group of ICMR and FAO/WHO recommended 0.4-0.5 mgs of thiamine per 1000 kcals. For infants 5.9 mgs/kg and from 1 year onwards proportionate kcals requirements are recommended.

During deficiency state 5 mgs of thiamine orally three times a day is recommended. It is readily absorbed from the small and large intestines. It is not stored in the body.

Sources : Rice polishings, dried yeast and wheat germ are rich sources of Bl. Whole cereals like wheat, oats, legumes, oil seeds and nuts are good sources. Milled cereals, vegetables, fruits, milk, meat and fish are fair sources. On milling, thiamine is lost from cereals. Commercial preparation of Bl is extracted from natural sources or it is synthesised. Nowadays cereals are enriched with thiamine. Flours are enriched and 'abelled to show the amount of enrichment.

Riboflavin : Scientists from various parts of the world reported that a growth promoting factor substance remained even after the beriberi preventive factor is destroyed from yeast A yellow green fluorescent compound was known to be in the whey from milk. In 1932 the German scientists Warburg and Christian discovered the first flavoproteins from yeast. It was first known as 'Yellow enzyme'. In 1933 Kuhn and Gyorgy isolated riboflavin and Booker recognised the growth promoting property of it. It was synthesised by Kuhn and Karrer in 1935.

Chemistry : Riboflavin is dimethyl-iso-alloxazine 'attached to ribose.' It is slightly soluble in water and stable in acid medium. In alkaline medium it is easily destroyed. Bright light also destroys riboflavin.

Functions : Riboflavin is part of food co-enzymes as riboflavin monophosphate or flavin mononucleotide (FMN) and flavin adenine dinucleotide. Riboflavin is closely related to biological oxidations in cells. Flavoprotein enzyme system is present in different parts of our body. It takes part in cell respiration. Cellular growth cannot evolve in the absence of riboflavin. Riboflavin through the enzyme system possesses the ability of transferring hydrogen in chemical reactions. Thus, it takes part in carbohydrate and protein metabolism. Riboflavin is essential for the health of skin, mucous membraneous structures of alimentary canal and eyes. For normal vision riboflavin is essential. The retina has free riboflavin which is converted by light to a compound that stimulates optic nerves.

Deficiency : In mild deficiency of riboflavin weakness, anorexia, apathy, burning at the angles of mouth, in the eyes and over the skin occur. Normal metabolism of carbohydrate and protein is affected. Riboflavin deficiency symptoms are generally known as ariboflavinosis. Ariboflavinosis consists of angular stomatitis, with cheilosis, glossitis, skin changes, seborrheic dermatitis and vascularisation of the cornea.

The angles of the mouth become pale and crust is formed in the corners. Fissures occur at the corners of the mouth and it is known as cheilosis. Crust is formed on the surface of lips and removal of it makes it red in colour. Papilae of the tongue are peeled off and the tongue shows a 'magenta' or red tint on the surface. This is referred to as glossitis. On the nasolabial folds, ears, vulva in the female reproductive organs and on the scrotum in the male, scaly desquamation of the skin takes place. These are known as nasolabial seborrhoea and scrotal lesions. The lesions are cured by the administration of riboflavin.

Even in a mild deficiency state burning and watering of the eyes, sensitivity to light and dimness of vision take place. Vascularisation of the cornea and abnormal pigmentation of the iris are the other symptoms manifested. Vision is blurred and conjunctivitis and itching may occur in riboflavin deficiency.

Requirements : Various studies on the requirement of riboflavin showed that 1.4 mg per day for a sedentary man, 1.7 mg for a moderate worker and 2.3 mg for a heavy worker are essential. For females the requirement is 1 mg, 1.1 mg and 1.5 mg, respectively. During pregnancy and lactation additional 0.2 mg and 0.3 mg riboflavin are required. During infancy (0-6 months) 0.71 mg/kg and 6-12 months, 0.65 mgVkg. For 1-3 years 0.7 mg, 4-6 years 0.9 mg, and 7-9 years, 1.00 mg. of riboflavin are required. During adolescence 15 mg in the earlier stages, that is, 10-12 years and 1.6 mg during later adolescence for boys had 1.4 mg in early adolescence and 1.4 during later adolescence for girls are recommended.

Sources : Fleshy foods like liver, egg, skimmed milk powder and other milk products are rich sources of riboflavin. Milk, meat, fish, whole cereals, legumes, dais, oil seeds, nuts and leafy vegetables are good sources of riboflavin. Milled cereal and flours, roots and tubers and other vegetables are fair sources of riboflavin.

Riboflavin is readily absorbed from the intestine and the excess is excreted through urine.

Niacin (Nicotine Acid) : Niacin is also known as nicotinic acid or nicotinamide. In 1725 a Spamsh physician Casal described die occurrence of red sickness. In the eighteenth century, the Italia physician Frapolt named the disease as pellagra, meaning rough skin. In 1911 Funk isolated niacin from rice polishings. In 1937 Elvegem and co-workers showed that the deficiency disease (black tongue in dogs) was cured by an isolated substance from liver. Goldberger cured and prevented pellagra by adding meat, vegetables, fruits and eggs in the diet. Thus, pellagra preventing factor (P-P-factor) was identified from foods. Further investigations showed that nicotinamide was equally effective as nicotinic acid in curing pellagra.

Chemistry : Niacin is a white crystalline compound soluble in water, stable to heat, light, acids and alkalies. In the body niacin is converted into niacinamide.

Physiological Functions : Like thiamine and riboflavin niacin takes part in the metabolism of carbohydrates, proteins, and fats through the enzymatic action. It is part of co-enzyme I and co-enzyme II which are essential for the conversion of lactic acid to pynivic acid. They are otherwise known as NAD and NADH. More than 40 biochemical reactions depend on these enzymes. Dehydrogenation reaction in Kreb's cycle, oxidation of alcohol to aldehyde, of glucose to gluconic acid, of malic acid to oxaloacetic acid, of lactic acid to pyruvic acid, glucose -1-6-phosphate to phosphohexouric acid and reactions in the synthesis of fatty acids, cholesterol and for the conversion of phenylalanine to tyrosine are some of them. It takes part in tissue oxidations. Nicotinic acid is essential for the normal functioning of the skin, intestinal and nervous systems.

Effects of Deficiency : In a mild deficiency state weakness, loss of weight, loss of appetite, lassitude, irritability, burning tongue, and constipation occur. Pellagra is the characteristic disease resulting from niacin deficiency. It is otherwise known as three 'D's disease—dermatitis, diarrhoea and dementia. The dermatitis of pellagra is different from other skin diseases. It occurs mainly on the exposed portions of the body. Lesions are seen on the face, neck, surfaces of the hands, elbows, feet and other parts of the body which are exposed to sunlight. Another peculiarity of dermatitis is that the skin lesions are symmetrical in shape, and they always occur on both sides of the body simultaneously. The skin is cracked and scaly. Burning sensation and itching are common. The lesions leave a reddish, dark brown colour. The mucous membrane of the alimentary canal is affected and glossitis, abdominal pain, and loss of appetite are some of the symptoms. Nausea, vomiting and diarrhoea are the other symptoms of niacin deficiency. Delirium is the common mental disturbance in pellagra. Earlier symptoms of nervous disturbances are depression, dizziness and insomnia. It leads to death.

Pellagra is common among jowar eaters. Jowar has a very high leucine content and this excessive leucine interferes with the utilisation of nicotinic acid. Leucine precipitates mental changes but isoleucine promptly reverses them to normal.

Dr. Gopalan and his co-workers in India have done extensive studies for a decade (1960-69) in this connection. Among maize eaters pellagra *is* common because of the high leucine content in them. Leucine interferes in the conversion of tryptophan to niacin and also the conversion of niacin to NAD and NADP in the tissues.

Sources : Cereals, especially whole cereals, dried yeast, liver, groundnut, legumes, fleshy foods and fish are good sources. Milk, eggs, and vegetables are fair sources.

Tryptophan-niacin Relation: Tryptophan is an essential amino acid and it is a precursor of niacin. Vitamin B1 or thiamine, B2 or riboflavin and pyridoxine are essential for this conversion of tryptophan to niacin. Another essential amino acid leucine interferes with the conversion of tryptophan to niacin and niacin to co-enzymes. Milk is a rich source of tryptophan. Animal foods rich in protein are good sources of niacin.

In a healthy person's body 60 mg of tryptophan is equivalent to 1 mg of niacin. Thus the dietary allowances are usually made in terms of niacin equivalents.

Niacin equivalents (mg) = Niacin content (mg.) + tryptophan content (mg)/60

Requirements : The FAO/WHO expert group has recommended an allowance of 6.6 mg niacin equivalents per 1,000 kcals. For all age groups this requirement pattern is suggested.

Pyridoxin (B6) : Three naturally occurring compounds with similar activities are together known as Vitamin B6. They are pyridoxine, pyridoxal and pyridoxamine. It is also known as adermin. In 1934 Gyorgy found a factor leading to a type of dermatitis in rats. In 1938 it was isolated and named pyridoxine.

Chemistry : It is a white crystalline substance soluble in water, quite stable to heat and acid. In an alkaline medium it is easily destroyable.

Physiological Function Vitamin B6 acts as a co-factor for several enzymatic reactions in all the three metabolisms. In protein metabolism, transaminauon takes place with the help of this enzyme system. In carbohydrate and fat metabolisms, there are intermediary reactions where vitamin B6 acts as a co-factor.

It is part of the transaminase system which has a number of enzymes, e.g.,

1. Glutamic acid + oxalo acelate - alphakete-glutarate + aspartate. Alanine + alpha ketoglutarate = glutamic acid + pyruvic acid.
2. Amino acid decarboxylase- Amino acids to amines

   e.g., Histidine = Histamine + CO_2

   Tyrosine - Tyramine + CO_2

Tryptophan conversion to niacin has a series of biochemical reactions. In pyridoxine deficiency this conversion step will not materialise and niacin is not formed.

Removal and transfer of sulphur groups from sulphur-containing amino acids like cysteine requires pyridoxine.

In carbohydrate metabolism pyridoxal phosphate is essential for the action of phosphorylase enzyme which converts glycogen to glucose.

Vitamin B6 helps in energy transformation in brain and in nervous tissues. In its deficiency, convulsive seizures occur in animals and in human beings. Especially infants fed on evaporated milk develop nervous irritability and convulsive seizures.

Deficiency : Separate deficiency of B6 is not common. But as a part of general deficiency of B complex, vitamin B6 deficiency also occurs. In experimental cases anorexia, nausea, restlessness, lethargy, seborrhoea, hyperpigmentation, pellagra like dermatitis, cheilosis, glossitis and angular stomatitis were observed. Changes in the periphery of nerves with sensory impairment and motor abnormalities were also observed. In infants convulsions are seen. Immune response is impaired in B6 deficiency.

Sources : Dried yeast, rice polishing and liver are rich sources. Milieu like cholam, cambu, pulses, beetroot, cabbage and fleshy foods are rich to fair sources, respectively.

Pantotnenic Acid : In 1933, R.J. Williams reported that a factor present in food is essential for growth in yeast. It is named pantothenic which is derived from a Greek word 'Panthos' meaning everywhere. In 1939, it was isolated and synthesised in 1940.

Chemistry : It is soluble in water and it contains the amino acid alanine. It is stable to heat but prolonged exposure to dry heat destroys it. It is also destroyed by acid and alkali.

Functions : Pantonenic acid forms a complex compound co-enzyme A (CoA) and acyl carrier protein and thus it takes part in the metabolism of carbohydrates and fats. It is also necessary for the formation of acetyle choline, which is a precursor of heme which in turn is essential for haemoglobin synthesis. CoA is also essential for the synthesis of cholesterol and sterob.

In our body pantotnenic acid is present in the liver, adrenal gland, kidneys, brain and heart.

Deficiency : Deficiency of pantothenic acid is rare. Experimental production of its deficiency showed loss of appetite, fatigue, weakness, headset, insomnia, nausea, emotional instability, dizziness, imparted motor coordination, muscle cramps, burning sensations in the feet and changes in the heart beat rate and in blood pressure.

Burning feet syndrome is associated with neurological and mental disturbances and it is cured only by pantothehic acid. Gopalan (1946) suggested the treatment for burning feet as administration of capantothenate. He also found that other

B-complex vitamins like thiamine, riboflavin and niacin were not effective in its cure.

Sources : Dried yeast, liver, rice polishings and wheat germs are the rich sources. Whole cereals, legumss, nuts, fleshy foods, eggs and fish are fair sources.

Requirements : For adults 10 mg/day, for adolescents 8-10 mg/day, children 5-8 mg/day and for infants 1.5-2.5 mg/day was recommended by the FAO/WHO expert committee. During pregnancy and lactation 10-15 mg/day was recommended.

Vitamin B_{12} : Minot and Murphy in 1926 reported that pernicious anaemia could be controlled by feeding liver in large quantities to patients. Castle found that beef muscle is also effective in controlling pernicious anaemia. He established that two factors were involved in curing pernicious anaemia, the intrinsic factor of normal gastric secretion and the extrinsic factor vitamin B12. In 1948, two British scientists, Smith and Parker, and Dickes from the USA isolated B12. Its structure was determined in 1955.

Chemistry : Vitamin B12 contains the mineral cobalt which gives its characteristic dark red colour. Cyanocobalamin and Hydroxy cobal are the two forms of B12.

Functions : B12 is essential for the maturation of red blood corpuscles and it combines with globulin. B12 takes part in the enzymatic reaction. Especially co-enzyme B12 is very prominent in the body. Niacin, riboflavin and magnesium are essential for this conversion of B12 to co-enzyme B12. B12 co-enzyme is required for the action of dehydrase reaction.

Vitamin B12 acts as a necessary substance for the lobo protein action in the nervous tissues. It is also necessary for the synthesis of myelin. Within the bone marrow B12 cc-enzyme participates in the synthesis of DNA.

Co-enzyme B12 is essential for carbohydrate, protein and fat metabolism. In experimental animals vitamin B12 has been found to stimulate growth. B12 co-enzyme is also essential for the synthesis and transfer of single carbon units such as the methyl group, eg., methionine and choline synthesis.

Deficiency : Due to the deficiency of vitamin B12, pernicious anaemia occurs. To prevent anaemia two factors are required: an intrinsic factor produced by the gastric parietal cells and the extrinsic factor of vitamin B12. If due to any cause enough intrinsic factor is not secreted, B12 is not absorbed and utilised properly. 312 is otherwise known as extrinsic factor. Congenital intrinsic factor deficiency and low absorption of B12 or its deficiency leads to pernicious anaemia. In pernicious anaemia the life span of red blood cells reduced to 60 days instead of the normal 120 days and their size and shape also is changed. Skin changes show a yellowish pallor. Soreness and shine on the tongue are other manifestations of skin lesions. Gastric secretion is reduced. Stomach mucosa shows atrophic changes. Nervous tissues are degenerated, especially the posterior column of the spinal cord is affected. Peripheral neuropathy and later on mental changes are common. Tingling, numbness, loss of sensation in the limbs, exaggerated knee jerks and ankle jerks are the other symptoms. Depression and other psychosis symptoms are common in later stages.

Sources : Organ meats like kidney, liver, brain, meat, poultry, egg, fish and milk are good sources of B12. Vegetable foods lack B12.

Requirements : B12 requirements for a normal person is 1 mgm (microgram) per day. During pregnancy and lactation it is increased to 1.5 mgm (microgram). Infants and children up to 10 years require 0.2 microgram and adolescents need 1 microgram.

Choline : In, 1934, Best and Huntsman discovered fatty liver in choline deficiency among rats. Choline is a component of phospholipid lecithin arid it forms acetyl choline with alcohol and takes part in biochemical reactions. It is involved in the transmission of parasympathen'c and other nerve impulses to effective organs. It promotes growth in animals.

In its absence neutral fats and cholesterol esters accumulate in the liver which produce fatty liver. Choline is otherwise known as a lipotropic factor as it prevents fatty liver. Choline enhances the oxidation of fatty acids and promotes the formation of lipo proteins in the liver. Phospholipids take part in the transport of fatty acids and cholesterol in the blood and from the liver and their deposition and removal in adipose tissues. It is essential for the transfer of nerve impulses. It takes part in the formation of methionine from homocystine. In its deficiency production is reduced.

Sources : Egg yolk, wheat germ, organ meats like kidney, brain, heart, meat, yeast, soyabean and peanut and skimmed milk are good sources of choline. The amino acid methionine acts as a donor of methyl group to convert cltanolamine to choline.

Requirements : Choline requirements of human beings is not known and its deficiency is also uncommon.

Biotin : The importance of biotin was first observed by Boas in 1927. He found mat raw egg white produces dermatitis, loss of hair, and muscular incoordination in experimental animals. Therefore, it was first called anti-egg white injury factor. In 1931, Gyorgy gave die name vitamin H or Haunt, which means skin. It is sparingly soluble in cold water and alcohol Biotin is present in human tissues and is necessary for bacterial growth. It is also necessary for normal gestation and lactation in sals. Biotin plays an important role in maintaining the skin soft and for the health of nervous system. Biotin also acts as a co-enzyme for carboxylatioa of acetyl-CoA to form malonyl-CoA in fatty acid metabolism.

Egg White Injury Factor (Avidin) : Rats kept on a diet of uncooked egg white developed dermatitis, muscular incoordination and spastivity of legs and loss of hair. This is produced by a protein present in egg white called avidin. Avidin combines with biotin which is unavailable to the body. This avidinbiotin compound is excreted in the stcol. Since avidin is present in egg white it is also known as egg white injury factor. Since heat destroys avidin, cooking of egg destroys it.

Deficiency of biotin among human beings is not common. Sydenstricker and co-workers in 1942 produced biotin deficiency in man by giving 30 per cent of the total calorie intake from fresh egg white. Dermatitis of varying types on the necks, hands, arms and legs were seen. It was followed by lassitude, muscle pain, anorexia, nausea and anaemia. Injection of biotin brought rapid cure.

Sources : Dried yeast, rice polishings, liver, groundnut, soyabean, whole cereals, legumes and fleshy foods are good sources of biotin.

Requirements : A normal balanced diet supplies the requirement of biotin. Its deficiency is not common. The exact requirement of biotin is not known.

Inositol : Wolley and co-workers in 1940 reported that deficiency of inositol produced alopaecia in the mouse. Inositol is a component of phospholipids.

Inositol is essential for the growth of fibroblasts. It is present in phospholipids and it exerts a lipotropic effect. It is present in different tissues, especially in heart and skeletal muscles. It also enables peristalisis of the intestine.

Effects of Deficiency : In experimental animals deficiency of inositol produced retarded growth, loss of hair or alopaecia. It also produces spectacled eye or loss of hair around the eye.

Sources : It is widely distributed in nature. Fruits, milk, vegetables, nuts, meat, wholegrains, yeast and bacteria are good sources of inositol.

Para-amino-benzoic Acid : In 1941, Ansbacher found out that para-amino-benzoic (PABA) is essential for normal growth of rats. For normal lactation and for the prevention of greying of hair para-amino-benzoic acid is necessary. It is present in folic acid.

Para-amino-benzoic acid inhibits the bacteriostatic effect of sulphonamides. It also inhibits the production of thyroid hormones.

Requirement of PABA for human beings is not known. Average diets provide adequate amounts of PABA.

Vitamin C : In the eighteenth century James Lind, a British naval surgeon, demonstrated that citrus juices cured scurvy among sailors. Hoist and Frolich produced scurvy in guinea-pigs. The study of ascorbic acid was thus actually initiated in 1907 by these scientists. Later in 1928, Szent-Gyorgy isolated a factor from orange and cabbage juice. It was hexuronic acid which prevented scurvy. Wangh and King in 1932 showed that hexuronic acid was identical with vitamin C and they isolated it from lemon juice.

Chemistry : Ascorbic acid is a white crystalline substance readily soluble in water. It is present in two forms: L-ascorbic acid and D-ascorbic acid. L-ascorbic acid has anti-ascorbulic property while D-ascorbic acid has no such property. It is easily destroyed by oxidation, heat and alkalies. Even when it is exposed to air it is oxidized to dehydroascorbic acid. During cooking much of it is destroyed. Iron and copper act as catalysts and cooking in these vessels increases the loss of vitamin C. Ascorbic acid is also lost by the action of an enzyme, ascorbic oxidase, present in vegetable oils. When vegetables are cut into fine pieces more enzyme is released and it causes more loss. This can be minimised by immersing vegetables in boiling water as heat destroys the enzyme. Adding sodium bicarbonate to preserve the colour of vegetables destroys vitamin C. As it is readily soluble in water washing vegetables after cutting, cooking in too much water and cooking in cold water enhances the loss of vitamin C.

Ascorbic acid is readily absorbed from the intestine.

Physiological Functions : Due to its great affinity for oxidation it takes part in oxidation and reduction reactions in the tissues. Ascorbic acid is essential for the formation of collagen present between cells. It is otherwise known as the cementing material or intercellular cement of the body. If collagen is defective tissues in the capillaries, teeth and bone matrix are net formed properly. Ascorbic acid is also necessary for the formation of osteoblasts. Vitamin C is essential for cholesterol metabolism. It is also necessary for the conversion of phenylalanine to tyrosine and for the oxidation of tyrosine. It takes part in the conversion of tryptophan to 5-hydroxy-tryptophan. Ascorbic acid reduces the ferric iron to ferrous iron and iron is absorbed only in this form. It is also essential for rapid healing of wounds as it helps in the formation of connective tissues. Vitamin C is also essential for the utilisation of folacin.

Deficiency of Vitamin C : Prolonged deficiency of Vitamin C produces scurvy. In mild deficiency fatigue, weakness, irritability and frequent infection occur. Pain in bones is common.

Scurvy among infants is known as infantile scurvy. The infant cries on moving its legs and arms. Early symptoms include listlessness, fatigue, fleeting pains in the joints and muscles, bleeding gums, internal haemorrhages and shortness of breath. In severe deficiency of vitamin C scurvy occurs. Swollen gums, loose teeth, soft or malformed weak bones, anaemia, degeneration of muscle fibres including heart muscles are some of the symptoms of vitamin C deficiency. Separation of ribs through beading is also observed. This can be distinguished from rickets from its shape. The skin becomes rough and dry. Pyrexia, rapid pulse and susceptibility to infection are common in scurvy.

In adults, weakness, spongy bleeding gums, loose teeth, swollen joints and haemorrhages in various tissues are the symptoms.

Propylaxis for infantile scurvy include 50 mgs of vitamin C with orange juice three or four times a day. In the case of adults 100 mgs of ascorbic acid twice daily is recommended. Citrus fruits, sprouted pulses and soaked cereals must be liberally used for vitamin C.

Dietary Sources : Citrus fruits like lime, orange, pineapple, ripe mango, papaya, cashew fruit and tomato are good sources. Amla or Indian gooseberry is the richest source of vitamin C. One hundred grammes of amla contain 700 mg of vitamin C. Guava is also rich in this vitamin C. Leafy vegetables are good sources of vitamin C. But this vitamin is readily destroyed by oxidation during cooking and storing. About half of the vitamin C in raw food is lost during cooking.

Requirements : The ICMR (1984) recommended 40 mgs of vitamin C as sufficient for an adult considering its cooking loss. During pregnancy and lactation the requirement was reviewed and the committee recommended that no additional allowances were needed during pregnancy but 80 mgs of vitamin C is required during lactation. For infants 20 mg per day and for children 40 mg per day is enough.

Folic Acid : Folic acid has been called folacin, pteroylglulamic or vitamin Bc. In the early 1930s a search for anaemia-preventing factor was done by many scientists. Dr. Lucy Wills (1934) showed that there was a factor in yeast and crude liver extracts which cured tropical macrocytic anaemia. This factor was known as "Wills factor". In 1946 Keresztesy and Silverman isolated folic acid from liver. The word folic acid is derived from the Latin word folium (leaf) as it is found in green vegetables. Many bacteria synthesize folic acid but lactobacillas grows well only if folic acid is supplied. So folic acid is otherwise known as Lactobacillas.

Chemistry : Folic acid is a yellow crystalline compound widely distributed in nature. Both folic acid and folinic acids occur in foods as conjugates of glutamic acid. It is slightly soluble in water, relatively unstable to heat and easily destroyed by exposure to light. Folacin occurs in different forms. Most of it occurs as pterolypolyglutamates. It is sensitive to light and cooking loss is reduced by anti-oxidant agents like vitamin C.

Physiological Functions : Folic acid is converted to active folinic acid. Folic acid acts as a co-enzyme. Co-enzyme takes part in metabolic process when it transfers single carbon units from one compound to another. Folacin co-enzymes are involved in many methylation reaction. Folic acid is essential for nucleoprotein synthesis and it is required for cells division and also for maturation of erythrocytes. The role of folacin in protein synthesis is made in cancer therapy. Folacin inhibitor like methortrexale are used in chemotherapy to inhibit tumor growth. It is essential for reproduction in animals. Folic acid also helps the hair growth and health of skin.

Deficiency : Lack of folacin results in megaloblastic anaemia. It is otherwise known as macrocytic anaemia. This mainly affects pregnant women. A study in Madras showed that 20 per cent of maternal deaths occurred due to anaemia. Poor vegetarian diet supplies very little folacin. This type of anaemia is due to inadequate formation of nucleoproteins. This prevents the megaloblasts in the bone marrow from maturing into erythrocytes. In megaloblastic anaemia the number of red blood cells produced in the bone marrow is reduced. The haemoglobin level is also reduced to 6-9 per cent and R.B.C. count to two to three million per mm. Along with inadequate diet it also results from malabsorption syndromes like tropical sprue. Often B12 deficiency after a surgical removal of some part of the stomach also produces megaloblastic anaemia.

Requirements : 100 mgm for adults and 300 mg during pregnancy are recommended by the expert committee. Breast-fed infants will receive about 25 to 30 mg of folate daily. Requirement for children is not known.

Sources : Organ meats like kidney, liver and dark green leafy vegetables, soyabean and groundnuts are rich sources of folic acid. Other vegetables, legumes, eggs, wholegrain cereals and fruits are good sources.

16

Vitamin Sources

Vegetables and Fruits are the classes of foods reflect the origins and methods of storage of the principal vitamins in these natural sources.

The water soluble factors are synthesized in the first place by plants and the quantities produced in excess of a plant's requirements are laid up in its own normal storage organs, such as underground roots and tubers, fruits and seeds. These are subsequently the foods to which we look for our supplies of the B vitamins, vitamin C and carotene. The oily fractions of seeds also contain vitamin E.

Vitamins A and D are found in foods of animal origin, having reached their final form in the animal body.

We would like to study first vegetables as sources of supply, then fruits and cereals, and finally dairy products, meat and fish.

Vegetables

Vegetables refer to edible parts of plants other than fruits and seeds. There are certain anomalies in any classification and this is no exception. Tomatoes, for example, cucumbers and marrows are structurally fruits, but are often eaten with meats and savouries, and are frequently included in the term 'vegetable.' By contrast rhubarb, which is eaten in the same way as fruits, as a desert, is the coloured stalk of the rhubarb leaf and is no more a fruit than is celery.

Vitamins synthesized by plants remain stable as long as they are in living undamaged cells, but injury to the cell walls exposes the contents to air and permits oxidation and vitamin destruction to occur. During harvesting, transport and storage, vegetables are subjected to conditions that may considerably lower their vitamin content. Gathered freshly from the garden they not only look and taste more attractive but are of greater nutritive value.

In present day civilizations vegetables have to be transported sometimes for long distances by rail, road or water and measures have to be taken to preserve as much as possible of their nutritive value.

Careful packing is essential to avoid bruising and mechanical injury and this is even more important for delicate leafy vegetables such as spinach than for firmer green vegetables such as cabbage or root vegetables such as parsnips and potatoes. We generally see green vegetables packed fairly tightly in open work sacks. This avoids overheating in the interior of the pack, and also protects them from being jostled and bruised.

Exposure to the sun causes not only oxidation but also loss of water and consequent wilting, which is another factor introducing loss of vitamins.

Finally, high temperatures reduce the content of certain vitamins to a considerable degree. Vegetable produce carried by sea is generally packed at refrigeration temperatures.

After travelling for hours or even days, vegetables are often exposed in markets or shops for another period of varying duration before they are taken home to be eaten or cooked.

***Storage of Fresh, Unprocessed Vegetables*:** Even green vegetables have to be kept for short or long periods and most seasonal crops are stored in order to be available after the season has ended.

Root crops can be stored in bulk, as for example, in potato caves (or clamps) for several months. Home storage takes place in cellars, lofts, cupboards, larders and refrigerators. Large-scale storage is also essential in catering establishments and ships.

The two vitamins which are found in highest concentration in vegetables are A (as carotene) and C. Of these carotene withstands storage better than vitamin C.

Carotene in Raw Vegetable Foods : We cannot assume that a yellow colour always indicates the presence of carotene because there are other yellow pigments, e.g. xanthophyll in some vegetables. Carrots contain a high proportion of carotene; swedes do not. Carotene is present in its highest concentration in green leaves, but is masked by the more dominant green of the chlorophyll. It is generally true that the greener the leaf, the higher the concentration of carotene. So good sources are broccoli leaves, watercress and spinach, and among animal feeding stuffs — grass. The outer deep green leaves are better sources than the inner pale leaves.

Vegetables fail to yield all their carotene during digestion and this fact must be taken into consideration while estimating the vitamin A values of diets where the vitamin A is provided from vegetable sources. From 40-50 per cent of the carotene of green leafy vegetables is available, but only about 25 per cent of that in carrots. Utilization of the carotene is best when it is present, as in lettuce, along with vitamin E — another example of vitamin interactions.

The following table shows the carotene content of some vegetables which are generally eaten raw.

Carotene in Raw Vegetables (i.u. per ounce)

Name	*Up to 200 i.u./oz.*		*200-1,000 i.u./oz.*		*Over 1,000 i.u./oz.*	
Celery	Peas	60	Tomato	720	Watercress	1,200
Cucumber	Onion	160	Lettuce	900	Carrut	3,000-4,000
Radish	Cabbage	180				

These days in agriculture it is a common practice to cut young grass instead of making hay. The grass, that is very green and contains a good supply of carotene, is finely divided and dried rapidly by a special process. It is used as a source of carotene in animal foods. It is bright green when fresh but may lose as much as 40-50 per cent of its potency during storage, or more if permitted to become damp.

Vitamin C in Raw Vegetable Foods

Vitamin C in raw vegetables (mg. per ounce)

1-4 mg.oz.		*5-9 mg./oz.*		*10-20 mg./oz.*		*Over 20 mg./oz.*	
Celery	1	Onions	5	Cabbage	1	Mustard and Cress	23
Cucumber	2	Tomatoes	7	Watercress	17		
Carrots, Lettuce, Peas	3						
Radish	4						

Please note that these are all values in *one ounce* of each vegetable. While it is easy to eat several ounces of raw tomatoes or celery, shredded raw cabbage cannot be eaten in *quantities* similiar to those of cooked cabbage and both watercress and mustard and cress are very light in weight. Thus it generally happens that a liberal helping of a vegetable with a low vitamin C potency supplies more than a small helping of one with a high potency.

Effects of Short Periods of Home Storage : Cabbage, left whole and stored in a cool, dry place, loses only little of its vitamin C in a week and only about one-third of its content in a month, where as minced or finely cut cabbage may lose much of its vitamin C in a matter of minutes. This is important and means that raw cabbage for use in salads should be shredded as near as possible to the time of serving, spinach and asparagus, both of which wilt quickly may lose 40-80 per cent of their content if kept for several days before being used.

The following table shows recorded vitamin C values in stored vegetables. From these, losses in 4 days and in 1 week can be read.

Vitamin C (in mg./oz)

Vegetables	*Fresh*	*Stored 4 days*	*Stored 1 week*	
Brocolli tops	49	37	30	room temperature
Brussels sprouts	38	34	—	" "
Cauliflower	24	—	17	" "

Contd..

Vegetables	*Fresh*	*Stored 4 days*	*Stored 1 week*	
Cabbage	14	—	13	0-3°C.
			11	room temperature (20°C.)
Spinach	14	—	13	0-3°C.
			2	room temperature (20°C.)
Asparagus	3	0.5	—	" "
Peas	3	—	1.5	" "

The green vegetables are better sources of vitamin C than root vegetables. Potatoes/parsnips, onions and carrots contain relatively little, but as the quantities eaten are greater, they contribute what may be quite a valuable quota to the daily intake. In many diets, potatoes are the chief source of vitamin C.

These vegetables of hard texture are generally better stored at low temperatures. Potatoes may lose up to 50 or 60 per cent of their vitamin C in six months. Onions may lose as much in three months.

Methods of Preservation of Vegetables : From olden times the housewife has practiced methods of preservation to have vegetables available when out of season. Her methods: pickling, salting and drying have been followed by modern procedures such as canning, freezing and dehydration. All these, both old and new, are still used.

Salting : This is, the oldest method of food preservation that is still in use. The domestic preservation of vegetables like beans, by packing them in jars with salt is very destructive to vitamins. Little or no vitamin C remains after three months' storage and 50-80 per cent of the original carotene is lost.

Pickling : Methods of pickling differ much and figures available indicate that the effect on vitamin C is also variable. Homemade chutney is considered to retain in some instances as much as 94 per cent of the vitamin C in the fresh ingredients — other samples contain only 40 per cent. Sauerkraut has a good vitamin C value when freshly made but loses it rapidly on storage.

Canning : Carotene is relatively stable to canning but vitamin C losses vary. It may be destroyed by the preliminary blanching of vegetables and is not well retained by non-acid vegetables like beans, peas and spinach, which lose about half their original content. Canned foods are best kept in a cool dry place. Storage of canned vegetables at 50°F. Causes losses of 5-15 per cent of the vitamin C content, while at 80°F. This loss is increased to 20-35 per cent.

Freezing : In 1929 a quick method of freezing foods was developed and it has proved effective for the preservation of vitamins. Although preliminary blanching has to be used to destroy oxidative enzymes in vegetables, ultimate losses in vitamin C are only from 20-24 per cent. The best temperature for storage seems to be about 0° F.

These 'quick frozen' foods do not readily lose vitamins on thawing but do on being allowed to stand at room temperatures. They ought therefore to be used as soon as they have been thawed.

Cooking : Whatever the method of harvesting, storage or preservation, most vegetables are eaten cooked, and the vitamin value of the food as it arrives at the table is the figure that is of the greatest interest to those engaged in the planning of diets and the preparation of meals.

Carotene is stable to cooking and there is very little difference between estimates made on raw and cooked vegetables, but vitamin C, one of the least stable of all the vitamins, is readily lost and every possible measure should be taken to preserve its activity.

The following points should be kept in mind :

1. Vitamin C is readily soluble in water. It tends, to pass into solution in the water in which the vegetable is being cooked. This happens more readily if the vegetable is already cut into small pieces or shredded. If the cooking water is to be used, the leakage of vitamin C into it does not mean loss, but in those cases where the cooking water is not to be used, it should be kept to as small a quantity as possible.
2. Vitamin C is also lost by oxidation. Vegetables should therefore to be plunged into boiling water which has been

actively boiling for some little time so that air has already been boiled off. Rapid boiling does less damage than slow simmering.

3. Vitamin C is less stable in alkaline than in neutral or acid solutions. The practice of adding soda to leafy vegetable so as to retain the green colour is therefore not good for the preservation of vitamin C.

Pressure Cooking : This is the method where by the largest quantities can be vitamin C retained, the losses being only about 10-40 per cent. This is, of course, not the only consideration when a method of cooking is being chosen, but is important in view of the increasing use of pressure cookers.

The following diagram shows a comparison between the percentages of vitamin C and carotene retention in three different methods of cooking vegetables.

The following table gives what information is available on the average values of cooked vegetables.

Carotene in Cooked Vegetables (i.u. per ounce)

None or Very Little		
Beetroot	boiled	
Cauliflower	boiled	
Celery	boiled	
Marrow	boiled	
Mushrooms	boiled	
Parsnips	boiled	
Potatoes	boiled	
Swedes	boiled	
Turnips	boiled	
Up to 200 i.u./oz.		
Artichokes	boiled	28 (estimated)
Asparagus	boiled	37
	press cooked	40
Beans	baked	47
French beans	boiled	162 (estimated)
Runner beans	boiled	197
Brussels sprouts	boiled	85 (estimated)

Contd...

None or Very Little			
Cabbage.	boiled	161	
	press, cooked	173	
Leeks	boiled	90 (estimated)	
Lentils	boiled	57 (estimated)	
Onions	boiled	150 (estimated)	
Peas	boiled	51 (estimated)	
	processed	48	
Over 200 i.u./oz.			
Carrots	boiled	3,000	(approx.) of which 750 available for conversion
	press cooked	3,200	800 available for conversion
Kale.	boiled	3,300	1,320 available for conversion
Spinach	boiled	2,300	1,020 available for conversion
Turnip tops	boiled	5,700	2,280 available for conversion

Vitamin C in Cooked Vegetables (mg./oz.)

None		
Beans	baked	
Lentils	boiled	
Marrow	boiled	
Peas	dried	
	processed	
	split, boiled	
1-4 mg./oz.		
Asparagus	boiled	2
Broad beans	boiled	2.5 (estimated)
French or Runner beans	boiled	1.6 (estimated)
Beetroot	boiled	1.7
Carrots	boiled	up to 1
	press cooked	2
Celery	boiled	1
Leeks	boiled	3 (estimated)
Onions	boiled	1

Contd...

None		
Parsnips	boiled	1.5
Peas	boiled	2.0
Potatoes, old	boiled	1.4-3
	press, cooked	4.5
	roast	2-3
	in jackets	3.3
	fried	2
Pumpkin	boiled	0.6
Tomato	fried	1.5-3.0
	canned	4
Turnips	boiled	2.8
5-9 mg./oz.		
Artichokes, globe	boiled	5.6 (estimated)
Broccoli, tops	boiled	6.3
Brussels sprouts	boiled	8-14
Cabbage	boiled	3-10
Potatoes, new	boiled	6.4
	press, cooked	6.9
Kale	boiled	5
	press, cooked	
Spinach	boiled	7.2
	press, cooked	8.6
Swedes .	boiled	6.5
20-20 *mg./oz.*		
Cabbage	press, cooked	13
Cauliflower	boiled	11
	press, cooked	15
Turnip tops	boiled	11.6
Over 20 mg./oz.		
None		

The usual practice of warming up vegetables or cooking them a second time is thus destructive of vitamin C and is justifiable only when there are abundant alternative sources of the vitamins in the diet.

In large scale cookery, vegetables are frequently prepared some hours or even a day before they are cooked and are left standing in water. This is not recommended for green vegetables but potatoes suffer merely slight losses.

It is clear from the above discussion that vegetables can be good sources of both carotene and vitamin C. The following measures help in keeping the values as high as possible.

1. Using as soon after gathering as possible.
2. Storing in a cool dry place, away from direct sunlight.
3. Rapid cookery with the use of as little water as possible.
4. Avoiding cooking too much and having to re-cookleft-overs.

Fruits

Various considerations about vegetables also apply to fruits. These, too, are sources chiefly of carotene and vitamin C. Although small quantities of the B vitamins may be present, these reach a higher concentration in seeds.

Hard fruits, such as hard vegetables, withstand transport better than the softer varieties. Many fruits that are imported from abroad are packed while still unripe to prevent damage from bruising and crushing during transport.

Carotene in Raw Fruits : Apricots, peaches and tomatoes are the only fruits having a carotene content comparable with that of vegetables. Raw apricots contain approximately 200 i.u. per ounce, peaches slightly less and tomatoes 280-300 i.u. per ounce.

The following values may be useful for reference. Carotene in Raw fruits (i.u. per ounce)

50-200 i.u./oz-Apricots	
Blackberries	196
Gooseberries	85
Peaches	57
Plums	185
Tangerines	100

Other fruits contain less than 50 i.u. per ounce and are not in general worth considering as sources of carotene. They are important, however, for their vitamin C content.

Vitamin C in Raw Fruits (mg. per ounce)

1-4 mg./oz		*5 9 mg./oz.*		*10-20 mg./oz-*		*over 20 mg./oz.*	
Cherry		Apple		Gooseberry	11	Black currant	57
Grape	1	Mulberry	5	lemon (juice)	12		
Pear		Blackberry	6	Red currant	13		
Plum		Grape-fruit	7	Tangerine			
Banana	2	Loganberry		Orange	16		
Apricot		Raspberry	9	Strawberry			
Melon	3						
Pineapple							
Cranberry	4						

Among such, apples, pears and bananas, that bear transport well and can be stored without mechanical damage, are not good sources of vitamin C. Those a high value, which can be stored are the citrus fruits. These may lose up to 20 per cent during one month's storage. Black currants, which exceed all the others, have to be preserved so as to be available when not in season. Rose hips that are also very high in their vitamin C content are not eaten raw.

Methods of Preservation of Fruits : The different methods used for preservation of fruits are drying, bottling, canning and jam making.

Dried fruits contain no vitamin C. Currants, sultanas, raisins, prunes, dried bananas may be valuable for their content of mineral salts and sugars but no reliance can be placed on them as sources of vitamins.

Jam : This, is a useful domestic method of preserving fruit with its vitamin C. The rapid boiling required does not destroy the vitamin C and, although there is progressive loss on storage, jams kept for reasonable periods still contain measurable quantities, particularly those made from fruits like black currants, strawberries and gooseberries that have high initial values. Orange peel also contains vitamin C, so that marmalade, if made by modern methods (e.g. pressure cooking), is a fairly good source of vitamin C. The old-fashioned custom of leaving the cut-up orange to stand for twenty-four hours before being cooked exposes the fruit to air and to consequent loss of vitamins.

Vitamin C in Jam (mg. per ounce)

Black currant	1
Strawberry	6
Gooseberry	3
Raspberry	2.5
Marmalade	2
Plum	1

Canning and Bottling : Exclusion of air is more efficient in canning than in bottling and this helps in the retention of vitamin C. However, home-bottling is a useful method and, if the bottled fruit is kept away from light, the vitamin value is retained moderately well.

Losses in canning or bottling fruit juices and sliced fruits are about 25 per cent.

The addition of ascorbic acid in weights not more than 0.2 per cent of the weight of the fruit is said to stabilize the colour of canned fruit without affecting its flavour and the vitamin is retained for long periods in 30-60 per cent of sugar as syrup.

Little vitamin C is retained in canned fruit. Black currants may contain as much as 26 mg./oz. and strawberries, grapefruit and gooseberries from 5-7 mg./oz. Other fruits contain less than 4 mg./oz.

Cooking **:** Fruits cooked in pies, stewed or baked contain very little vitamin C- in most instances less than 1 mg. per ounce. Exceptions are those fruits which have very high initial values, e.g. black currants, strawberries, raspberries and gooseberries. These, even with a loss of about 50 per cent, still contain appreciable amounts.

On the whole, vegetables are better sources of both carotene and vitamin C than fruits. The latter, with one or two notable exceptions, need to be eaten raw and as fresh as possible if they are to contribute an appreciable amount of vitamin C to the daily food intake.

Cereals and Nuts

Though the term cereal should be applied only to grain products but in practice other foods of a starchy nature like sago (from a stem) and tapioca (from a root) are included because of their similar uses in cookery.

Nuts, too, may be either seeds, whole fruits or parts of fruit, but for the purpose of considering them as sources of supply of vitamins they will be grouped together.

All these are storage organs of plants. They are not structures where in we find vitamin C. Carotene likewise is found in only one or two exceptional instances like yellow maize and some samples of rye and millet. Cereals and nuts are, however, the storehouses for vitamins of the B complex and therefore they form a group of very interesting products — interesting because of their geographical occurrence, regional use and widespread consumption.

They are eaten as whole grains, as flour or meal from whole grains or as products prepared from parts of grains or other plant structures.

To understand the composition of grain products it is essential to be quite familiar with the structure of a single grain of any cereal, e.g. wheat, oats or rice. All these are members of the grass family and every grain is alike in being a single-seeded fruit where in the wall of the seed is fused with that of the fruit forming a single outer layer or integument. This is generally very hard, consisting of cells with a thick tough wall. Inside is a compact white mass the 'endosperm.' The cells of this part of the seed have thin walls but are packed so tightly with starch grains that the cell walls can be ignored. At one end of the grain, generally the more pointed, is the embryo or germ, the part that would grow into a new plant if the seed were sown. This is very small and can be made out as a slight indentation on the otherwise smooth surface of the grain. It occupies only about one-fiftieth part by weight of the whole grain but a section examined by microscope shows it to be made up beginnings of a shoot with rudiments of leaves and the growing point of a root. This embryo is separated from the endosperm by a layer of cells, the scutellum through which nutrients from the endosperm pass to the embryo when growth begins.

The starchy endosperm is totally devoid of vitamins but members of the B group are present in the outer cells close to the integument, in the scutellum and in the embryo itself. Whole grains therefore contain fair quantities; products made from the germ contain more still, but foods prepared solely from endosperm are poor sources.

Whole wheat was previously ground between stones to produce flour for bread making and a few stone-grinding mills still exist and work in Great Britain. The meal produced by the first grinding is very coarse and is sifted to produce fine whole-meal flour. About the year 1870 steel roller mills were introduced. These can be worked so as to extract from flour whatever proportion of the grain is desired. If only about 70 per cent (chiefly endosperm) is used, the flour, which is white, is said to be of 70 per cent extraction. The highest extraction rate used is about 90-92 per cent and includes most of the grain with the germ but without some of the hard integument or husk that goes to form bran. Various grades are also milled between 70 and 90 per cent.

Other grains used whole are oats, barley and maize, but these are coarsely ground to produce meal rather than flour. Finely ground flour from maize endosperm (with a low extraction rate) is cornflour and this is also used for custard powders.

Pearl barley and rice are freed from their husks before being offered for sale as food, and this process removes not only husk but embryo with the result that the final product contains very little vitamin B.

In primitive regions like parts of the Sudan, sorghum (Kaffir corn) is prepared by village women. They soak it until the grains are soft enough and then pound these to a coarse pulp that can be made into flat cakes and baked. The resulting cakes or biscuits are called Kisra and contains most of the vitamin B of the fresh grain.

All the B vitamins are not distributed in the same way in whole grains; for example, oats, which has a high content of B_1 is low in nicotinic acid and the converse is true of barley.

The values of grains which are used whole are listed in the following table.

Whole grains (raw)	*Microgrammes per oz.*			
	B_1	*Riboflavin*	*Niacin*	B_6
Barely	30	10	700	130
Maize	94	37	430	200
Millet	130	10	460	90
Oats	180	31	340	—
Rice, polished	18	20	300	92
Rye 80	30	340	100	
Sorghum	130	23	740	—
Wheat	135	60	1,070	110

The Preservation

Grains being hard and dry and not easily damaged, grains withstand storage well. Loss of vitamins occurs if they are allowed to become damp or overheated. Moreover they are liable to be attacked by parasites — larvae of moths, weevils and moulds which may cause deterioration. But grain kept dry and clean retains its B vitamins for long periods of storage.

Most cereals are used in food as meal or flour, and in this form they provide the staple foods to people in many countries, the choice depending on climate, soil, custom and economics. The various rye breads of the European continent, the polenta and spaghetti of Italy, the oatcakes and porridge of Scotland, and the wheaten bread of Australia, the North American continent and Great Britain are examples. In India and China whole rice is still eaten to greater extent than bread made from any kind of flour.

The Storage

Ground products of whole grains keep only for limited periods because they contain in addition to starch, small quantities of fat (in the germ) and moisture. In conditions of temperature conducive to enzyme action, the fats are split producing fatty acids and the presence of these makes the entire meal rancid.

It was this tendency of whole meal flour to become rancid that made the roller mill such a welcome innovation to millers, because the white flours thereby produced were for the first time free from the oil of the germ and kept for long periods without rancidity. They were also whiter and this pleased public tastes, satisfying a desire that can be traced back to the time of the Roman occupation of Britain. At the time when modern white flour first became readily available, i.e. in the latter half of the 19th century nothing was known of the effects that might be produced by a resulting shortage of B vitamins. Thus white flour became popular because it could be stored without deterioration and made a loaf light in colour and in texture.

Vitamin-B Values of Meals, Flours and Pellets

Microgrammes per oz.

	B_1	*Riboflavin*	*Niacin*	*Pyridoxin*
Arrowroot	—	—	—	—
Barley meal	—	—	—	—
Cornflour \	—	—	30	—
Custard powder	—	—	30	—
Wheat flour				
70 per cent	24	10	300	72
81 per cent	68	23	450	—
92 per cent	130	60	800	104
Lentil flour	—	—	321	—
Maize meal				
60 per cent	14	9	140	—
96 per cent	90	370	430	16
Millet meal	130	10	460	90
Oatmeal	128	40	300	70

Contd...

Microgrammes per oz.				
	B_1	*Riboflavin*	*Niacin*	*Pyridoxin*
Rye flour	77	30	340	98
Sago	—	—	60	—
Tapioca	—	200	330	—
Taro (Hawaiian)				
Corms	60	8	170	—
Leaves	66	132	460	—

These are not all prepared from cereal grains.

Arrowroot is obtained from the underground stems of a species of Maranta. The starch is extracted and grated to a very fine powder. With boiling water or milk it readily forms a smooth jelly-like paste and was previously used in invalid cookery, but now it is clear that it has no vitamin content and any value it possesses is by virtue of its calories and digestibility only.

Custard powders consist chiefly of cornflour, coloured and flavoured.

Lentil flour, from a leguminous plant, contains some protein but little vitamin value except from the presence of some nicotinic acid.

Maize meal is unique among ground cereals in that it contains carotene. Whole yellow maize contains 28 i.u. per oz. Italian polenta is made by boiling yellow maize meal with a small amount of water to form a thick paste. It is eaten either alone or with cheese.

Sago and *tapioca* resemble each other in being composed chiefly of starch, with no vitamin B_1 Tapioca contains fair quantities of riboflavin and nicotinic acid but sago very little. Tapioca is prepared from starch found in the tuberous roots of the cassava, and sago is obtained from the starchy pith in the stems of several species of palm.

Taro is a plant of some interest. Both its corms and leaves are used in the preparation of food in Hawaii and both contain good amounts of the B complex.

These are other cereals are mostly eaten cooked or at any rate processed on a large scale.

Processed Cereals **:** A great many cereal foods are cooked during large scale manufacture but not further cooked in the home. These include all those generally grouped as 'breakfast cereals.' They vary widely in their vitamin B content according to the grain from which they are formed, the part of the grain used and the temperatures to which they are subjected during preparation.

The values of some of these are listed below.

Cereal Products, Processed but not Further Cooked

	Vitamin B_1	*Riboflavin*	*Niacin*	*Pyridoxin*
Bemax(wheat germ)	450	200	1,700	450
Bran, processed	180	140	5,000	?
Cornflakes	trace	–	—	—
Puffed wheat	—	30	1,300	?
Shredded wheat	73	40	1,400	?
Wheat flakest	38	50	1,800	?

Manufacturers' data. Lampitt and others, 1945.

Bicknell and Prescott, 1947.

Bread **:** The domestic cooking of bread has mostly given way to large-scale preparation. Breads are thus intermediate products between processed cereals and cooked cereals. They cannot be stored as long as the cereals which are specialty packed but can be kept for several days and in this respect are unlike cereal dishes that are eaten immediately after being cooked. The various crisp-breads properly packed can be kept in the same way as breakfast cereals and perhaps starch-free breads like Energen rolls should be included here also.

Wheaten bread has been eaten in Great Britain since the latter part of the 18th century. Before that, most of the bread was made from barley or rye and rye bread is still popular on the continent of Europe.

By the middle of the 19th century the consumption of bread was very high and has been estimated as about 21 ounces per head daily in 1840. This would all have been from stone-ground flour. By 1934, when most of the bread was white, the daily consumption had fallen to 11.6 oz. per head and during the period when bread was rationed the level permitted was 9 oz. Since rationing ceased, the level has again risen and in 1951 was 12.3 oz. per head. The effect of this difference in quantity consumed and type of flour is seen in the variation caused in vitamin B_1 intake from bread.

Year	*Estimated mean consumption of bread per day oz.*	*Type of bread*	*Vit. B_1 mg./oz.*	*Vit. B_1 eaten as bread mg.*
1840	21.3	Stoneground wholemeal	o:13	2.8
1934	11.6	White, 70 per cent extraction	0.024	0.28
1945	9	90 per cent extraction (ration)	0.13	1.17
1951	12.3	81 per cent extraction	0.048	0.59
1954	?	white, restored	0.048	
	?	or 81 per cent extraction (National flour)	0.048	?
		or wholemeal, 90-92 per cent extraction	0.09	?

The bread cannot be exactly equivalent in nutrients to bread of a high extraction rate. There are many known and no doubt still some unknown factors in wheat germ, all of which are not added.

Nothing has so far been said about the vitamin E of grains. It is all situated in the oil of the germ and is, removed with the other germ factors in the preparation of white flour.

We can argue that there are no signs of vitamin E deficiency in man and therefore removal of vitamin E is of no importance, but this is by no means unquestioned. All these large-scale changes in foods consumed by any population are in the nature of nutritional experiments and time alone shows their effects.

Among breads should perhaps be considered the crispbreads and various forms of biscuit. Most of these are subjected to temperatures high enough to make the final products brown and crisp throughout. This is destructive of vitamin B_1 but not of the heat-stable fractions. Bread contains more B_1 in the central uncharred portion than in the crust and slices of bread, toasted, lose some B_1 in the charring that occur on the outside.

Vitamins of Breads and Biscuits

	Microgrammes per oz.		
	B_1	*Riboflavin*	*Niacin acid*
Biscuits, water type	37	30	300
Biscuits sweet	18	10	200
Bread, wheaten 70 per cent	24	10	200
Bread, wheaten 81 per cent	48	17	330
Bread, wheaten 92 per cent	90	50	600
Bread rye crisp	70	—	—
Sorghum, Kisra	—	12	800

Cooked Cereals : Cereals, cooked before purchase or else in the home form part of almost every meal and contribute the bulk of the calories in any daily diet. It is therefore easy to see that the intake of 3 vitamins is largely controlled by the type and quantity of the cereals eaten.

The cooked polished rice of the Far East is still responsible for the prevalence of beriberi in India and China, and white bread, unfortified as it was in Great Britain before 1940, doubtless casued minor degrees of deficiency, for example, digestive and cardiac disturbances.

A great many cereals in various forms are now available for domestic cookery—flours for making into cakes and pastries; meals such as oatmeal, maize meal ground rice, semolina and ground almonds for the making of porridge and puddings — grains, whole and split such as barley, rice, lentils and peas for puddings, soups and savouries, pellets such as tapioca and sago for puddings and other dishes, and finally, made-up cereal preparations such as macaroni and spaghetti.

These can be classified into two main classes, those mixed with different amounts of fat and only small amounts of water and usually baked into pastry and cakes, and those cooked with larger amounts of water which is afterwards drained off, sometimes to be used, sometimes not.

In all those processes the B_2 vitamins withstand heat well, but being water-soluble may be dissolved out into the cooking water and lost if this is discarded.

Vitamin B_1, being less stable, is retained in the inner portions which are not oxidized but is gradually lost in prolonged or excessive heat or by surface charring.

It is important, too, to remember that cereals cooked in water take up large quantities and that while this increases the bulk, the weight of the cooked food is to a large extent that of its moisture content.

Thus it requires only 2 oz. of rice to make 24 oz. of rice pudding and only 1 oz. of oatmeal of make 6 oz. of porridge.

These points are illustrated in the following comparisons between the vitamin B_1 content of equal weights of raw cereal and pudding or other dish prepared from it.

	Raw Cereal			*Cooked*	
	Moisture per cent	*B_1 mg./oz.*		*Moisture per cent*	*B_1 mg./oz.*
81 per cent extraction flour	12.3	.068	Bread from 81 per cent extraction flour	36.0	.048
Oatmeal	8.9	.128	Porridge . .	89.1	.015

Other Vitamins : The newly discovered members of the B-complex are present in most cereals but values are not yet available.

Vitamin E is obtained mainly from cereals, the oils of which are rich sources, although some is present in leafy vegetables and in animal products. Values are known for a limited number of foods and some are listed below:

Vitamin E in Some Cereal Foods

	mg./oz.	
Wheat and wheat products	whole wheat	0.3-1.0
	wholemeal flour	0.6
	white flour	0.3
	macaroni	0.3
	wheat germ (Bemax)	8.0
Oats	whole oats	0.3-0.6
	oatmeal	0.6
Other grains	barely	0.9-1.4
	maize	2.8
	rice	0.1-0.2
	rye	2.9

Vitamins in Nuts : The only vitamins in nuts are those of the B-complex and vitamin E, but some nuts are relatively high in these.

The following table provides the information at present available.

	Microgrammes per oz.			
	B_1	*Riboflavin*	*Niacin*	*mg./oz.*
Almonds, whole	24	90-152	334	—
Brazil	—	—	—	—
Chestnuts, raw	80	—	—	—
roasted	—	—	60	—
Hazel	168	—	60	—
Coconut, fresh	10	30	100	0.1-0.3
desiccated	trace	—	—	—
Peanut	179	80	2,600	3.0
Walnut	51	37	320	—

Green Walnuts also Contain Vitamin C : Since the B vitamins and particularly vitamin B_1 are obtained mainly from cereals, there is a real danger of deficient intakes when large quantities of refined cereals are eaten because no foods of any other kind will make up for what these lack.

When the population can choose freely it is essential that everyone should have some knowledge of these facts so that the choice of food should be governed not only by preference but also by knowledge of nutritional value.

Dairy Produce : Modern methods have resulted in the production of farm animals very different in type from their wild ancestors: cows which yield milk when it is not needed by their calves, and hens which lay more eggs and at most seasons of the year. Man no longer has to hunt wild animals for his food. He can to a great extent control the production of both vegetable and animal food for his own use.

The cow eats carotene in grass, changes it into vitamin A and passes this into its milk from which it is available not only in the form of liquid milk but as butter and cheese. Hens eat grain and poultry mashes and pass different vitamins, e.g. A, D and riboflavin, into the yolks of their eggs.

Dairy produce is thus a source of vitamins very valuable in human nutrition and particularly so in relation to vitamin. A because it becomes available as the performed vitamin rather than as the less readily assimilated carotene of vegetables.

Vitamin D is also present in eggs and butter though in much smaller quantities than in fat fish.

These dairy foods are available in the fresh state/when they can be eaten raw or cooked, .They are also treated to make them suitable for storage, after which again some may be eaten without further cooking but some, e.g. dried eggs, are suitable only for use in cooked dishes.

Dairy Foods : Truly speaking, only fresh milk, whole or skimmed, raw cream, and fresh eggs are the dairy foods which are freshly used.

Milk

Fresh milk may be:

(a) T-T, or *Tuberculin tested,* when it is from cattle regularly tested and proved to be free from tuberculosis and kept from contact with other herds,

or

(b) Accredited, when it is from cows subject to veterinary inspection every three months. This milk must pass a certain standard indicating freedom from bacterial contamination.

(c) *Cream* : Raw cream is made up of the milk fats that rise to the surface when milk is left to stand. It contains less water than whole milk and a higher concentration of vitamins.

The vitamin content of milk and cream depends partly on the food of the cow and partly on its breed. Animals fed on young spring grass yield milk with the highest concentration of vitamin A in the fat.

Eggs : The concentration of vitamins in eggs depends upon the hen's food, the vitamin — A content being at its highest when the hens are fed on grass or green leaves and that of vitamin D when they are given cod liver oil or some other concentrated source of vitamin D in their food or when they live in sunshine. Laying hens kept intensively are sometimes exposed to ultra-violet light so as to increase the quantity of vitamin D in the eggs.

The colour of the yolk is no guide to its vitamin content. A pale yolk may be relatively high in performed vitamin A. The colour can be made deeper by the feeding of foods rich in carotene or even by feeding various pigments. Deeply coloured yolks are preferred in some areas and pale yolks in others. The vitamins of eggs, except riboflavin and vitamin B_6, are present only in the yolk.

Milk can be subjected to certain processes while still fresh, the most important of these being pasteurization, to that destroy bacteria that are specially injurious to health, e.g. those responsible for diphtheria, typhoid and tuberculosis. The milk is subjected to a temperature of 145°-150°F. (62.8-65.6°C.) for 30 minutes or else to 162° F. (72.2°C.) for 15 seconds and cooled.

Pasteurization does not affect the content of vitamins A, D or riboflavin. *Cream* may be clotted by being heated to

Vitamins in Fresh Milk and Eggs 1 oz. of each

	A	*D*	*E*	B_1	*Ribo-flavin*	*Niacin*	B_6	*C*
	i.u.	*i.u*	*mg.*	μg.	μg.	μg.	μg.	mg
Milk, whole	20-40	3	0.03	13	40	—	20	0.4
skimmed	—	—	0.01	13	40	—	—	0.4
Cream, raw.	450	14	?	13	40	—	?	0.3
Egg whole	250	17	0.6	37	170	—	56	—
yolk	200-800	upto 50	1.5	120	100	—	200	—
white	—	—	—	—	100	—	30	—

A better comparison a half pint of milk (10 oz.) is perhaps between and one egg (2 oz.): the vitamins of

	A	*D*	*E*	B_1	*Ribo-flavin*	*Niacin*	B_6	*C*
	i.u.	*i.u*	*mg.*	μg.	μg.	μg.	μg.	mg
½ pint milk	300	30	0.3	130	400	—	200	4
1 egg	500	34	1.2	74	340	—	112	—

scalding' temperature and then being cooled in shallow pans. This causes the fat to be caught in the meshes of clotted protein and the whole mass thickens to a stiff paste.

The vitamin content remains unaffected.

Yoghourt is milk soured by the culture in it of a specific lactic-acid-producing bacillus. The temperature of the culture is not high and there is nothing in the process to destroy the natural vitamins of milk. Vitamin C may be slightly reduce but is low in milk even when this is fresh.

Preservation of Dairy Foods : Milk can be preserved by concentration or by being converted into cheese; cream can be canned and the fats of milk can be made into butter that keeps much longer than liquid milk.

Evaporated milk is prepared by subjecting whole milk to heat to destroy any bacteria that are present and then concentrating it to one-quarter or one-fifth of its original bulk. It is treated to reduce the size of its fat globules, canned, and sterilized.

Condensed milk is concentrated rather less, down to one-third of its original bulk and may be sweetened by the addition of cane sugar. This helps to keep it sterile.

Dried milk is a fine powder, produced either by spraying or by passing liquid milk over heated rollers, thus reducing its moisture content to under 3 per cent. Dried milk may be full-cream, half cream, or skimmed. Dried skim-milk does not contain the fat-soluble vitamins.

Butter

Butter is produced by separating the milk fats from the rest of the milk. The separation is more complete than in the preparation of cream, butter containing about 85 per cent of the milk fat and only traces of moisture. The natural colour of butter depends on its carotene content, this being derived from green foods, but some pale butter may have a high concentration of performed vitamin A.

Butter thus contains the fat-soluble vitamins of milk in higher concentration but not the water-soluble vitamins. Its keeping properties depend on the amount of moisture present. Water in butter causes the development of rancidity.

Cheese

Cheese contains most of the solids of milk, a gallon of milk being required to produce about a pound of cheese of the Cheddar type. The milk is soured either by the action of its natural enzymes or by rennet coagulation after which the whey is extruded and the cheese is permitted to ripen. Different moulds and yeasts are used to produce cheese of characteristic type and flavour. Most of the common cheeses are made from full-cream or partly skimmed cow's milk but sheep's and goat's milk are also used and in some countries, the milk of the reindeer, buffalo and llama.

Cheeses contain only small quantities of vitamin D, but are rich sources of vitamin A and contain also some riboflavin. They form the best means of storing the vitamin A of milk. The riboflavin may even increase in the surface layers during ripening, probably owing to synthesis by micro-organisms.

Eggs

Eggs can be preserved whole in water-glass (Sodium silicate) or by being dipped in a cellulose varnish. Dipping them in oil also preserves them for short periods. The riboflavin quite stable during storage but 50 per cent of the pyridoxine may be lost in 12 months. Eggs are preserved commercially by drying, the whites and yolks being beaten together first and then sprayed through warm air forming a powder.

They are reconstituted by the addition of water. Some vitamin A may be lost by prolonged storage.

Other Foods

Other foods that are similar to dairy foods must be included here, i.e. margarine, lard, cooking fat and dripping.

Margarine was first produced during the Franco-Prussian war from emulsified purified beef suet. Modern margarines can be obtained from animal fats, hardened vegetable oils or mixtures of both, and in Great. Britain margarine is now fortified by the addition of concentrates of vitamins A and D.

Neither lard nor cooking fat contains any vitamins but dripping may contain small amounts.

Vitamins of Prepared Dairy Produce per ounce

	A	*D*	*E*	B_1	*Ribo-flavin*	*Niacin*	B_6	*C*
	i.u.	*i.u*	*mg.*	µg.	µg.	µg.	µg.	mg
Milk, evaporated condensed	105	1	.09	18	100	100	?	0.6
Sweetened	105	1	?	30	100	100	15-30	0.9
Dried	300	3	.014	85	330	200	90	2.0
Butter Cheese	800-	17	0.5	—	—	—	—	—

Contd..

	A	*D*	*E*	B_1	*Ribo-flavin*	*Niacin*	B_6	*C*
	i.u.	*i.u*	*mg.*	μg.	μg.	μg.	μg.	mg
Cheddar	370	4	0.3	9	140	100	18	—
Roquefort	700	?	?	8	100	350	?	—
Processed	310	?	?	3	110	40	?	—
Eggs, dried	850 or less	?	?	10	—	—	—	—
Margarine	560	56	15	—	—	—	—	—
Lard	—	—	—	0.7	—	—	—	—
Cooking fat	—	—	—	—	—	—	—	—
Dripping	28	?	9-12	—	—	—	—	—

The Effect of Cooking on the Vitamins in Daily Foods : Milk is extensively drunk raw but is also added to hot drinks such as tea and is itself heated in the preparation of beverages like coffee and cocoa. It is also an ingredient in many cooked dishes like custards, milk puddings, soups and sauces.

The vitamins that would be destroyed by these cooking processes are fortunately present only in small amounts in milk, e.g. vitamins B_1 and C.

Vitamin A and riboflavin are both stable to heat and are therefore reduced only slightly if at all in cooking.

This stability of vitamin A and riboflavin is important and means that cooked dairy produce can be relied upon as sources of these vitamins. Individuals who are advised to take one or two pints of milk daily sometimes have difficulty in consuming this quantity and are apt to assume that it must all be taken as liquid milk. Provided that vitamin B_1 is adequately supplied by other foods there is no reason why milk should not be taken cooked in a variety of ways, e.g. half a pint in coffee for breakfast or at a mid-morning break, another half-pint at lunch or dinner partly in soup, partly in pudding, and some in an evening drink.

Eggs also retain their stable vitamins when used in the making of cakes and puddings.

Meat and Fish

Animals, whether mammals or fish, store fat-soluble vitamins in their body fat and also in glandular organs, e.g. the liver. Small amounts are also found in muscle. Thus when these parts of the animal body become human food, they provide not only protein or fat but also vitamins to the diet.

Vitamin A is stored in the liver and the human liver may contain more than a year's supply that is one reason why gross vitamin-A deficiency is not common in Great Britain (Tomaszeski, 1947). It occurs only when individuals have been in short supply for fairly long periods.

Vitamin D is found not only in the livers of fish but also in the body oils so that fat fish can be a source of supply of this vitamin for those who do not take fish liver oils.

Vitamin E is stored partly in muscle but more in the liver. Its concentration in animal foods is, however, much lower than in cereals.

Of the various water-soluble vitamins, vitamin C is found in the animal body in only very small quantities, the only organ containing appreciable amounts being the adrenal gland. This is very small and is not generally eaten. Vitamin B_1 is found in the liver and smaller quantities in muscle, but except for pork and bacon, muscle meat contains very little. The pig is exceptional in its ability to store vitamin B_1 in its muscles.

Some of the B_2 vitamins are, by contrast, found in large quantities — riboflavin in liver and kidney and nicotinic acid in these glandular organs and also in muscle.

Most of the vitamins obtained from meat and fish are heat-stable and are thus available when the food is cooked.

Preservation : Curing of meat is an old process. Both curing and corning are effective for the preservation of the B vitamins other than B_1. Corned beef retains only 20-50 per cent of its vitamin B_1 but as much as 90 per cent of its riboflavin.

Hams prepared commercially by being placed in pickling solution for 12 days and then being smoked for 18 hours were found on test to contain 73 per cent of their original vitamin B_1 92 per cent of their riboflavin and 84 per cent of their nicotinic acid.

Generally nitrites and nitrates are used in the curing of meat and, if the same meat is used with vegetables in stews, these salts will rapidly destroy the vitamin C of the vegetables.

Canning is a satisfactory method of preservation particularly for fish containing vitamins A and D. Unless subsequent storage is unduly prolonged, there is very little loss. Canned fish such as herrings and sardines are thus valuable sources of vitamin D.

***Cooking*:** During cooking, vitamin is B_1. may be lost. In stewing as much as 50 per cent may be lost and in roasting 40-60 per cent. But by the ordinary methods of cooking meats: roasting, grilling, frying, braising or stewing, the losses of riboflavin and nicotinic acid amount to less than 15 per cent.

Comparing these as sources of the B vitamins with foods in the cereal class, we find that one slice of bread made from 81 per cent extraction flour provides 3 times as much vitamin B_1 as one ounce of roast beef but only one-third as much nicotinic acid.

Vitamin A in Meat and Fish (i.u. per oz. approx.)

Less than 20 i.u. loz.		*20-100 i.u. loz.*		*100-500 i.u. loz.*		*Over 1,000 i.u./oz*
Veal	11	Bloater	32	Oyster	120	Calves' liver
Beef		Herring		Sheep's kidney		1,100-1,200
Mutton	about 14	Mackerel	42	250-300		
Cod	about 14	Kipper	50			
Haddock	about 14	Sheep's heart	60			
Hake		Canned salmon sardine	70-80			

Vitamin D in-fat Fish (i.u. per oz. approx.)

Cod roe	fried	20
Herring	fried	250
Kipper	grilled	250
Mackerel	fried	100
Salmon	canned	170
Sardines	canned	280

Vitamin E in M 250 Meat and Fish (mg. per oz. approx.) (only a few values available)

Bacon	15
Beef lean	15
Chicken	.06
Calf liver	.04
Mutton, lean	.22
Pork, lean	.20
Cod roe	1.5-2.2
Haddock, fresh	11

Although there the B vitamins, foods is great variation in the distribution of do fall roughly into groups.

Foods poor in	*(Microgrammes per oz. approx.)*		
B vitamins	*Vitamin* B_1	*Riboflavin*	*Niacin*
Beef, corned roast or stewed	10-15	40 60-70	50 1,000
Sweetbreads Tongue }	12-20	30-60	up to 800
Chicken, roast	30	20-70	1,000-2,000
Mutton, roast boiled }	20-25	40	1,600
Better sources of the B vitamins			
Bacon, fried	120-200	100-150	1,000
Ham, boiled	110	30-100	500
Pork, roast	300-400	35-50	1,200.
Cod roe	200	200	400
Salmon, canned	300	60	1,800
Liver	110	890	3,600

Miscellaneous Foods

Sweets and Preservers : The only sweet that may contain appreciable quantities of performed vitamin A is *ice-cream,* good samples of it may contain as much as 400 i.u. per ounce. Small quantities of carotene are present in chocolate (20 i.u./oz.), lemon curd (13 i.u./oz.) and jams (6 i.u./oz.).

Chocolate contains no vitamin C, but enough of the B complex (B_1 30 ug., riboflavin 70 ug., niacin 300 ug. per oz.), carotene 20 i.u. per oz. and traces of vitamin D.

Molasses, that is obtained from the crude liquor from which sugar is crystallized, contains the B vitamins in addition to cane sugar, glucose, fructose and mineral matter. Cane molasses contains about:

14 microgrammes of vitamin B_1	per oz.
55 microgrammes of riboflavin	
and rather over 400 microgrammes of nicotinic acid	

During further purification these are destroyed and neither sugar nor golden syrup contain any vitamins.

Honey contains small quantities of:

vitamin B_1	5 microgrammes	
riboflavin	20 microgrammes	per oz.
nicotinic acid	100 microgrammes	
vitamin B_6 up to	80 microgrammes	
and vitamin C	1-6 mg.	

Jellies contain no vitamins but are convenient for 'setting' fruit. The flesh of a large orange or grape-fruit can be removed with very little damage from half a fruit and can be set in jelly in a small container. This is an easy way of carrying the fruit for picnic lunches and obviates the sticky fingers caused by peeling an orange. Although geltine contains no vitamins, it is not without value. Eaten with bread, the amino acids of gelatine together with those of wheat contribute the equivalent of good protein to the diet.

Puddings and Deserts : These depend for their value on the ingredients used and the method of cookery.

The value can be estimated easily by reference to the values given for cooked cereals, fruit, eggs and other ingredients.

Spices, Condiments and Flavouring Agents : These are all eaten in small amounts and, with few exceptions contribute little of nutritive value to the diet.

Vitamins in 1 Teaspoonful (½ oz.)

	A	B_1	*Riboflavin*	*Niacin*	C
	i.u.	μg.	μg.	μg.	*mg.*
Paprika (green)	—	—	—	—	4
Paprika pods.	—	—	—	—	2-100
Marmite	—	112t	225	2,150	—
Bovril	—	9	60	2,000	—
Ham & beef paste	—	4	5	17	—
Salmoon & Shrimp paste	—	3.5	3	35	—
Dried thyme	—	40	—	—	—
Dried mint	—	—	—	—	1-14
Ginger	—	—	—	25-30	—
Sage	trace (carotene)	—	—	—	1

Manufacturer's data.

Freshly prepared. [There is progressive loss on keeping.]

Booth and others, 1944.

Fish Liver Oils

	per oz. Vit. A i.u.	Vit. D i.u.
Halibut liver oil	600,000	60,000
Cod liver oil	30,000	3,000

Beverages : Drinks prepared with milk contain the vitamins of milk according to the amounts used. This may be so little as to be negligible in the case of tea and may be very little in cocoa. White coffee contains more. Coffee contains no vitamins and tea only a trace (6-11 ug./g. of riboflavin) but cocoa contains traces of vitamin A and of the B complex.

Fruit drinks and squashes vary in their content of actual fruit juice and neither flavour nor colour are reliable indicators. Where, however, the makers state the proportion of genuine fruit juices present, rough estimates of vitamin value can be made.

Special fruit juices and purees are now available where in the content of vitamin C is stated on the label. These are generally preparations of black currants, rose hips and orange.

Black currant puree	18 mg. vitamin C/oz.
Black syrup (Ribena)	20 mg. vitamin C/oz.
Rose-hip syrup	43 mg. vitamin C/oz.
Concentrated orange juicet	45 mg. vitamin C/oz.
Orange and rose-hip syrup combined (Rosena)	56 mg. vitamin C/oz.

Manufacturers' data. Min. of Food, 1945.

These can all be diluted.

Alcoholic beverages contain no vitamin C but some contain members of the B complex.

Microgrammes per oz	B_1	*Riboflavin*	*Niacin*
Ale, strong	3	10 or more	400-700
Beer, mild	1	10	40
Beer bitter	1	8	100-200
Stout	1	10	150-300
Wine	less than 1	2-8	?

Min. of Food, 1945.

Randoin, 1949.

Cailleau and others, 1949.

The vitamin B_1, even in a pint (20 fl. oz.) of beer is thus negligible but the riboflavin content may be as high as 150-200 ug. and nicotinic acid 2,000 or more ug.

The absence of vitamin B_1 in alcohol coupled with the fact that alcohol contributes a considerable quota of calories to a diet is one reason why individuals who consume large quantities of alcohol, and often consequently fail to take enough of the protective foods may suffer a shortage of vitamin B_1 This deficiency is in some cases severe enough to cause symptoms.

17

Vitamin Measurement

The isolation of pure vitamins led to further work on their chemistry and the development of chemical methods of analysis which had previously been impossible. With the evolution of these techniques we have become less dependent upon the use of expeimental animals, but a difficulty exists in that many chemical methods of estimation give accurate results only when no interfering substances are present and are not applicable to food analyses, since many other factors are always present in foods together with vitamins.

A brief description of some of the commoner chemical methods is given, followed by outlines of the biological and microbiological techniques which are still used.

Chemical and Physico-Chemical Methods of Estimation

Vitamin A : Vitamin A activity is measured either chemically or spectroscopically.

Chemical Method

This depends on the fact that a solution of vitamin A in

chloroform when mixed with a 25-per-cent solution of antimony trichloride in chloroform produces a blue colour, whose intensity depends on the quantity of vitamin A present. The colour develops gradually, reaching a maximum after ten seconds and then gradually fades. Its intensity is measured by a colorimeter.

Spectroscopic Method

This depends on the production of the same blue colour, but it is measured by a spectrophotometer instead of visually.

Vitamin B_1

Chemical Estimation

The method is based on the fact that vitamin B_1 can be converted by alkaline oxidation into another compound, thiochrome, that gives a blue fluorescence in ultra-violet light. The food or other material containing the vitamin is first extracted with water and then heated to get an adsorbate on Kaolin. This is oxidized to thiochrome and the intensity of the fluorescence produced, that is measured by special apparatus, is proportional to the amount of vitamin B_1 present.

The method used for testing pharmaceutical preparations is suitable only for high concentrations. It involves precipitating the vitamin by an alkaloid reagent, silicotungstic acid and determining the weight of vitamin precipitated.

Vitamin C : This is the vitamin that is most easily measured chemically. The method is based on the fact that vitamin C is a reducing agent, so it readily takes oxygen from other compounds and thereby reduces their oxygen content (or its equivalent) and becomes itself oxidized. A dye, methylene blue and another compound derived from indo-phenol are blue in the oxidized state but colourless when reduced. By means of a chemical titration it can be found out how much vitamin C (ascorbic acid) is required to reduce a given quantity of one of these dyes. This is compared with a solution of the substances under test and the strength of the latter can then be calculated. The practical details are a bit difficult. Various modifications of the test solution have to be made before titration, so as to remove any other reducing compounds.

Vitamin D : A chemical procedure has been published for the estimation of vitamin D, but unfortunately this can be used only for highly concentrated solutions. It is therefore still necessary to use biological methods in spite of the fact that these are less accurate than chemical analyses.

Biological and Microbiological Methods of Estimation : The methods used now, have been developed from the earlier procedures used but, the careful design of each test and the selection of animals has led to greater accuracy, and a further advance has been the discovery that micro-organisms can be used in some tests instead of rats, mice or guinea-pigs.

Yeasts, moulds, and bacteria need vitamins for their multiplication and fail to develop if even one vitamin essential for their multiplication is absent. These organisms cannot, of course, be observed singly but a colony can be used to replace a single animal of higher type.

Growth of a colony occurs either in a liquid containing the necessary nutrients in solution or on the surface of a nutrient jelly. Generally test-tubes or flasks are used for liquid media and shallow dishes for jelly cultures. The obvious saving of space by the use of such test organisms is a great advantage. Compare, for example, the space required for two or three dozen tubes or flasks with that required for cages of rats —> and the time involved in giving the necessary attention to each.

Whatever the organism used, whether rat, chicken, mould or bacterial colony, the principle underlying the assay is the same and the methods are comparable.

The commonest feature used for measurement is growth-rate, but there are others such as, for example, the heart-rate of rats, and the production of carbon dioxide by yeasts.

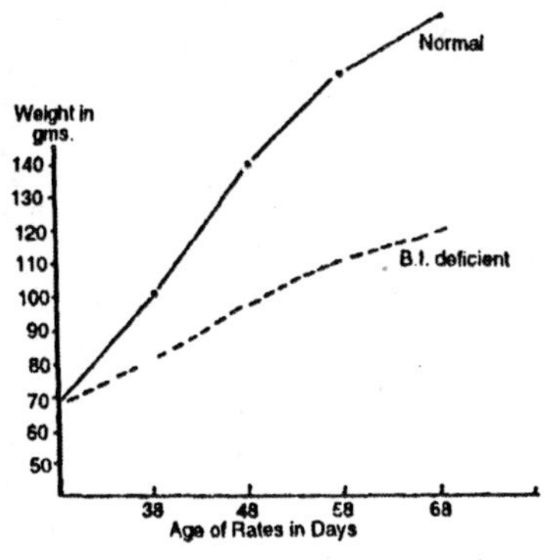

In a test, groups of animals or cultures of micro-organisms are given graded doses of the pure vitamin and their responses give a standard with which test substances can be compared.

Simultaneously, other groups of cultures similar to the ones on standard are given graded doses of the test substances and their responses are measured in the same way.

Since no two animals and no two cultures of bacteria are ever exactly alike, there is always some variation in response, even to the same dose. This is why *groups* of animals and two or more tubes or flasks of bacteria are used at each dose level.

There is a definite mathematical relationship between the response and the dose, but this is not a simple ratio and has to be worked out by statistical methods that are beyond the scope of this book.

However if care is taken to use experimental material which is as uniform as possible and if conditions are the same for all groups of material then the results generally show the graded responses to be expected from graded doses. Having measured the responses obtained by given doses of standard and the responses from given weights of test substances, it is not difficult to calculate the amount of vitamin in the dose of test substance used and hence the weight in say 100 g. or 1 oz.

Vitamins which are estimated by biological or microbiological methods are those of the B group, vitamin D and vitamin E.

Vitamin B_1

Biological Assay

The first methods to be used depended on growth rate or on the cure of head retraction in pigeons, but a more accurate method now in use makes use of the fact that rats deprived of vitamin B_1 develop in contrast to man a slower and slower pulse rate (a slow pulse rate = bradycardia: a rapid pulse rat = tachycardia).

Small animals such as rats, mice and birds have, when healthy, very rapid heart-beats, the normal speed for a rat's heart being 500 or more contractions, or beats, per minute. Rats are used for test when they have been on a vitamin B_1-deficient diet long enough for their pulse rate to be reduced to 350-400 beats per minute. Doses of either pure vitamin B_1 or foods containing the vitamin cure" this bradycardia, but the cure is only temporary and its duration depends on the amount of the B_1 given. When this feature of vitamin B_1 deficiency is used in the assay of foods, the calculation is based on the fact that the duration of cure bears a mathematical relationship to the weight of vitamin given.

The pulse rate is too fast to be measured by hand and is accurately recorded by an electrocardiograph.

Microbiological Assay

The microbiological method for vitamin B_1 is not as satisfactory as that used for some of the other vitamins of the B group, but it is being developed and gradually perfected. One procedure depends on the fact that the organism, lactobacillus fermenti, depends for its growth on the presence of vitamin B_1.

Other Vitamins of the B complex Microbiological Methods : Most of the other vitamins of the B group are measured satisfactorily by microbiological methods commonly. Two types of organisms are most used: bacteria that during multiplication produce lactic acid and moulds that produce an extensive mycelium. The lactic acid bacteria are permitted to develop in tubes of liquid media to which graded doses of, for example, riboflavin are added. When the cultures have been incubated for a long enough period for acid to be produced, definite volumes of the liquid cultures are titrated against alkali of known strength to determine how much acid has been formed. From this it is possible to calculate the weight of vitamin present.

When moulds are used, the mycelium develops on the surface of the medium, where it can clearly be seen showing the gradation between none or very little at one end of the series to a dense growth at the other end. These mycelial growths can be lifted off the surface of the medium, dried and weighed and again we get a graded series of weights according to the amount of vitamin present in the culture.

The following is a list of vitamins which can be estimated by mecrobiological methods and some of the many organisms which may be used.

Vitamin	Organism
Biotin, Nicotinic acid, Pantothenic acid	Lactobacillus arabinosus
Riboflavin	Lactobacillus helveticus
Vitamin B_1	Lactobacillus fermenti
Vitamin B_{12}	Lactobacillus leichmanii
Folic acid	Streptococcus faecalis
Para-aminobenzoic	Neurospora crassa (mutant 1633) acid
Choline	Neurospora crassa (mutant 34486)
Inositol	Neurospora crassa (mutant 37401)
Vitamin B_6	Neurospora sitophila (mutant 299)

It may be noticed that four of these use mutants of neurospora. These are varieties produced by exposing the fungus (mould) to X-rays, a procedure that makes the organisms sensitive to lack of these particular vitamins.

Vitamin D

Biological Estimations

Methods for the measurement of vitamin D are less accurate than those for most of the other vitamins. Test animals still have to be used, pending the perfecting of chemical procedures and, in spite of great improvement, the error is still considerable.

Rats and chicks are both used but the method involving the least error is that wherein bone calcification in chicks is considered the criterion.

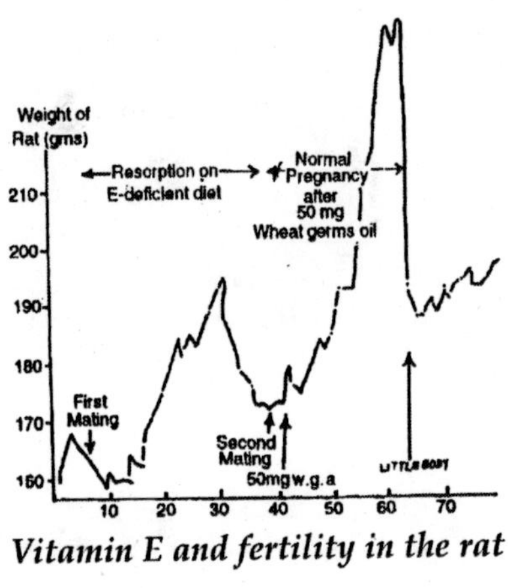

Vitamin E and fertility in the rat

Chicks grow quickly and X-ray negatives of their bones show the effects of lack of vitamin D before any gross signs of rickets can be detected. The method involves feeding to groups of chicks a basal diet containing no vitamin D, to which has been added graded quantities of standard oil or test oil and taking X-ray photographs of the legs at the end of a six-week period. The joint chosen for examination is the hock joint where the tarsal bones form a junction with the metatarsus. At this point there is a very easily seen centre of ossification (bone formation) separated from a very clearly defined metatarus. The distance between these two is very definite and can be measured to a fraction of a millimeter by means of a micrometer guage and magnifying eye piece. This is called the tarsometatarsal distance (or T.M.T.) and is found to differ with the vitamin D supplies given to the chicks, being narrow in normal chicks but wider in chicks with poor calcification. In this case there is an *inverse* ratio between the distance and the amount of vitamin D in the ration.

Because of the variation between chicks even on the same ration, large numbers are required for the test but where this is possible results obtained are better than by any other method.

This is the procedure used to measure the vitamin D potencies of cod liver and halibut liver oils used in infant feeding.

Vitamin E

Biological Method

This depends on the fact that the female rat, when fed a diet lacking in vitamin E fails to carry her young to term. They are formed, but part way through pregnancy the maternal weight begins to drop due to the resorption (re-absorption) of the embryos. No abortion and no birth of young occur. Vitamin E is measured by finding the least dose of any test food which will prevent resorption and allow the birth of living young.

There are also chemical and physical methods but the details of these are rather beyond the scope of this book.

Chemical Combination

Vitamins are chemically unrelated compounds that are grouped together because each is essential in the diet in minute amounts and is required for specific metabolic reaction within the cell. Several of the vitamins though conveniently considered *as* single substance actually are a group of structurally related compounds which tends to behave alike physiologically. In general however physiological function tend to be limited to one active form to which the related forms are converted.

The discovery of the vitamins unhered in a new era in nutrition. It was one in which *small* things in the diet assumed *great* importance. The percentage of body weight attributable to vitamins is minute. Nevertheless, the amount, even though small, is indispensable for normal functioning.

Early evidence of the existence of vitamins came from the search for treatments of certain human diseases that persistently plauged people. One of these diseases was scurvy. In was one of the hazards explorers and sea voyagers encountered. In the year 1747 relief from scurvy was enjoyed by a few of the British navy. Freedom from the illness came about in this way. Limited amounts of lemons and oranges were aboard ship and were given to some of the sailors; the recovery was remarkable. In another part of the world, another disease was making the sailors of the Japanese navy ill. This disease was beriberi. The diet of the men consisted mainly of white rice. With the addition of vegetables, meat, and fish, a change instituted in 1884, the incidence of beriberi decreased.

Despite the improvement in health brought about by changes in the diet, the realization that a lack in the diet was responsible for scurvy and beriberi was slow. In fact it was almost 150 years after the experiment at sea with lemons and oranges before there was acceptance of scurvy as being caused by a nutritional deficiency; and it was not until the early 1900's that scientists and physicians accepted beriberi as the result of a deficiency in the diet. That deficiencies in the diet could be responsible for disease was a new concept, which was adopted with reluctance.

Another avenue of approach to the study of the dietary deficiency diseases was the use of animals in the experimental laboratory. In the final analysis, it was the well-controlled laboratory experiments that eventually gave conclusive evidence that certain diseases are caused by the lack of dietary nutrients.

In the experimental studies a diet of pure chemical compounds was fed, pure protein (such as casein or albumen), pure fat (such as lard), pure carbohydrate (such as dextrin), and minerals. The purified diet was really crux of the experiment, for the natural food more than likely would have provided some of the unknown factors. As it was young animals on the purified diet failed to grow; mature animals failed to maintain body weight.

Mc Collum in his study of the history of nutrition found Jean Dumas (1800-1884) to be the earliest man of science to question the adequacy of a diet composed only of protein, carbohydrate, fat and salts. The experience that led Dumas to such a prophetic belief was not with experimental animals, however, but with citizens of France. At the time of the siege of France by the Germans (1870-1871) food was scarce; there was no milk for children. Scientists attempted to prepare artificial milk using protein, fat, carbohydrates, and salts. The combination was disastrous to the young. It led Dumas to conclude that there was something lacking in a combination of the then known nutrients essential for life.

Animal experiments that led to the descovery of vitamins included those of Pekelharing in 1905 at the University of Utrecht, who made the important observation that a diet of caseim, egg albumen, rice flour lard, and salt kept his rice alive only for four weeks. When the added milk to the mixture the rice assumed healthy.

Sir Fredick Hopkins (England) is considered as one of the most advanced thinkers of his time as animal nutrition in 1960 he made an observation,"... no animal can live on mixture of carbohydrates, proteins and fats, and even when the necessary inorganic material is carefully supplied, the animal still can not flourish."

Mc Collum in 1907, believed with many others of his time that the failure to survive on purified relation of protein, fat, carbohydrate and minerals, was due to unpalatability of the mixture and the consequent refusal to eat a sufficient amount of the ration. He tried to improve the flavour by adding sugars and other variables in rations, but the animal still failed to survive. Even though his experiments appeared to be a failure, it was successful in that it led to his discovery of vitamin A a few years later.

McCollum and Davis discovered that there was a substance present in butterfat or egg-yolk fat that made difference between moderate success in the nutrition of young rats on certain diets and prompt failure in its absence. Osborne and Mendel at the Connecticut Experiment Station found cod liver also to contain this essential substance. Thus the first vitamin was discovered and called fat-soluble vitamin A.

Historical Background

The discovery of vitamins is a long story that started with observations made centuries ago in China where the disease now recognised as beriberi was certainly known. It is probable also that liver was used in very early times for the treatment of diseases which may have been related to a vitamin deficiency. Experience in the Middle Ages showed that certain foods were definitely connected with the prevention and cure of scurvy. All these and other observations led ultimately to the scientific investigations of the present century.

Many a names are associated with this history of exploration into the field of nutrition, but in general our knowledge today is the result of the careful observations, experiments and experiences of many individuals in different countries.

Vitamins have existed in foods as long as the plants that produce them and the animals that eat them, but our knowledge of them as such dates only from the beginning of the present century. It is now recognised that without vitamins in their daily food both man and animal become unhealthily and at length succumb to various diseases. These have come to be called the 'Vitamin Deficiency Diseases' and several are familiar by name, for example *rickets* in which the bones of a child or young animal become soft and bend under the weight of the body, and *scurvy* in which the gums bleed and the joints become swollen and painful on account of internal bleeding into them. These and other deficiency diseases have been known for many years but it is only within living memory that we have learnt of their cause and of how to prevent theme.

Descriptions two to three thousand years old have been found in China of the disease beriberi, now known to be due to the lack of an essential vitamin, and in the Middle Ages many sailors died of scurvy because the food they were able to carry on long voyages was short of vitamins. Vasco da Gama who sailed from Portugal and found a route to India and the East via the Cape of Good Hope in 1497, lost 100 men out of his crew of 160 from scurvy.

One hundred years later, when Drake and other seafaring men from Great Britain were exploring the seas and finding new sea-routes, Sir Richard Hawkins was able to write in his *Observations on his Voyage to the South Sea* published in 1593, this description of the disease and its remedy:

"The signes to know this disease in the beginning are divers, by the swelling of the gummes, by denting of the flesh of the legs with a man's finger, the pit remaining without filling up in a good space; others show it with their lasinefse, others complain of the cricke of the backe ... all of which are for the most part, certain tokens of infection. That which I have seen most fruitful for this sicknefse is sower oranges and lemmons."

This is a very good description of scurvy. Hawkins knew that it was caused by lack of fruit and vegetables and he knew mat he could prevent it by giving his crew oranges and lemons.

In the 18th century Captain Cook was also careful to take a supply of fruit and vegetables with him to protect his crew from the same serious disease, and in 1804 regulations were introduced into the British Navy providing all ratings with a daily ration of lemon juice. These measures were all taken as a result of experiences and observations. Nothing was known about vitamins and we were as ignorant in 1900 as Hawkins was three hundred years before, about *why* oranges and lemons were effective remedies for scurvy.

Scientific discoveries are sometimes made by a deliberate and planned search, by experiments designed to give solutions to definite problems, but sometimes the 'discovery' follows years of careful observation and recording; one could cite many instances, of which the use of fruits and vegetables for scurvy is one, in which the practical application came first and the underlying reason was discovered later.

Towards the end of the 19th century, Dr. Eijkman, a military medical officer stationed in the Dutch East Indies, noticed that poultry, when fed on rough unpolished rice remained healthy but that when fed on the polished rice which was usually reserved for the men in the camp, they developed a paralytic condition in which they were unable to stand and could not hold up their heads (head retraction). Soldiers at the station were also suffering at the time from beriberi, a disease in which great weakness was accompanied by a condition resembling paralysis of the limbs and frequently causing death. Dr. Eijkman found that he could treat his men in the same way as he treated the hens by giving them either rough unpolished rice or rice polishings and that this was effective (Eijkman, 1897). He did not, at the time, recognize that beriberi was a deficiency disease. He thought there was some toxic substance in the starch of the polished rice and that there was an antidote in the polishings and it was not until several years later that another Dutch scientist, Grijns, put forward the suggestion that beriberi was caused by a lack of some vital ingredient which was present in the polishings of rice but not in the starchy part of the grain.

The existence of such 'accessory food factors' was finally established by Frederick Gowland Hopkins at Cambridge. Between 1906 and 1912 he was carrying out of series of experiments on proteins and other known nutritional factors in foods. He fed groups of rats on whole milk, and others on all the known constituents of milk and found that while the first group grew normally and were quite healthy the second group soon began to lose weight and became ill. When, however, he gave milk to these also, they recovered and began to grow normally. Hopkins concluded that there must be, in whole milk, unknown substances necessary for life and growth but in such small quantities that they had so far defied chemical analysis. He quoted the well known diseases, scurvy and rickets, as examples of such deficiency diseases and foresaw that there were 'many more nutritive errors' affecting health, but at that time quite unknown.

There was a need for a name for the new factors in nutrition. The name which became *vitamin* originated in this way. Funk, a Polish biochemist, coined the term "vitamine" in 1911 to designate the anti-beriberi factor that he was investigating at the time. He combined *vita,* to designate "essential for life," and *amine* to indicate the chemical structure. The term was accepted and used for all of the unknown factors. In 1919 the final "e" was dropped, for by then it was abundantly clear that there were several unknown dietary nutrients, and that the known ones did not all have the amine structure. The change in spelling avoided any implication of chemical for the as yet unidentified dietary factors.

There seemed to be only one factor needed to be supplement the purified diet at the time of discovery of vitamin A. In a couple of years, however, it was recognised, that not one, but two factors were lacking in the purified diet for normal growth, one soluble in fat, the other soluble in water. Later of course, a number of new fat soluble and some more water soluble factors were identified.

Here we give a summary of the isolation and synthesis of various vitamins:

The scientists who made these first discoveries about vitamins were wise in not attempting to give each one a name as nothing was known of their chemical nature. All that was certain was that there were several different factors in a number of different foods and that each played its own part in the maintenance of health and the protection of the body from particular diseases.

Year	*Isolation or Identification*	*Synthesis or Structure*
1925	vitamin D by Irradition	
1926	Vitamin B1	
1925	Iusitol	
1931	Vitamin A	
1932	Vitamin C	
1933	[Riboflavin 1	Vitamin C
	Pantothenate	
1935	Biotin	Riboflavin
1936	Anti-egg-white factor	Vitamin D
	Vitamin E	Thiamine

Year	*Isolation or Identification*	*Synthesis or Structure*
1938	Nicotinic acid Pyridoxine	
1939	Vitamin K	Vitamin K
1940	P- Amino benzaic acid	Pentothenatic
1942		Biotin structure
1945	Folk acid	Folic acid
1948	Vitamin B_{12}	Folioric acid
1955-56		Vitamin B_{12} structure

McCollum and Davis, two other Americans, adopted the custom of referring to the vitamins by the letters of the alphabet (McCollum and Davis, 1915), and this proves so useful and easy that we still use the letters today.

A = the growth factor in butter fat

B = the anti-beriberi factor in rice polishings and wheat germ

C = the anti-scorbutic factor in fresh fruit and vegetables.

It was not until after the World War that vitamins D and E were discovered.

Mellanby (1918-19), showed by feeding tests on puppies that rickets was a deficiency disease and, moreover, that it could be cured or prevented by the same foods, i.e. butter and cod liver oil, which contained the growth-promoting vitamin A. It was thought at first that growth and the prevention of rickets were both due to the same factor, However, later work showed that the growth-promoting factor could be destroyed by heat and oxidation while the anti-rachitic property remained, and also that although butter was weaker than cod liver oil in both properties it had *relatively* higher proportion of the growth factor A. It was therefore concluded that there must be two factors: vitamin A controlling growth and vitamin D protecting against rickets.

By this time it was known also that vitamin A prevented a disease characterized by drying of the eyes leading to blindness and known as xerophthalmia.

It would be a mistake to pass on without mentioning the notable work of Herriett Chick and her colleagues, who went to Vienna in the years 1918-21 and did much to improve the health of children who at the time were dying of malnutrition. Supplies of cod liver oil were used for the infants who had both xerophthalmia and rickets, and the rather surprising discovery was made that rickets could be cured by exposure to sunlight without any cod liver oil or butter but that one or other of these was needed to cure the condition of the eyes.

This observation led to recognition of the fact that in sunlight the body can make its own vitamin D, and this is one reason why exposure to sunlight is beneficial. Vitamin D is sometimes called the 'sunshine vitamin.'

It will be noticed that both A and D occur in fats and oils of animal origin/and B and C in plant products. This led to an early distinction between fat-soluble and water soluble vitamins, the latter being found in the cell-sap of plants or in the non-fatty storage organs.

In 1922 came the discovery of vitamin E. Evans and Scolt (1922), found that rats, even when given a good diet containing protein, salts and all the other known essential factors, failed to breed unless certain natural foodstuffs were added. The foods they found effective were fresh lettuce leaves, wheat germ and dried alfalfa. These, therefore, were sources of the fertility or anti-sterility factor which was called vitamin E.

The best source of vitamin E was found to be the oil extracted from wheat germ. Vitamin E was therefore the first exan ple to be found of a fat-soluble vitamin occurring in a. vegemble oil.

Thus by the early 1920s five definite vitamins had been recognized and something was known of their occurrence in food and of the functions each.

This knowledge is summarized in the following Table:

Vitamin	*Solubility*	*Function*	*Best Sources*
A	Fat-soluble	Promotes growth Prevents or cures xerophthalmia	Cod liver oil, butter.
B	Water-soluble	Prevents or cures beriberi in man, polyneuritis in birds	Wheat germ, rice germ, yeast.
C	Water-soluble	Prevents and cures scurvy	Citrus fruits. fresh vegetables
D	Fat-soluble	Prevents and cures rickets	Cod liver oil, butter.
E	Fat-soluble	Prevents and cures sterility in female rats (not in male	Fresh lettuce leaves, wheat germ oil, dried alfalfa. rats)

During the second quarter of the present century there have been discoveries not only of other vitamins but of the chemical nature and distribution of all those known and the physiological actions of most. Many have been isolated from natural sources and methods have been evolved by which many can now be synthesized. The following is a brief outline of the various discoveries which have been made since 1925.

The first step of the importance was the recognition (Goldberger and others, 1926), that pellagra was associated with lack of a factor closely related to, but not the same as, the anti-beriberi factor. This he called P-P or pellagra-preventing. The following year a distinction was drawn between them on account of their stability to heat and they were renamed vitamins B_1 and B_2, the former or anti-beriberi factor being less stable to heat than the anti-pellagra factor B_2.

Shortly afterwards B_1 was isolated in crystalline form (Jansen and Donath, 1926), and it was shown that B_2 was not a single substance but was in itself a 'complex.'

The next step was the recognition, between 1920 and 1929, that carotene, the yellow plant pigment, was a precursor of fat-soluble A (Moore, 1929).

This discovery was followed in 1930 by the first preparation of vitamin D, calciferol (Askew and others, 1930).

In 1933 vitamin C was synthesized (Ault and others, 1933), and riboflavin was isolated from the rest of the vitamin B complex (Booker, 1933).

In 1935, a vitamin concerned in the coagulation of blood was discovered by a Danish scientist (Dam, 1935) who named it vitamin K (Koagulations vitamin) and the next year riboflavin was synthesized. In 1935, too, another factor was found in lemon rind and juice which appeared to be different from ascorbic acid (Szent-Gyorgyi, 1936). This was called vitamin P and its status as a vitamin is still uncertain. More will be said about this later.

In 1936, vitamin E (tocopherol) was isolated from wheat germ oil (Evans and others, 1936), and a series of researches began on its chemistry and very complicated behaviour.

In 1937, ten years after its isolation, vitamin B_1 was synthesized and given the name aneurin and about the same time (Williams and Clive, 1936), it was recognized that the so-called P-P or pellagra-preventing factor was nicotinic acid, another factor of the vitamin B_2 complex (Harris. 1937).

A second form of vitamin D was also distinguished: D_3 or irradiated 7-dehydrocholesterol, the form in which vitamin D is present in fish liver oils.

In 1938 came the isolation of another vitamin B factor — B_6 (Harris and Folkers, 1939), needed to protect rats from dermatitis. The terms vitamins B_3, B_4 and B_5 had been tentatively given to factors thought to exist, but these had not proved to be genuine new vitamins and the terms did not persist, B_6, however, was established as a real and necessary vitamin and it was synthesized in 1939. Vitamin B_6 is pyridoxin.

In the same year two other substances with vitamin activity were isolated: vitamin K or menaphthone and pantothenic acid. Choline and para-aminobenzoic acid were not newly isolated or synthesized but were for the first time found to be necessary for certain vital physiological processes and were therefore included among the vitamins.

In 1941, a factor, known since 1932 as H was identified as a newly isolated chemical compound biotin (Du Vigneaud and others, 1941). In the same year folic acid was isolated from the green grass and inositol, a known constituent of wheat germ, was recognized as a vitamin.

It took chemists until 1945 to discover the nature of biotin and to synthesize it and it remains one of the rare Vitamins.

The most recent vitamin to *he* added to the list is vitamin B_{12} (Rickes and others, 1948; Smith, 1948), isolated from liver after six years of intensive research and since proved to be the active factor which is used in the treatment of pernicious anaemia.

There are various factor whose nature and functions are still being studied, i.e. B_{12}, B_{14}, B_{15} and U, but in the summary which follows are included all those vitamins which are the present identified and about whose functions we have some definite knowledge.

A Few Dates in Vitamin History

B.C.	Beriberi known in China.
Middle 18th century	Scurvy common among sailors. Discovery made ages that oranges and lemons were protective. Capton Cook took vegetables with him to protect century his crew from scurvy.
1804	Regulations introduced into the British Navy providing for a daily ration of lemon juice for all ranks.
1897	Eijkman, in the Dutch East Indies, discovered that beriberi in man and polyneuritis in birds were both caused by eating polished rice and cured with rice polishings.
1906-10	Gowland Hopkins demonstrated the existence of Accessory Food Factors.
1912	The word 'Vitamin' first used by Funk.
1915	McCollum and Davis demonstrated two growth factors: fat-soluble A and water-soluble B. The anti-scurvy factor became water soluble C.
1918	Mellanby showed that the anti-rachitic factor was different from vitamin A and called it D.
1920	J.C. Drummond suggested the present spelling 'vitamin' for all these factors.

1922 Evans and Scott described vitamin E, the anti-sterility vitamin.

1926 Recognition of the P-P (pellagra-preventing) factor. Distinction between B_1 and B_2. Isolation of crystals of vitamin B_1.

1928 Carotene, the plant pigment, recognized as a precursor of vitamin A.

1930 Vitamin D, calciferol, prepared by the irradiation of ergosterol.

1933 Synthesis of vitamin C. Recognition of ribofiavin as part of vitamin B_2.

1936 Isolation of vitamin E- tocopherol.

1937 Discovery of vitamin K. Synthesis of riboflavin.

Synthesis of vitamin B_1- aneurin. Identification of nicotinic acid as the P-P factor. Recognition of vitamin D_2 as the anti-rachitic factor on fish liver oils.

1938 Isolation of vitamin B_6.

1939 Synthesis of vitamin B_6- pyridoxin. Preparation of menaphthone, a form of vitamin K. Pantothenic acid isolated.

1940 Choline suggested as a member of the B_2 complex.

1941 Isolation of biotin and folic acid. Recognition of inositol as a vitamin.

1945 Synthesis of biotin.

1948 Isolation of vitamin B_{12} from liver.

From the historical development of vitamins presented in the preceding pages it is obvious that vitamins were discovered not by their presence in the diet but because of their absence. A lack of most, but not all of vitamins known to be required by results in typical clinical syndromes or symptoms that can be cured by adding necessary vitamin to the diet. This negative approach to the functioning of vitamins is a useful tool in research. The investigator obtains clues to the metabolic rate of the nutrient when he observes the functional failure that results nutrient is absent from diet.

As a preface to the discussion of individual vitamins an explanation of the general meaning of the term should be made. Vitamins are organic compounds, necessary for growth and maintenance of life, which must be provided in the diet. The body is not able to synthesize them, at least in amounts sufficient to meet its needs. An exception is vitamin D; exposure to ultraviolet rays develops the vitamin from its precursor present in the skin. Vitamins are regulatory substances, each performing a specific function. They are carried in the blood stream to all parts of the body.

18

Essential Diet

Vitamins are essential in animal diet though they are needed only in minute quantities. Most of the vitamins are essential for the metabolism of all living organisms. Some fat soluble vitamins may have metabolic roles unique to animals and they all synthesized from plants and micro-organisms. Some vitamins can be synthesized from pro-vitamins. A large proportion of vitamin requirement of humans and higher animals is supplied by the intestinal flora.

Whether a substance must be supplied intact to the cell or can be synthesized by the cell depends upon the assortment of enzymes peculiar to the cell species. A substance may therefore be a vitamin for one species but not for other. The difference lies in whether the substance is required in the diet or whether it can be synthesized by the animal. The number of substances supplied by cellular synthesis tends to vary inversely with the complexity of the organism and appears to reflect an evolutionary adaptation as more complex organisms evolved from a simple one cell origin. Thus man is dependent upon his environment for some nutrients that microorganisms and other lower form of life can synthesize for themselves.

Early Methods

Very soon after vitamins began to be studied seriously it became apparent that their distribution in foods was very uneven. Cod liver oil and butter, for example, both contained vitamin D and could be used to cure rickets but the amount of butter needed was much greater than the amount of cod liver oil. Similarly wheat germ and yeast both contained the anti-beriberi factor B but some investigators found wheat germ the better and others yeast and no one had reliable information. Various attempts were therefore made to discover which were the best sources and later to find some means of measuring the extent to which one food was superior to another.

The fact that animals failed to grow when deprived of either fat-soluble vitamin A or water-soluble vitamin B provided a ready method of testing foods for the presence or absence of these two factors. Rats, when fed on purified foods containing all the known dietary essentials except vitamin A, grew for a time, but their rate of growth gradually decreased until finally they began actually to lose weight. At this stage foods to be tested for vitamin A would be given. If the effects were a restoration of growth, then it could be concluded that the food added to the deficient ration contained the missing factor A, but if no growth took place, then the food was deficient in vitamin A.

Foods found by this test to contain vitamin A were cod liver oil, halibut liver oil, butter, milk, liver, egg yolk.

The same type of experiment was used for testing the presence of vitamin B in foods, because by this time it was known that vitamin B not only protected from beriberi but was also a growth factor.

Experiments of this sort gave information not only on the presence or absence of these vitamins in a wide variety of foodstuffs but also an indication of their relative concentration. If a very small quantity of food were needed to restore or maintain normal growth, it was considered that this food was 'rich' in the vitamin; if a large amount were needed, then the concentration was taken to be 'poor.'

Other tests besides the growth test were used for some of the vitamins, for example, the cure of head retraction in pigeons for vitamin B, the prevention of scurvy in guinea pigs for vitamin C, the cure of rickets in rats for vitamin D, and the cure of sterility in rats for vitamin E.

Cure of Head Retraction : It had long been known that pigeons fed on polished rice developed avian polyneuritis, the condition in birds corresponding with beriberi in man. In this disease, the muscles of the neck and limbs become paralysed, the bird can no longer stand upright but sinks down on its hocks and the head cannot be held up. It usually flops backwards in the position known as head retraction.

In the test, birds were fed on polished rice until they developed this condition and were then given doses of the food to be tested until, by trial and error, the weight was discovered which was just enough to effect a cure. A good source of vitamin B would be indicated by a small curative dose and vice versa. Vitamin B was found in the following :

Wheat germ

yeast

oats

whole wheat

egg yolk

liver

peas and beans

Prevention of Scurvy

Animals were found to vary in their sensitivity to lack of vitamin C. Some, like rats and birds, seemed able to live and thrive without foods containing it, but guinea pigs were very sensitive to a shortage and on diets consisting of oats, skimmed milk powder and butter fat without any vegetables, their growth would begin to fait after about three weeks and afterwards they would develop scurvy. Their joints would become tender and sometimes bleeding under the skin could be detected. The test consisted of adding to a scurvy-producing diet small quantities of test foods and finding out (again by trial and error) the least quantity needed to prevent the onset of scurvy.

By this test it was discovered that most fresh fruits and vegetables contained vitamin C, the best sources being citrus fruits and fresh green leaves. Other.foods, like meat and milk, contained only very small quantities.

Cure and Prevention of Rickets : Rickets does not readily develop in rats if they have access to sunlight, but if they are reared in darkness (a condition quite normal for a rat) and are fed a diet containing no cod liver oil or other source of vitamin D and if the calcium and phosphorus of the diet are suitably unbalanced, rickets will develop in three or four weeks. This is shown by swollen joints and a disinclination on the part of the young rats to run about. The joints, if X-rayed at this stage, show swollen ends of the long bones and poor calcification indicative of rickets.

The test consisted either of feeding minimum curative doses to rats with rickets or of finding the smallest daily dose needed to prevent the onset of rickets and permit normal growth.

Fish liver oils were found to be the best sources,.others were animal fats like butter.

Cure of Sterility in Rats : The same principle was applied in this test. Female rats were kept on diets known to lack sources of vitamin E. On such diets they failed to produce any young, but normal fertility could be restored by feeding daily doses of wheat germ oil or another source of vitamin E.

This vitamin was found in some green foods as well as in the oils of wheat germ and cotton seed. Lettuce and grass both contain it.

The Categories

A great deal was learnt by these methods about the distribution of vitamins in foods and the two main groups were distinguished: the ıat-soluble vitamins, namely, A, D and E, and the water-soluble, B and C.

But still very little was known of the actual quantities present and a great need was felt for a method of accurate measurement.

An attempt was made to express vitamin activity numerically by giving an arbitrary value of 100 to some well known and reliable source and then assigning other appropriate values to other foods. For example, if food X were chosen as the standard of comparison and given the value 100, and the dose of food Y needed to produce the same result as a given weight of X were twice as great, then Y would have half the value, i.e. 50.

Comparisons such as these were made and lists of values published. They were useful as preliminary tests but owing to variation in potency from sample to sample and variation in response from animal to animal, the resulting figures tended to be confusing and even sometimes contradictory.

The next step was the adoption of 'units' of activity. Several such units were suggested. For example, a unit of vitamin B was taken as the amount of vitamin B which, when fed as a daily allowance, resulted in maintenance of weight of a standard rat over an 8-week period.

There were obvious disadvantages in such units as these. There is, for example, no such thing as a 'standard' rat any more than there is a 'standard' baby. Moreover, 'units' failed to take into account the different conditions under which the tests were performed and thus the standards themselves varied.

It was therefore a significant advance in technique when, in 1931, the Health Organization of the League of Nations suggested the adoption of 'International Standards' for the four principal vitamins A, B, C and D.

A suitable highly active preparation was chosen for each vitamin and it was agreed that a given weight contained so many units. Except in the case of vitamin C which is unstable, a bulk supply sufficient for several years' use throughout the world, was put into storage and from this, allocations were made to scientific institutions where estimations were being carried out.

This measure helped greatly to dispel, at one stroke, all the difficulties caused by the use of as many standards as there were countries and workers.

Vitamin A : On account of the unstable nature of vitamin A, the material chosen as a provisional standard was its precursor P-carotene, which occurs as a yellow pigment in vegetables such as carrots. One unit was the activity of 0.6 mg. of pure P-carotene.

Vitamin B : This standard was specially prepared from a single bulk supply of rice polishings. The material was extracted with water and then absorbed on kaolin. The resulting powder was highly active. One unit was the activity of 10 mg.

Vitamin C : On account of the impossibility of keeping fruit-juices without loss of vitamin C activity from oxidation, fresh lemon juke was chosen as the standard substance and a unit was 0.1 ml. of this. It was assumed that this was relatively constant which of course it was not, since lemons, like all products, vary among themselves.

Vitamin D : For this, a good quality cod liver oil was selected. The unit was the activity of 1 mg. of this oil.

Present-day Units : Later on with advances in the chemistry of the vitamins all these were modified but in the meantime between the years 1931 and 1938 much very valuable work was done by their use.

People were familiar with units of length, weight, temperature, heat. The vitamin units were not given names corresponding to inches, grammes, degrees or calories, but were called merely units of activity. The ideas, however, were the same and just as there is a standard yard against which all other yardsticks are measured, so these international units of vitamins became standard and all other estimates were measured against them.

Before vitamins were chemically synthesized in a state of purity estimations of vitamins in foods were carried out by biological methods. Chemical procedures were greatly simplified when the chemical nature of the vitamins was discovered.

International Standard : The general method was to select a group of experimental animals (usually rats) all from the same stock and arrange them in several groups with eight or ten in each so that their weights were as nearly as possible distributed evenly and fairly between all the groups. They were fed with a basal diet containing everything needed for health except the one vitamin to be measured and were weighted regularly until they either ceased to grow or began to develop typical signs of deficiency. Each rat was then fed an accurately weighed dose either of the international standard or of the food being tested.

Both the standard and the test would be fed at several dose levels and from the weight measurements obtained a graph was drawn showing what was known as a 'Dose-response curve'.

From this curve it was possible to read off the weight of test substance which produced a response equal to that produced by a given number of units of standard and hence to calculate the potency of the test food in terms of international units per gramme or ounce.

The mathematical procedures were at first very simple and to attempt was made to measure how accurate they were. It v as, however, accepted that methods depending on the use if living animals were subject to considerable error but for the time being, even these simply performed tests using the international standards were a great advance and many foods were subjected to estimation or 'assayed' for their vitamin content.

From values published it became possible to calculate the approximate total vitamin content of any diet and hence the next logical step was the carrying out of dietary surveys.

The international standards now in use (Timmerman, 1951) as established by the World Health Organization (1951) are as follows:

International Standard Preparations	*One International Unit*	
Vitamin A acetate	0.000344	mg.
Provitamin A (beta-carotene)	0.0006	mg.
Vitamin B_1 (Pure synthetic vitamin B_1)	0.003	mg.
Vitamin C (1-ascorbic acid)	0.05	mg.
Vitamin D_3 (crystalline)	0.000025	mg.
Vitamin E (alpha-tocopherol acetate)	1.0	mg.

The Functions

The function of each vitamin in living processes is studied in the first place by excluding it from the diet and observing the effects either on experimental animals or on man. But the study of deficiency states does not reveal the whole story because no vitamin acts alone. In the living body there are interactions between all the essential nutrients and lack of one may interfere with the actions of others.

However, to arrive at an understanding of how the vitamins work, it is logical to consider first what we know of each one individually, and then to study their relationships one with another.

The Role

Vitamins are concerned in the development of the body from its beginning until it has reached full maturity and with the maintenance of its structure throughout life. They are not structural constituents in the way that calcium phosphate is of bone or protein is of muscle but they are concerned in the formation and maintenance of the whole body of the various organs and tissues.

It is important to recognize that the body, while retaining its identity and its individual characteristics throughout life, is in a continuous state of change. The outer layers of the skin are shed and cells of the deeper layers constantly renew the covering of the body. This is a process which we can all observe. The same kind of renewal is all the time taking place even with structures as hard and apparently permanent as bone.

The initial growth and continued renewal of the bodily structures depends in the first instance on food taken. That is the source of supply. But although food consists of proteins, fats, carbohydrates, mineral salts and water, and the body consists of the same there is a vast difference between bread and butter, meat and vegetables and the human or animal body into which they are transformed. This transformation to be effective, needs the essential vitamins and each one plays some part in the camplicated process. Some have functions still only very imperfectly understood. Others we know more about.

The deficiency diseases through which vitamins were first discovered showed how lack of a single vitamin caused a disturbance of function and also structural change in organs or tissues peculiarly sensitive to that particular factor, e.g. the connection between vitamin A and the skin, vitamins B and the nervous system, vitamin C and capilaries, vitamin D and bone, but as soon as observations and experiences were followed by experiments, it was found that in all these deficiencies growth was affected. Vitamin A and Vitamin B were, in fact, known as growth factors for experimental animals.

The first investigations into the ways where by these factors actually functioned in the body and thus brought about the effects on growth, were made on vitamin B_1 which was found to be essential at certain stages in the breakdown of carbohydrates and the production of energy.

Vitamins and Carbohydrates : Carbohydrates (starches and sugars) are basically fuel foods. They supply the day-to-day needs of the body for the maintenance of body temperature and for the performance of all physiological processes, whether involving muscular activity, glandular secretions or even mental and nervous functions. Carbohydrates do not become part of the body structure except in so far as they are stored as reserves. Glycogen is the form in which storage occurs in muscle and liver and carbohydrate eaten in excess over the daily needs may be stored as body fat, but in general these foods are eaten to provide energy and this is liberated when they are chemically broken down (or metabolized) to produce at the final stage of the process, carbon dioxide and water.

Various chemical stages are controlled by enzymes (compounds produced within the body, which accelerate or regulate chemical reactions). Lactic acid and pyruvic acid are intermediate products produced during the breakdown of carbohydrates and in deficiency of vitamin B_1 is associated with the enzyme which is concerned in the oxidation of pyruvic acid. It is, in fact, a co-enzyme and is responsible for the control of chemical oxidative processes which take place within the living cell in connection with its energy relations. Since vitamin B_1 performs such a vital function as this, it is not surprising that shortage affects many physiological processes in addition to inhibiting growth.

Vitamin B_1 Functions : Being readily soluble in water, vitamin B_1 is quickly absorbed from the digestive tract and enters the blood stream to be carried to the tissues of the body. It is used continuously in the vital physiological processes and the tissues are not able to store it except for short periods. Consequently regular supplies are required for health.

The effect on growth is seen in children, who make only subnormal gains in weight if their food is short of vitamin B_1. At the same time their appetite may be poor, the slackness of their intestinal muscles may cause constipation and they may lack the mental alertness of normally healthy children.

The same apathy and lethargy affect adults, abnormal fatigue may be because of excess of such breakdown products as lactic acid and pyruvic acids in the tissues. All stages occur between such general signs of debility and the symptoms and signs characteristic of beriberi.

In this disease, the loss of appetite and consequent reduced food intake may produce extreme emaciation, but this may be masked by the presence of fluid in the tissues causing swelling of the limbs. The heart gets always affected in beriberi, being considerably enlarged and its rate is increased, so that when death occurs, this is frequently from failure of the circulatory mechanism.

The nervous system changes that causes abnormal sensations in the limbs, extreme weakness, wrist-drop and loss of knee-jerks.

Relations with other Vitamins : Riboflavin and nicotinic acid also participate in the same enzyme systems concerned in the intermediate breakdown of carbohydrates. It is possible in experiments with animals to make diets deficient in one vitamin only and to study the effects of deficiency of one factor at a time but deficiency diseases in man rarely occur singly. The, effects of one deficiency could be more marked than those of another but even in extreme cases like beriberi and pellagra, several factors are involved.

Riboflavin : Although riboflavin is more concerned in carbohydrate oxidation, deficiency is shown by effects quite different from those caused by vitamin B_1 deficiency. The chief characteristics are cracking of the lips with sore, red patches, scaliness with raw cracks at the corners of the mouth and a sore, red tongue. There may also be scaly red patches across the nose and lips and over the eyelids and sometimes redness of the eyes.

These signs may appear in individuals even in Great Britain if their diet has been poorly supplied with such riboflavin-containing foods as milk, eggs and liver.

Nicotinic Acid : Nicotinic acid is also fundamentally a tissue co-enzyme and is essential, along with vitamins B_1 and riboflavin for the breakdown of carbohydrate, but extreme deficiency of nicotinic acid causes not beriberi but pellagra, the disease that is still prevalent in countries where maize is the staple cereal. The word is derived from pelle=skin, agra=rough, and was used first to describe this disease in 1771 in Italy. It was Funk, who had coined the word 'vitamine', that suggested in 1914 that pellagra was a deficiency disease. Its exact cause has given rise to much controversy but can probably be explained as follows: Maize, like any other cereal grain, consists of both carbohydrate and protein and the protein is made up of a collection of simpler nitrogenous compounds, the amino-acids. Some of these are necessary for life and health; others can be dispensed to be very poorly represented in maize. It has been known that the occurrence of pellagra was associated with the consumption of foods poor in protein. The connection with nicotinic acid was recognized when it was found that tryptophan was a precursor of nicotinic acid. The disease disappears when food protein containing tryptophan or when nicotinic acid (or amide) is given.

Pellagra is characterized by soreness of the tongue and mouth, presistent diarrhoea and the development of pigmentation of the skin leading to dermatitis. At the same time patients become mentally depressed. The pigmentation of the skin occurs in exposed parts : hands, face and neck. The skin is dark and brown, later becoming red and scaling with raw patches. The depression may develop into a severe dementia.

Pellagra is fairly common in parts of America and in the Middle East : Mild degrees of nicotinic acid deficiency is found in other countries if individuals fail to get their quota of nicotinic acid and protein foods, or if some other condition prevents either absorption or use by the body of these food substances. The chief sources of nicotinic acid in human foods are meats and fish. Hence when people eat only a little of these and large quantities of maize, they lack both the protective factors, tryptophan and nicotinic acid, and pellagra is the result. In animals other vitamins, and in particular B_1 and B_6 are needed for the full utilization of nicotinic acid and also for its formation from tryptophan - another example of vitamin interactions.

B Vitamins and Proteins : Riboflavin and nicotinic acid are not only concerned with carbohydrates and energy relations but also with protein and the muscular structures of the body.

All proteins are made up of amino acids and there are some twenty of these latter, which, combined in different ways and in different proportions account for the characteristics of the various proteins as we know them. For example lean meat is protien but we know from our own experience that there are differences between beef and mutton and between mutton and pork; and even greater differences between meat and kilk, cheese or eggs. But the body can use all these proteins as sources of the amino acids which are to be built up into human muscle. Protein eaten in any form is first digested and absorbed and then its amino acids have to be re-arranged and built up again either for the growth of the human body or for its maintenance when it is fully grown.

In this process both riboflavin and nicotinic acid play a part. Moreover, riboflavin, when eaten in food, is transported to the liver where it is temporarily stored, to be released again when required during the digestion and assimilation of food. It has been shown that for this release from the liver vitamin B_1 is needed.

Thus these three vitamins are very intimately connected not only in foods where they are found in nature, but also in their functions in the animal body, and we can see why the absence of any one of them interferes with growth and with general well-being.

B Vitamin and Fats : Just as certain vitamins have a special function in the metabolism of carbohydrates and others in that of protein, so a few seem to have close association with the body's use of fats in connection with either absorption or subsequent use.

Fats are digested in the small intestine and afterwards pass through the cells lining the tract into the lacteals and thence to the Lymphatics.

After absorption, fat is carried first to the liver and thence passes to other parts of the body, being either broken down for the production of heat and energy like carbohydrates or laid down in various fat depots for storage.

Choline : Choline is concerned in the use of fats by the body. It is required first of all for their absorption through the cells lining the intestinal wall and it is required subsequently by the liver for the conversion of fats into substances suitable for passing in the circulating blood to different parts of the body. Fats do not circulate as such but as compounds with phosphorus, phospho-lipids, and part of the choline molecule is required for this conversion. For this reason choline is known as a lipotropic agent and in this respect is similar to inositol and vitamin B_{12}.

Lack of choline affects the functions of the liver, but in the absence of choline, an amino acid, methionine, can take on some of its functions. Because of this reason some doubt has been cast on the justification for considering choline a vitamin since it is not *absolutely* essential for life. However, at present it is usual to include choline among the vitamins and its role in metabolism is certainly similar to that of other vitamins.

In the absence of choline, fat accumulates in the liver, interfering with the proper functioning of the hepatic cells.

This causes dyspepsia, and if the process continues, the liver may become permanently damaged.

Since normal metabolism of fat depends on supplies of choline, animals on a choline-deficient diet grow badly and various signs of deficiency develop, for example, in adequate lactation in the rat and slipped tendon in chicks.

Inositol : The function of inositol in the animal body is only partly understood. Like; choline, it helps in the metabolism of fats and it may have some effect on the skin. It has also been suggestesd that with vitamin E it affects the health of the muscles.

Inositol itself is soluble in water, but in plant products it is often found as phytin, a compound with phosphorus, the calcium salt of which is insoluble. This fact gave rise to the opinion that cereals containing inositol caused rickets by preventing the absorption of calcium. It has been found however, that this only happens in severe deficiency of calcium and that with normal supplies of calcium and vitamin D, no apprehension need be felt on this score.

B_6 or Pyridoxin : The role of this vitamin is not yet clearly understood Animals on a deficient diet develop eczema of the skin especially of the feet, but nothing corresponding closely with this seems to occur in man. However, the vitamin has been found useful in medicine for the treatment of several conditions, e.g. nausea and vomiting in pregnancy and radiation sickness and is one example illustrating the fact that a vitamin may be found of value before an explanation is forthcoming of exactly how it works. One fact that is known at present is that it is associated in the body with protein metabolism.

Biotin : Experiments with animals on restricted diets have demonstrated that this is essential for the growth of almost every animal and also of yeasts, bacteria and fungi, but because of its almost universal presence in human foods, little is known of deficiency in man or of human requirements.

We know, that raw egg white combines with it in the intestinal tract rendering it insoluble. The factor in egg white is responsible for this action is referred to as avidin due to its avidity of biotin.

There is thus a very slight danger that individuals consuming excessive amounts of raw egg white may become biotin-deficient. Actually this is a rather remote possibility, but sick children in hospital, given abundant quantities of albumin water, have been reported as developing a form of scaly dermatitis that has cleared up with small doses of biotin. These are, however, only isolated observations and have not been confirmed in general.

Pantothenic Acid : Rats and chicks fed on diets deliberately made to contain no pantothenic acid develop various signs of deficiency: e.g. dermatitis in chicks and growth failure, haemorrhages and changes in the adrenal glands in rats.

Deficiency in man is not likely to occur because the vitamin is so widely distributed in all kinds of foods.

It has been suggested from time to time that pantothenic acid affected greying of the hair but this has not been confirmed. It has been used in medicine in doses higher than those which would normally be taken in foods. These relatively high doses have been found effective in the treatment of several conditions including the sore tongue which some patients with pernicious anaemia develop. Pantothenic acid is not of any value in the treatment of the anaemia itself.

Para-aminobenzoic Acid : As a dietary factor this presents no problem because there is little risk of shortage in foods, but like several other vitamins, it has found a place in medicine given in high doses and has been used in the treatment of various forms of eczama and also of rheumatic fever.

Vitamin D and Bone : Among animals with a bony skeleton, the form, shape and size depend on the growth and perfection of this skeletal framework. The bones are chemically of calcium phosphate, but each bone has its own characteristic architecture and internal structure. The bones begin as relatively soft cartilage and gradually the hard calcium phosphate is deposited in them, being carried there in the blood stream.

The formation of bone from the calcium and phosphorus brought by the blood is controlled by vitamin D. The body cannot make good bone (or teeth) without vitamin D and conversely the vitamin cannot produce good bone without the materials from which bone is made. Calcium and phosphorus, together with traces of other elements such as fluorine are the building materials; vitamin D regulates the process.

The supply of these minerals and vitamins reaches its greatest importance during the period of rapid growth and it is when the bones are young, cartilaginous and relatively soft that deficiencies may cause rickets with the characteristic bending of the bones of the limbs and widening at the ends. It may be noted, however, that since bones are hard, they are not of unchanging structure like a piece of wooden or metal furniture. Every bone is supplied with blood vessels and nerves and is as much a living part of the body as muscle or skin. Moreover the blood is continuously bringing new nutrients to the bones and removing waste products. There is thus a state of balance or equilibrium between the calcium and phosphorus of bones and of circulating blood and this continues throughout life.

The amount of vitamin D required during active growth is definitely greater than that needed when the full size of the body is reached, but some vitamin D continues to be required to maintain the bones in a healthy condition. Old people, whose consumption of calcium and vitamin D has been neglected, sometimes develop very brittle and fragile bones which fracture easily — and in some cases the bones of the spinal column lose their firm hard texture and collapse, causing great pain and weakness.

Vitamin D is the one vitamin that can be produced by sunlight, vitamin D is formed in the oil on the surface of the skin and in its deeper layers. This is absorbed and in sunny climates makes it less necessary for sources of vitamin D to be taken by mouth.

Vitamin D is also formed of the fur and feathers of mammals and birds and is consumed when they lick themselves or preen their feathers.

The fact that vitamin D is formed when the skin is exposed to sunlight has led to its use in high doses for certain diseases formerly treated by ultra-violet irradiation. One such disease is lupus vulgaris (tuberculosis of the skin) in which results have been encouraging.

Bones and other Vitamins : Concertration on the importance of vitamin D in the development of bones and teeth led to neglect of other factors possibly concerned. However, vitamins A, C and E all seem to play some part in promoting normal bone formation.

In vitamin A deficiency the structure becomes disorganized — the bone becomes cancellous (less hard and compact) and overgrown, losing its fine moulding and outline.

In infantile scurvy, the ends of the long bones are abnormally brittle and fractures frequently occur. Vitamin E deficiency is thought to lead to imperfect calcification of the skull, possibly through its effect on vitamin A or on the phosphorus 'turn-over' in bone.

Vitamins A and the Skin : All parts of the body whether external or internal are covered or lined with skin. The type of skin differs according to situation, being tougher and thicker on the outside of the body and more delicate when lining internal cavities, but carrying out important functions throughout the life of any animal.

Vitamin A has special functions in relation to the skin, both that covering the outside of the body and also the mucous membrane lining the various parts which communicate with the exterior, e.g. the mouth, nose, trachea and bronchial tubes, the conjunctiva of the eyes and the linings of the eyelids.

The eyes are peculiarly sensitive to severe deficiency and the condition of xerophthalmia (dry eyes) was one of the first signs of vitamin A deficiency to be discovered.

The mucous membranes of the respiratory tract become less healthy if there is a shortage of vitamin A and due to this fact vitamin A was considered 'antiinfective.' We know now that it has no direct effect on infective organisms but it may help to prevent respiratory infections by maintaining the mucous membranes in a healthy condition.

The outer skin of the body shows little change in mild degrees of deficiency but subjects suffering exterme deprivation for long periods may develop a horny, dry condition of the skin — keratomalacia.

Other Functions of Vitamin A : Much has been discovered about the connection between vitamin A and vision in dim light. Deficiency causes a condition known as night-blindness in which individuals experience difficulty in adjusting their sight to distinguish objects in a poor light especially after being in a bright light or looking at bright objects. Normal persons can adapt themselves readily to changes in light intensity but vitamin A-deficient individuals have poor dark-adaptation. It can be explained as follows: seeing in semi-darkness depends on the presence in the retina of the eye of a pigment, visual purple. When bright light falls on the eye, the visual purple is bleached. If there are ample supplies of vitamin A in the body, the pigment is re-formed, but if supplies are poor, this re-synthesis is delayed with the result that the clarity of vision in a dim light is reduced. Visual purple is actually a chemical derivative of vitamin A itself.

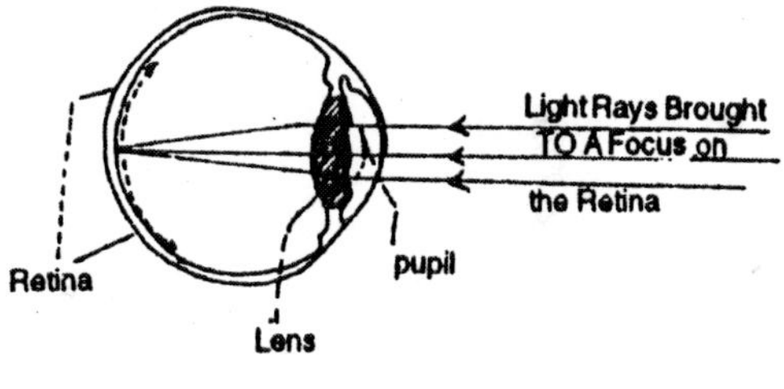

Above facts from the basis of tests used to find out the state nutrition of groups of people in relation to vitamin A. By means of it, subjects not otherwise unhealthy have been detected as in a state of mild deficiency and it has been possible to correct the faults in their diet before any serious consequences have developed.

Vitamin E and Body Tissues : The function of vitamin E are less well understood than those of vitamins A and D there is still much conflict of opinion about its role in human nutrition.

It is absorbed from the digestive tract along with fats, and anything which interferes with fat absorption may reduce the quantity reaching the tissues of the body.

In experimental animals, e.g. rats, it has been proved that vitamin E is necessary for normal fertility and this fact is still used in biological methods of estimation, but because of this, it cannot be assumed that the same is true of other species.

Although the role of vitamin E in connection with fertility is still not understood and seems to vary from species to species, there is no doubt at all about its importance in the physiology of some of the tissues. Just as vitamin A is concerned with the skin and mucous membranes and vitamin D with bones, so vitamin E is concerned with muscle and connective tissues. Deficiency in man does not produce such well-marked and easily recognized diseases as xerophthalmia and rickets but in animals such as rats, the young develop a paralysis of the limbs due to a muscle defect or dystrophy. This is not unlike a rather uncommon human disease but attempts to cure the latter with vitamin E have so far met with very inconclusive results.

One function of vitamin E that appears to be well established is its role as an 'anti-oxidant.' When present in the tissues or in foods with readily oxidizable constituents of fats, including such substances as vitamin A and carotene, it prevents their destruction by oxidation. It is thus said to 'save' vitamin A. This effect can occur both in the intestinal tract where vitamin A in food may be oxidized, or in the liver where vitamin A is stored, and this fact means that if vitamin A is given *with* small quantities of vitamin E, less A is needed to achieve a given result. It has been shown, for example, that vitamin E increases the growth-promoting power of vitamin A.

It may be partly because of this property and partly for other reasons not fully understood that vitmain E is as essential for the maintenance of tissue integrity and is thought to have an effect in delaying the process of ageing.

Vitamin E is one of the best examples to illustrate the complications with which a study of vitamins may be involved. There is no simple deficiency here. The vitamin itself is made up of a complex of at least four components. Their biological activity differs; their distribution varies from source to source; they are known to interact within the body with proteins and protein constituents, the amino acids, with vitamin A and probably vitamins of the B complex and to be concerned in the metabolism of some of the most complex tissues, the muscular and fibrous tissues, of the body. It is not surprising in the circumstances that there are still many gaps in our knowledge of this particular vitamin.

Vitamins, Blood and Blood-Vessels : Vitamins C, K, P, folic acid and B_{12} are all concerned either with blood or with the smallest blood-vessels, the capillaries, which bring blood into contact with all the tissues.

Blood is made up of a fluid, plasma, of highly complex composition where in are suspended cells or corpuscles: red cells containing the pigment hemoglobin and white cells which are unpigmented.

The blood, by circulating through the vessels comes into contact with all the tissues, even the hardest bone. Holes can be found in the shafts of the long bones and in the skull, through which these vessels pass to the interior.

As the fluid (plasma) circulates, it carries the cells with it and together they form the distributing and collecting system which functions for the whole body. Oxygen is carried from the lungs to all parts; carbon dioxide, formed in tissue respiration, is removed from its sites of formation and taken back to the lungs. For this, the red cells are largely responsible.

The plasma transports nutrient materials from the alimentary tract to the organs or tissues where they undergo further breakdown or assimilation. The waste products of physiological processes are similarly removed by the plasma, which may contain at any one time proteins, inorganic and organic substances in solution, nutritive and excretory materials, antibodies, hormones, vitamins and possibly compounds still unidentified.

By studying the effects of deficiencies of vitamins C, P and K, we can see how they affect the blood. All three are associated with bleeding into the tissues.

Vitamins C and P : In scurvy, there are haemorrhages into the joints and into the tissues. These appear as purplish-red patches in dependent parts, i.e. around the shoulder blades and lower part of the back when the patient is lying down. The gums may also bleed, and there is swelling due to an increase in the amount of fluid in the tissues. It is evident that the capillaries have allowed blood to escape.

The walls of these minute vessels are made up of a single layer of very delicate cells, held together by a cement substance and surrounded by a sheath of very fine fibrils.

Bleeding into the tissues occur when blood-vessels of any size are damaged. This can happen in mechanical injuries and the severity varies according to the size of the vessels which are damaged. Bruising causes injury to the walls of the capillaries and the discolouration that follows is because of the presence of blood in the tissues.

When similar escape of blood takes place without mechanical injury it means that the walls of the capillaries have become unduly permeable. This is what happens in scurvy and in another condition : purpura — a term applied to a number of conditions in which spontaneous bleeding occurs beneath the skin. The capillary sheath and the cement substance holding the cells together give way and allow the blood to escape.

The haemorrages seems as purplish spots of various sizes and the colour changes which occur are similar to those of an ordinary bruise.

Purpura occurs in diseases of different kinds — many quite unrelated to scurvy, and it should not be looked upon as a disease in itself but rather as a disease sign. In the most serious varieties of measles, scarlet fever and smallpox, purpuric spots appear and the diseases are then described as 'haemorrhagic.'

Purpura may also take place in conditions of irritation of the digestive tract and associated with weals in some forms of urticaria (nettle-rash).

Scurvy can be arrived and treated by means of vitamin C; purpura by vitamin P.'

Vitamin C (ascorbic acid) is concerned not only with the prevention of tissue bleeding as in scurvy, but in the maintenance of a healthy state of equilibrium between the fluid plasma in the capillaries and the fluid in the body tissues (Hoskin, 1948). Where excess body fluid tends to collect as in diseases of the heart and circulation, vitamin C helps to reduce this.

Vitamin K : This is concerned not with the capillary walls but with the process of clotting, which is slowed down in deficiency of the vitamin.

Clotting is a complicated process. It does not normally occur within the intact, healthy blood-vessels but when the vessel or the tissues are damaged the blood which escapes, clots within a few minutes. The clot is made up of fibrin, a jelly-like protein that forms in the plasma and that entangles the blood cells in its meshes. This fibrin sets as a jelly only in the presence of another constituent of plasma, thrombin, and this in turn is formed from a precursor prothrombin, that is liberated into the circulating blood from the liver where it is formed. And this is where vitamin K comes into the picture. Without it, there is a failure in the maintenance of prothrombin in the liver and if the amount liberated into the plasma is much reduced there develops a tendency for injuries, which would otherwise be negligible, to continue to bleed because no clot forms to seal them.

In the blood of infants, the prothrombin is low for the first few days of life and consequently haemorrhage may occur from undetectable small injuries to blood-vessels. These haemorrhages may be in the digestive tract when their existence is detected by the passing of discoloured stools, or in the nose or palate or under the skin, hence the descriptive term 'haemorrhagic disease of the newborn.'

The low levels of prothrombin in the blood of infants may be due to the fact that vitamin K is obtained only to a small extent from food but chiefly as a result of bacterial synthesis in the digestive tract. Only relatively few bacterias are associated with disease; some inhabit the digestive tracts of man and of other animals and their presence is an advantage rather than the reverse.

Although vitamin K is eaten with certain foods, there is no doubt but that much of a man's needs for this vitamin are supplied by these bacteria in the interstines. Vitamin K cannot, however, be synthesized by the intestinal bacteria until these are established and this does not happen until several days after birth, so that during this short period there is a real danger of vitamin K deficiency and subsequent haemorrhage. To prevent this, it is now a usual practice to give vitamin K either to the mother before the child is born or to the infant after birth and this has done much to reduce the incidence of this type of haemorrhagic disease.

Vitamin K, whether formed by bacteria in the intestine or eaten in foods, is fat-soluble and is so absorbed along with fats. Fat-absorption, in turn, depends on the presence of bile in the intestine since bile salts act on fats converting them into a form in which they will pass through the cells lining the intestine. Anything which interferes with the presence of bile salts in the intestine will therefore prevent the absorption of fats and with them of vitamin K. The commonest such condition is an obstruction in the bile duct, through which bile passes from the gall-bladder to the intestine. A gall-stone in this duct is a common form of obstruction and a construction from a growth can produce the same sort of result.

When the bile can no longer pass into the intestine it becomes re-absorbed into the blood stream and the pigments in it circulate in the blood and turn the skin yellow producing the condition known as obstructive jaundice.

Sometimes surgical operations are required to remove the obstruction and these operations formerly attended with some danger from haemorrhage because, with poor vitamin K absorption from the intestine, the liver had failed to produce adequate quantities of prothrombin. Now, vitamin K is given to such patients before operation and this danger is greatly reduced.

A secondary deficiency of vitamin K may occur in any disease where in there is defective absorption of fats. Actually individuals may have low prothrombin levels in the blood with no signs to show this until some minor damage to the tissues occur such as a tooth extraction, after which bleeding may be excessive. The practice of giving vitamin K before such minor operations is becoming more widespread.

The other two vitamins in this group effect the cells of the blood rather than the plasma. Red blood cells are formed in the red marrow of bones and go through stages in development before they are fully mature and ready to be released into the circulation. They remain in circulation carrying out their various functions for about four months after which they tend to become fragmented and are disposed of by the spleen and new ones take their place.

Primitive cells in the marrow contain no haemoglobin and they stain with basic dyes, so that in stained films a cell seems blue or purple with a large nucleus. Gradually haemoglobin is formed within the cell, and as this stains with acid dyes (usually pink) the cells appear in films with both pink and bluish colouring. These cells are larger than circulating red cells and are called megaloblasts. They gain more haemoglobin, lose their nuclei and become smaller before being liberated into the circulation as normal red cells or normocytes.

In those cases where this process for some reason or other goes wrong, various things can happen; cells may be liberated with less than the normal quantity of haemoglobin, or before they are fully mature, or are of uneven size.

These abnormalities of blood cells or the amount of haemoglobin is reduced. All these are anaemias of that many varieties are known. Those types that are characterized by abnormally large red cells (macrocytes) and by a reduction of haemoglobin are the macrocytic anaemias, the most common of which are the so-called pernicious anaemia and the anaemias that occur in the disease, sprue. There are other forms of macrocytic anaemia.

Folic Acid and Vitamin B : It has never been proved that these anaemias are directly due to dietary deficiencies, but it is now known that a normal blood picture can be restored by both folic acid and vitamin B_{12} However, the mechanism by which this is brought about is still only vaguely understood.

The story of vitamin B_{12} has been, in recent years, as dramatic as the history of the earlier vitamin discoveries. For the beginning we have to go back to 1926 when Minot and Murphy found that a form of anaemia hitherto resistant to treatment and hence described as 'pernicious' responded to treatment with liver. From that time onwards chemists sought to isolate the active factor from liver and, after nearly twenty years, the goal was thought to be reached when folic acid was synthesized. This substance restored the blood of pernicious anaemia patients to normal but this disease is characterized not only by abnormal blood formation but also changes in the nervous system causing disturbances of sensation and these were not affected by folic acid.

In 1948, workers in the laboratories of pharmaceutical firms in the U.S.A. and in Great Britain simultaneously announced the isolation from liver of a crystalline compound several thousand times more active than folic acid. The crystals were dark red needles and contained, in addition to carbon, hydrogen, oxygen, nitrogen and phosphorus, 4 per cent of cobalt — the only vitamin known to contain this element.

The new substance, known as vitamin B_{12} was found to be active in pernicious anaemia in doses as small as 20 , micro-grammes and also effective in the treatment of the nervous system complications.

It was marketed in Great Britain in 1949 and is now widely used in doses of 50 to 100 micro-grammes given by injection. It is no longer extracted from liver, twenty tons of which are required to produce one gramme, but is produced from cultures of streptomyces griseus (an organism allied to primitive moulds).

It is stable in the solid state and also in neutral or slightly acid solution. It exists in several forms — all physiologically active. The solution, which is pale pink, is supplied in ampoules for medical use.

19

Vitamin A

Vitamin A was the first of the accessory food factors to be identified as a component of specific foods. In 1913 McCollum and Davis reported that rats failed to grow as diets containing carbohydrates, protein, lard, and salts. Addition of an other extract of butter or eggs promoted growth and established the missing factor was a fat soluble substance.

In two laboratories experiments were in progress that led to the discovery of vitamin A in 1913. McCollum and Davis at the University of Wisconsin fed young rats a purified diet considerd to be adequate at that time. It contained protein, carbohydrate, mineral salts, and fat (lard), all of the nutrients then recognized. They used milk sugar in this experiment, which added enough impurities of nutritional value for the young to grow fairly well for a while. However, after a period the animals ceased growing. Adjustments were made in the relative amounts of the nutrients in the diet in an effort to prevent the cessation of growth but to no avail. When butterfat or egg yolk was added to the diet, growth was resumed. The addition of olive oil, on the other hand, had no effect. It was apparent that there was a difference in the nutritive value of olive oil and lard, and butterfat and egg yolk. At the Connecticut Experiment Station in the same year, Osborne and Mendel fed Young rats a purified diet consisting of protein, lard, starch, and protein-free, fat-free milk that provided the minerals, some lactose, and other water-soluble substances. As in the Wisconsin experiments, after a period growth failed. With replacement of lard by butter, the rats grew and matured normally. Eggs yolk and cod liver oil in the diet gave the same response as butter; these fats provided the *something* that was lacking in the purified diet.

It was evident from the studies of McCollum, and Osborne and Mendel, that the essential nutrient missing in the purified diets was soluble in fat. Knowing the solubility of the substance helped to speed the discovery of vitamin A; many compounds were thus excluded from consideration. Further help came from Steenbock's studies (1919) at the University of Wisconsin in which root vegetables were fed as a source of the-soluble vitamin. Yellow sweet potatoes and carrots supported normal growth and even supplied enough of the unknown substance for reproduction. Rutabagas, red beets, parsnips, potatoes, and sugar beets failed completely to provide the needed nutrient. Steenbock tested many plant materials, among them yellow corn and white corn. The animals fed yellow corn as a supplement to the purified ration grew normally and were able to reproduce; animals fed white corn failed to grow and developed deficiency symptoms. It is interesting that McCollum, in 1908, heard Wisconsin farmers who were attending a meeting at the University maintain that yellow corn was better feed for hogs than white corn; that was eleven years before Steenbock's studies confirming their contention. Thus in addition to solubility in fat, another characteristic of the nutrient was recognized, the association with yellow colour. It had been demonstrated that animals were unable to synthesize the yellow pigment carotene. This gave a lead that it must be provided in the diet. The question still remained, however, as to the relationship between the yellow plant materials and the animal fat substances, both of which supplied the fat-soluble vitamin. Ten years later, in 1929, crystalline carotene was demonstarted to have vitamin A value by a German scientist Hans von Euler-Chelpin.

The Properties

Vitamin A is an organic compound found only in the animal Kingdom. It is considered to be present in all species of fish, birds, and mammals. Yellow plant pigments of the carotenoid group *alpha-*, *beta-*, and *gamma-* carotene, and cryptoxanthin are precursor of vitamin A. The body has the capacity of converting these carotenoid compounds contained in the diet into vitamin A. Actullay the origin of all vitamin A in animal products comes directly or indirectly from the carotenoids of plants. Fish consume marine plankton; herbivorous land animals eat vegetable foods; and the carnivorous animals depend indirectly on herbivorous sources.

Two forms of vitamin A have been identified: vitamin A_1 which occurs in the livers of salt-water fish, and vitamin A_2 found in the livers of fresh water fish. Although the two vitamins differ slightly in chemical structure they are alike in physiological functions. Vitamin A_2 contains an additional double bond in its ring structure. In the present discussion they will be treated as one, only the term vitamin A will be used.

Vitamin A is a pale yellow substance composed of carbon, hydrogen, and oxygen. The chemical formula for vitamin A_1 is given in Fig. It is soluble in fat, not soluble in water, stable to heat, that is, to ordinary cooking temperatures, and destroyed by both oxidation and ultraviolet irradiation.

The carotenoid pigments are compounds composed only of carbon and hydrogen. The crystals of the carotenoid pigments are a deep red color but in solution intensely yellow. Their properties of solubility and stability are like those of vitamin A. Beta-carotene is more efficiently converted to vitamin A than alpha- or gamma-carotene, or cryptoxanthin. The chemical structure of beta-carotene is such that it oxidizes to form two molecules of vitamin A; the other provitamins form only one. However, for the sake of clarity, the carotenoid groups will be spoken of as *carotene,* as though it were one substance.

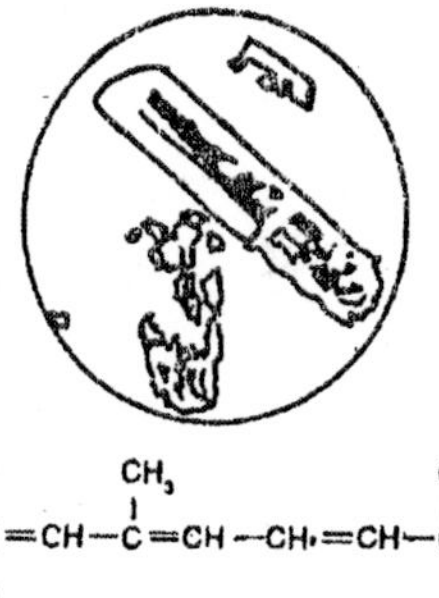

Measurement Unit

Vitamin A is measured in International Units (IU). The term itself indicates the universality of the unit of measurment. The Health Organization of the League of Nations proposed the universal unit for vitamin A to replace the many different units in use at that time. The International Unit of vitamin A activity as adopted by the current official body, the Sub-committee on Fat-Soluble Vitamins of the World Health Organization is equivalent to 0.3 *ug* of vitamin A alcohol. Equal biologic activity is afforded by 0.6 *ug* of the provitamin beta-carotene.

In the United States pharmaceutical companies use the United States Pharmacopeia (USP) Unit to express the vitamin A potency of their products. The USP Unit is exactly the same as the International Unit. The U.S. Pharmacopeia is a book describing drugs, chemicals, and medicinal preparations published under the official authority of the Food and Drug Administration.

The Demonstration

When a diet lacking vitamin A value but adequate in all other respects is given to a young growing rat, growth continues for a shorter or longer time according to the extent of the stores of this vitamin which the body possesses, after which growth ceases and there usually soon sets in a loss of body weight and a condition of decline in health. About when the body's reserves of vitamin A have been exhausted and growth ceases, a large proportion of the experimental animals begin to develop a characteristic disease of the eye. This usually begins with a swelling of the lids of one or both eyes or with indications that the eye is becoming unduly sensitive to light. Following this there commonly develops an inflamed and catarrhal condition of the conjunctivae, with a bloody or purulent discharge, the lids becoming scabby or sticky. This, with the swelling of the lids and the sensitiveness to light, sometimes results in the eye being found completely closed. If the eye condition is not treated, and the animal continues to live, the cornea may become affected and blindness results. On the other hand the eye condition if not too far advanced, usually improves quickly upon feeding the animal with any food of adequate vitamin A value.

Thus a demonstration of vitamin A value in a food may be made by feeding the food to be tested to a young rat which has recently ceased to grow and begun to develop eye trouble form vitamin A deficiency as just described. Under these conditions if the test food has significant vitamin A value it will usually induce resumption of growth and recovery from the eye disease.

And such testing of foods can be made quantitative by standardizing the conditions and measuring the rate of growth induced by the feeding of graded allowances of the test food and of a reference material of accurately known vitamin A value.

The Health Organization of the League of Nations prepared beta-carotene of which 0.6 microgram had, by definition, one International unit of vitamin A value. The U. S. Pharmacopeia unit of vitamin A value is the same; but the Pharmacopeia organization provides a reference material, more convenient for ordinary use than carotene, in the form of a cod liver oil whose value has been measured in comparison with the International Standard.

More recently, the World Health Organization has offered, as a new reference standard, crystalline vitamin A acetate. At present (1951-52) both this and the beta-carotene standard are in use. Description of the method should also include definition of the oil in which the carotene is dissolved.

The same test may be made both qualitative and quantitative by allowing development of the typical sore eye before the feeding of the test food is begun; but if this involves serious illness in the test animal it is apt to render the results more variable, so that for quantitative work the test feeding is usually begun before the test animal becomes seriously ill, while at least one member of each litter or set of test animals is continued on the basal diet alone as a *negative control* in which the typical development of the vitamin A deficiency disease can be observed.

Quantitative measurements of the relative amounts of vitamin A in food can be made by means of sufficiently numerous and carefully controlled feeding experiments with rats which are amply provided with all other nutrients which they require.

The general plan to start with normal young rats 21to29 days old, feeding them a basal diet good in all other respects but free from vitamin A. When any surplus store of this vitamin which the body contained at the beginning has been depleted (or sufficiently diluted by the growth of the animal) the body weight increases and usually remains nearly stationary for a few days. At this time, if not before, each of the animals is placed in a separate cage and kept under controlled conditions. One (or more) of each litter continues to receive the basal ration only, until it dies from lack of vitamin A. Thus serving as a negative control. Other animals are fed, in addition to the basal diet, systematically graded amounts of the reference material or of the food which is being tested.

Natural biological variability makes it probable that the results of individual experiments of the kind will vary more than should the data of chemical experiments conducted *in vitro;* but when sufficient numbers of animals are used and the results are averaged, the average weight curves show a fairly regular gradation according to the amount of vitamin A which the animal actually received.

The accompanying graph shows the results obtained by Batchelder in negative controls and in test animals fed at several different levels of vitamin A intake (allowance of test food), each curve being the average result obtained with nine test animals. The starting point of these curves is the end of the *depletion period* and beginning of the *test period.*

In the experiments of Todhunter, vitamin A values were determined by a method of single feedings (Sherman and Todhunter, 1934). The typical results summarized in Fig. are of interest in demonstrating the similarity of weight-curve resposne to equivalent vitamin A values whether in the form of the vitamin itself (as in cod liver oil) or its precursor (as in the case of carotene or kale) or a mixture of the two; and also in showing the quantitative relationship of weight gain and survival period to the amount of the single feeding of any of these typical sources of vitamin A value. The result of this relationship is that the area, bounded above by the weight curve resulting from a given feeding and below by the negative control curve and the base line, is (under sufficiently standardized conditions) quantitatively proportional to the vitamin A value of the weighed amount of test material used in the single feeding. This makes it possible to measure quantitatively the vitamin A values of materials which are only occasionally available, or of individual samples of fresh foods without subjecting them to the possibility of any decline in vitamin A value due to drying or storage.

Food Value

The carotenes which constitute the chief precursors of the vitamin A of normal human nutrition, as well as of ordinary farm animals are formed in plants. The functional interrelationship of the C250 tenes and chlorophylls in the nutritional processes of the plant is still a problem of active research. The action of light does not seem to be strictly essential to the formation of carotene, for MacLeod and coworkers have observed an instance of increase of vitamin A value in a root (sweet potato) on storage, and some small increases have been reported in etiolated seedlings. But, as shown by Drummond and Coward, the pronounced development of vitamin A value in a seedling accompanies the appearance of green color, so that carotene and chlorophyll formation normally go on in a more or less parallel way in young plants and green leaves. The green leaves of food and forage crops and of pastures serve directly and indirectly to supply us with vitamin A values. The high vitamin A values of typical pasture grasses have been accurately measured by Woods and Coworkers (1932,1935). When we eat green leaves, we absorb (with varying degrees of completeness) the carotenes which they contain; and our bodies convert these carotenes (again, with varying degrees of completeness) into vitamin A which thus becomes available either for immediate nutritional needs or for storage in the body. And the carotenes of a wider range of green leaves are brought indirectly into the service of human nutrition through farm animals—particularly milch cows and laying hens. The cow and the hen render important services to human nutrition by bringing into the readily digestible and assimilable forms of milk and eggs a significant share of the nutritive values of leaves of coarser quality than we care to eat or which would exceed the quantity that we readily digest.

In any broad view of the problem of food supply, the outstandingly important sources of vitamin A value are the green and yellow vegetables, eggs, milk, and such products of milk as contain a significant proportion of its fat (e.g., butter, cream, ice cream, and whole-milk or "cream" cheeses). There follows a brief discussion of these and other food sources of vitamin A as illustrated in Table.

The vitamin A values reported for green leaf vegetables vary rather widely, even for the same species. This is doubtless attributable in part to actual variability of the materials and in part to laboratory errors due both to use of methods still in process of quantitative development and to differences in the aptitude and experience of the laboratory workers.

The high vitamin A value of the fully green leaf foods, especially kale, is probably much understated by the figure 7540 I.U. for raw kale in *Agriculture Handbook No.8*. Evidently this does not take full account of the large numbers of determinations of vitamin A values of kale made by Dr. Todhunter in the chemical laboratory of Columbia University. Granted that it might be "undue particularism" to stress any one average from the work of Todhunter, the same criticism is doubtless applicable in at least equal degree to some of the figures which did enter into the averages now appearing in the *Handbook*. For instance, chard leaves, 8720; collards, 6870; dandelion greens, 13,650; parsley, 8230; spinach, 9420; turnip greens, 9540 I.U. per 100 grams of the edible portion.

Vitamin A Values of Typical Foods: International Units Per 100 Grams

Food	*Range Within Which The Final Average May Be Expected To Be found*
Green leaves such as kale,	
spinach, turnip greens	(10,000)a
Peas, fresh young green	600-1300
Mature seeds generally	nearly negligible
Mill products and breadstuffs	negligible
Sugar and sweets	none or negligible
Fruits: Apples	40-100

Contd...

Food	*Range Within Which The Final Average May Be Expected To Be found*
Bananas	160-450
Cantaloupes	400-3500
Oranges	50-400
Vegetables: Broccoli	3000-9000
Potatoes	20-50
Sweetpotatoes	1500-7700
Tomatoes	500-1200
Beef muscle	10-50
Milk	160-225
Butter	3300-5000
Eggs	1000-2000

The differences among these figures may very likely be accidental, for kale which appears among the lowest of the fully green leaves in the *Handbook,* is certainly among the highest in fact. The writer has therefore suggested the tentative average figure of 10,000 which is very near the *Handbook* figures for spinach and turnip greens (perhaps the most used greens of the North and South respectively), and as affording a part of the upward revision which is clearly needed in the case of kale.

The evidence now available on the green leaves used as food for man indicates that representative samples contain around 10,000 International Units per 100 grams; but it should be emphasized that such high values are to be expected only in leaves that are *fully* green. Leaves which are relatively colourless must be expected to be of low vitamin A value. Thus Kramer and co-workers found less than one thirtieth as much vitamin A value in the white inner leaves of head lettuce as in the green outer leaves of the same plants. And the white inner leaves of tight-headed cabbages are of very much less vitamin A value than the green leaves of the loose leaf cabbages, while the still greener leaves of the closely related kale have much higher vitamin A values.

Quinn, Burtis, and Milner (1927) demonstrated the correlation with greenness in plant tissues other than leaves when they found snap or string beans and green peppers to be of high vitamin A value. Crist and Dye extended this to the case of green *versus* bleached asparagus tips.

The vitamin A values of other parts of plants are in general much lower; but in some cases vary widely with the extent to which carotene is transported to, and stored in, some organs other than the leaf. Thus most roots and tubers are of very low vitamin A value, while the carrot with its high carotene content shows regularly a high vitamin A value, and in sweetpotatoes the value is high but varies widely with the depth of yellow colour. But not all yellow plant pigments are precursors of vitamin A; for example, xanthophyll is not. However, F.L. MacLeod has shown that in sweetpotatoes, and M.C. Smith has shown that in squash, the differences in depth of yellow colour of the flesh are indicative of corresponding differences in carotene content and thus in vitamin A value.

As already noted, the leaves of forage plants and pasture grasses are of high vitamin A value. The extent to which this value is conserved in hay depends largely upon the curing process. Grains (and therefore our cereals and breadstuffs) are of very low vitamin A value.

As milch cows may consume grains and grasses in quite different proportions, a question arises as to the extent of the resulting variations in the vitamin A values of milk. *In the long run* the nursing mother or the lactating animal is dependent upon her food for all the vitamin A value which she puts into her milk; but the storage capacity of the body for vitamin A permits a regulatory process by which the vitamin A value of the milk can be maintained during considerable periods of low intake by drawing upon the body stores accumulated in periods in which the intake was high. Thus the vitamin A value of the milk ultimately depends upon, but does not literally vary with and as, the vitamin A value of the food.

The vitamin A values of sea foods originate much as do those of the land. Diatoms, small algae, and other tiny aquatic plants which have thus formed provitamin A may then be eaten by small crustacea, these by fish, and these in turn by larger fish like the cod. Thus the cod, through perhaps two or three or more intermediaries, has obtained its vitamin A from sources ultimately analogous to those which supply the cow.

The U.S.P. standard for cod liver oil calls for not less than 850 units *per gram,* and halibut liver oil is much richer in vitamin A, having, according to McCollum, about 20,000 units per gram, while percomorph oil is still richer.

The Inter-relationships

Vitamin A is a colourless, fat-soluble substance. It occurs in relative abundance in the fats of milk, eggs, livers, and fish. It is found in high concentration in fish liver oils. It now appears that there are two closely related substances, vitamin A_1 (which predominates in mammals and salt water fish) and vitamin A_2(which predominates in fresh water fish liver oils). It is convenient and scientifically admissible to continue to use the term vitamin A as a collective singular to cover these two substances functioning essentially as if they were one.

At least four substances (alpha-, beta-, and gamma-carotene, and cryptoxanthin) are known to be precursors of vitamin A, and are readily changed to vitamin A in the body. Hence the presence of any of these precursors in a food adds to its vitamin A *value,* though not to its vitamin A *content.* The fish liver oils contain vitamin A and practically none of its precursors. Foods generally however, owe their vitamin A values at least partly to the precursors which they contain.

Green leaves have a high vitamin A value, but no vitamin A content; the value is wholly because of the presence of the precursors. This is true with the vitamin A values of all plant organs. Animal organs contain both vitamin A and some unchanged precursor, and this is true of milk and eggs also.

The accepted structural formulae for vitamin A and its two best known precursors are shown in Figure.

It will be seen that the structural formula of the beta-carotene molecule, $C_{40}H_{56}$ is symmetrical, with a "double bond" in the centre (Fig.). By dividing at this point, each half taking cn both an H and an OH, it is theoretically capable of yielding two molecules of vitamin A, $C_{20}H_{30}O$; while by parallel reactions the other precursors known would each yield only one molecule of vitamin A, corresponding to the half of the precursor molecule here shown on the reader's left, because only this half includes the vitamin A structure.

In practice, the body may get considerably less than the theoretical yields of vitamin A from these precursors as they occur in foods. Even so, the vitamin A values of such green leaves as those of kale are high and shown by feeding tests (Sherman and Todhunter, 1934).

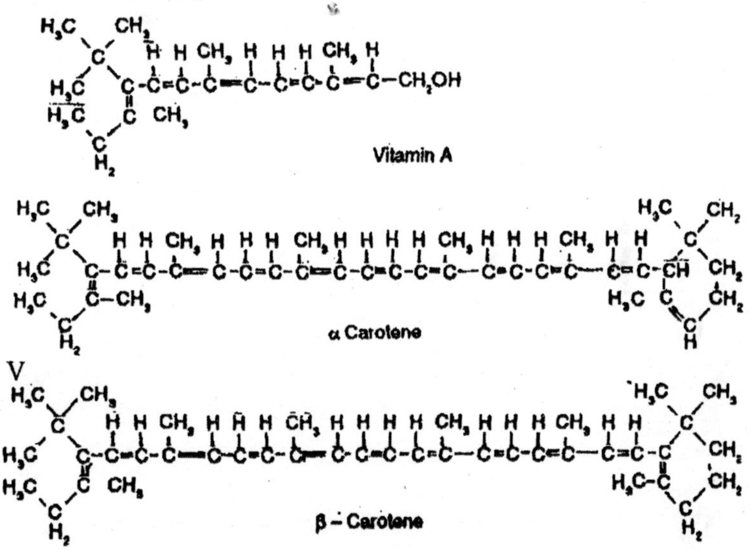

Vitamin A and two of its precursors

Nutritional Chemistry

It is of more than historical interest that the existence of vitamin A was discovered through experiments (made independently and almost simultaneously in 1913 by McCollum and Davis and by Osborne and Mendel) where in it was found that young animals on diets good in other respects would continue to grow and thrive or would cease growing and sicken, according as the sole fat in their food mixtures was butterfat or lard. Further work showed that egg fat and cod liver oil resemble butterfat in this respect while most commercial fats and oils are more like lard. The property, by virtue of which the maternal organism concentrates vitamin A in the material (milk or egg) provided for the offspring, seems to have a high *survival value* for both mammalian and avian species.

The importance of vitamin A to growth and development is illustrated in the experiments and observations, just mentioned, through which the existence of this factor was discovered. Its relations to the eyes are also of great importance.

Vitamin A is directly involved in the chemical changes upon which vision depends, and it is generally found that a diminished efficiency of adaptation of vision to a changed intensity of light (dysadaptation, hemeralopia, nightblindness) will be among the earliest observable effects of a shortage of vitamin A in any normal person.

The vitamin A which functions in vision is doubtless brought to the retina by the blood, a lowered concentration of this vitamin in the blood may be expected to accompany or precede the lowering of visual efficiency; but the evidence on this point is not entirely clear and consistent.

The physical evidence obtainable through biomicroscopy has been developed as a diagnostic method by Kruse (1941).

It shows that vitamin A deficiencies are considerably more prevalent than the chemical methods indicate. This apparent discrepancy may or may not be satisfactorily explained by the fact that what is found by chemical analysis of the blood is something quickly responsive to nutritional intake while the Bitot spots and perhaps other physical signs revealed by the biomicroscope are not so quickly changed by a change of intake. A question still remains as to whether a deficiency has been fully cured as soon as the blood chemistry has been brought to normal.

According to the methods and interpretations of biomicroscopy more vitamin A value should be provided, both for prevention and for complete cure, than has been thought necessary so far.

The Health Significance

The relations to health are clearer than the relations to disease.

Relatively early in the history of our knowledge of vitamin A, Dr. F. L. MacLeod found that in sufficiently long-term experiments great differences developed in family life histories of rats fed wheat and milk mixtures according as whole milk or skimmed milk was used (Sherman and MacLeod; 1925). The smaller mount of vitamin A (supplemented, no doubt, by reserves in the bodies of the animals at the beginning of the experiment) was sufficient for normal growth up to nearly average adult size. And the animals presented an excellent appearance during their youth and into adulthood. But this diet did not support successful reproduction, nor did the low-vitamin-A animals attain normal longevity, as did those getting more of this vitamin. In these early experiments, only two levels of vitamin A feeding were used, and these so far apart that the ultimate difference of outcome was too drastic for strictly quantitative measurement.

Dr. E. L. Batchelder (1934) fed animals alike in other respects upon four levels of dietary vitamin A and found that this stepwise increasing of vitamin A intake in the proportions 1, 2, 4, and 8 induced corresponding improvements in nutritional performance and family histories. The gains from extra feedings of vitamin A are summarized in Table. The number of animals were, however, not sufficiently large to allow of any quantitative comparisons of results. In later experiments at Columbia similar comparisons were made in larger numbers by Dr. H. L. Campbell and Mrs. H. Y. Trupp.

Here the basal ration was the Columbia Diet A (Laboratory No. 16) on which rat families have thrived throughout 70 to 75 generations so there is no doubt that the diet was adequate to all the nutritional needs of the experimental animals. This basal diet had 31.U. of vitamin A per gram of the dry food mixture, clearly adequate as just explained. When this allowance of vitamin A was doubled the result was a superior record including a longer average life. And when the vitamin A allowance was again doubled, making a four-fold increase over the original (about minimal-adequate) intake, the health record and length of life were still further bettered.

Thus vitamin A is one of the outstanding nutrients of which stepwise increased intakes induce correspondingly higher health and longer life up to intake levels about 4 times that of minimal adequacy. There is no doubt that in these experiments the addition of extra vitamin A to a diet which already contained "enough of everything" did positively and constructively build higher health and longer life. And we need have no doubt that this same thing may often occur with people.

The increase in the average length of normal adult life in these experiments was 10.4 per cent with males and 12.1 per cent with females. If one finds doubt about side questions of relation of vitamin A to disease, one can still be confident that it clearly promotes positive health.

Effects of Different Levels of Intake of Vitamin A (Based on Work of Batchelder)

In take	*Growth and Size*	*Adult Vitality*	*Length of Life*	*Vitality of Young*
At level of 1	Within normal range	Within normal range	Within normal range	Within normal range
Increased from 1 to 2	Slightly increased (within normal	Increased (within normal range)	Increased (within normal range)	Distinctly Increased range)
Increased from 2to4	Slightly increased (within normal range)	Slightly increased	Slightly increased	Distinctly increased
Increased from 4 to 8	No further change	No directly measurable	Slightly increased change	Distinctly increased

How human beings may best benefit by what has been learned through animal experimentation with different intake levels of vitamin A value, is a problem worthy of careful thought. And, to guide our thinking, we now have much more clearly and firmly established scientific research than existed even a few years ago, or than most people are probably yet fully aware of.

This animal experimentation upon the responses of the health and life history has been mainly with the rat because of its close nutritional similarly to the human. There is no reason to doubt and good reason to suppose that, in human as in rat nutrition, best result in long-term experience will require at least 2 to 4 times as much vitamin A value as suffices for minimal adequacy. If the minimal-adequate amount is taken as about 3000 I.U. for normal adult maintenance, from 6000 to 12,000 would be a scientifically more logical allowance (than the 5000 I.U. allowance of the National Research Council) to provide *both* for individual variations in requirement and for the maintenance of such a bodily reserve as has been found to be favourable to higher health and longer life.

To the extent that the vitamin A values of dietaries depend largery upon the precursor carotenes rather than on vitamin A itself and the doubts as to the completeness of availability of these precursors in digestion and assimilation, one will prefer to have his intake of vitamin A value nearer the top than the bottom of the 6000-12,000 I.U. zone above suggested.

Also it is scientifically sound in principle to provide liberal intakes of a nutrient which seems to be involved in so many different functions and in so many major and minor ills as is vitamin A, even if some of these involvements are not entirely clear.

The general principle, that nutritional shortages not of sufficient severity to be clearly diagnosed may nonetheless be reasons for dietary improvement, is clearly worthy of especially careful observance in the case of vitamin A values. This is a scientifically studied discrimination; not merely an undiscriminating open-handedness with food.

The Storage

The combination of properties which, makes vitamin A capable of storage in the body, in quantities sufficient to meet its nutritional needs for relatively long times, is of great importance to nutational well-being and health at all ages, and the clear recognition of the magnitude of this factor in the nutritional economy is highly essential to an adequate appreciation of the true significance of the intake of vitamin A values.

At some times and in some quarters, there has been failure of due recognition of the importance of the vitamin A value of the food supply, because of the very fact that bodily storage may protect from prompt penalty in periods of subsequent shortage.

Both for interpretation of experiments and for the everyday application of our new knowledge, we must keep in mind that the body does not quickly acquire nor quickly lose the full store of vitamin A which it is capable of taking on when habitually given a liberal intake. Cammack (Sherman and Cammack, 1926) found this a number of years ago in experiments with moderately liberal daily intakes, and the same fact appears to have been responsible for some confusing observation after Wald's administration of massive doses. Conceivably it may also have a bearing upon lack of pronounced correlation between bodily stores of vitamin A and either blood values or light-adaptation data in the work of Steininger, Roberts, and Brenner (1939).

Even at weaning time young animals (doubtless including babies) may already have a large store of vitamin A in the body. The body can also store vitamin A at later ages.

In one investigation (Sherman and Boyton, 1925) laboratory animals of known nutritional history were killed and dissected and graded weighed amounts of their tissues were fed as sole sources of vitamin A to young experimental animals. In this way it was shown directly that there is an actual storage of vitamin A in certain tissues, and it became possible to study quantitatively the distribution of the stored vitamin in the body and also to determine directly the extent to which the vitamin A content of the body is influenced by the intake of this vitamin in the food. If adipose tissue and skin be ignored, at least nine tenths of the total vitamin A in the body of a well nourished adult rat is found in the liver weight for weight, kidney appeared about 40 times, lung more than 40 times, and liver between 200 and 400 times as rich in this vitamin as the muscle of the same animal. Moderate differences in the vitamin A content of the food, such as are well within the range of variation likely to be encountered in human experience, resulted large differences in concentration of this vitamin in the liver, and distinct differences in the amount of it found in lung tissue. That the vitamin content of the lung tissue is dependent upon the abundance of this vitamin in the food is of interest because it has also been found that the higher intake of the vitamin is accompanied by a lower incidence of lung disease. Baumann, Riising, and Steenbock (1934) have also studied the body storage as affected by nutritional intake and the distribution in the body. They assigned about 95 per cent of the body's vitamin A content to the liver.

In another study of storage of vitamin A in the body (Sherman and Cammack, 1926) the placing of animals upon diet devoid of vitamin A was preceded by the feeding for different lengths of time of a diet bountifully supplied with the vitamin. This resulted in very great prolongation of life upon the vitamin-A-free food, showing that large surpluses of vitamin A can be stored under these conditions also. The maximum storage of vitamin A in rats of different ages was closely approached by feeding a ration containing 4 per cent of cold liver oil, which supplied about 80 times as much vitamin A necessary for adequate nutrition. However, the store does not increase in arithmetical proportion to the richness of the diet in vitamin A, because the animal uses the vitamin provided in the food more freely or less economically as the maximum store is approached.

The storage of vitamin A in the body has been shown to be an important factor, it seems probable that the supposed lesser need for this vitamin by the adult than by the young is apparent rather than real, and largely attributable to the fact that the adult has had more opportunity to provide himself with a bodily store of this vitamin which then carries him over periods of deficient intake.

Storage of Vitamin A in the Body May Complicate Interpretation in Research : Vitamin A has a combination of physical and chemical properties which makes it capable of long storage in the body, and up to qualities that are large in relation to the body's daily requirement. It is to our nutritional advantage that we can carry in our bodies such a valuable asset but its importance may be under-valued unless it is kept in mind that the storage factor makes long term experimentation particularly important in relation to vitamin A.

A distinguished pediatrician in the early history of vitamin A tried the experiment of feeding one group of children in an institution skimmed instead of whole milk for 9 months, which he judged would make it a quite thorough test, covering the whole winter season. Finding that this group apparently went through the winter as well as the others, he drew a conclusion which doubtless underestimated the importance of vitamin A for children; because he did not realize the significance of the storage factor, and that a nine months' experiment with children was only the equivalent of about a nine day experiment with rats.

Numerous experiments show that the body can store vitamin A at all ages. The very great extent to which bodily storage is influenced by the level of intake in the case of this vitamin is also now well established.

Site of Conversion of Carotene to Vitamin A : Following the work of Moore and of Olcott and McCann in 1931 the site of conversion of carotene to vitamin A was generally accepted to be liver. Conflicting results were reported by other workers but inconsistencies were attributed, at least in part, to species differences.

There has been renewed activity in the investigation of this problem. In 1946, Sexton, Mehl, and Deuel suggested that the intestine might be the site of the conversion, subsequent work in their laboratory showing that in the rat this is actually the case.

At the same time other research groups, including Glover, Goodwin, and Morton; and Thompson and Kon in England, using different techniques and, in addition to the rat, various herbivora and pigs and chicks, also reached the conclusion that the intestinal wall is the site of conversion of carotene to vitamin A, the efficiency with which the change occur varying in different species.

Absorption and Transport

Drummond, Bill and Palmer (1935) observed that the vitamin A content of the lymph increased after vitamin A or carotene had been ingested, the vitamin A in the lymph being esterified Popper and Volk (1944) found vitamin A in the facteals of the rat after oral during with vitamin A, absorption taking place from the small intestine, most rapidly between the upper and middle thirds. After oral dosing with carotene, Goodwin et. al. found the vitamin A level in the lymph of herbivora increased, and that no carotene appeared in the blood. Using a fluorescent method, Radice and Herraiz (1947) found in the rat that absorption of vitamin A took place into the lymphatics, most rapidly from the jejunum. Eden and Sellers found marked increase in the vitamin A content of intestinal lymph, especially from the duodenal region, after giving bullocks, sheep, and rats doses of the vitamin, but there was no significant change in the content of either portal or systemic blood, or of lymph from other regions of the body. Further evidence for the lymphatic route of transport of vitamin A was provided by the work of Goodwin, et al. (1948), in which increased levels of vitamin A were found in the lymph collected by cannula from the intestinal region in goats after feeding either vitamin A or carotene while blood values remained unchanged.

The work of Eden and Sellers confirm the earlier observation by Gray, et al. in 1940 that orally ingested vitamin A ester is hydrolyzed in the intestinal tract, the alcohol apparently being the form in which it is absorbed; but it is re-esterified in its passage through the intestinal wall and appears in the lymph almost entirely as the ester. It is also as the ester that it appears in the circulating blood.

The Metabolism

Conversion of Carotene of Vitamin A. After it was established that carotene is a source of vitamin A the next step was to ascertain where in the body the conversion of the pigment to the vitamin occurs. It was believed that the conversion occur in the liver. It was suspected because of the high vitamin A content of the organ. The hypothesis was questioned, however, when the provitamin administered intravenously was not converted to the vitamin. Further doubt was cast on that belief when blood vitamin A was elevated after feeding carotene to rats, with the liver taken out of circulation. The liver was cut off from all connections by trying off the veins, arteries, and common bile duct. It was under the conditions of this experiment, together with other evidence, that the intestinal wall was found to be the main site of the transformation in the rat.

The liver and other tissues have been shown to be alternative sites for the conversion of some carotene to vitamin A. Since man is capable of absorbing carotene from the intestine, other sites for transformation may be of importance. Education is required about the site and the conversion mechanism of the carotenoids to vitamin A.

The amount of conversion of carotene to vitamin A is variable and depends on the dietary source and efficiency of absorption. Pure crystalline beta-carotene was found to be more efficiency utilized than beta carotene from the green vegetables in studies with adults. In average diets carotene is estimated to be one-half as well utilized as performed vitamin A: 3000 IU of carotene in the diet would be expected to yield only 1500 IU of vitamin A in the body.

Absorption. Vitamin A and carotene are absorbed from the small intestine into the lymph system. The maximum absorption is reached three to five hours after consumption. The rate of absorption of vitamin A is more rapid than that of carotene.

The absorption of both vitamin A and carotene is influnced by other substances in the small intestine. Bile, dietary fat, antioxidants, and detergents enhance absorption. Bile is essential for the absorption of carotene. Mineral oil may interfere with absorption, particularly if taken simultaneously with food, or if a large amount of oil is fed with a less than adequate intake of the vitamin. Both vitamin A and carotene are soluable in mineral oil; since the oil is not absorbed, some of the vitmain A and provitamin A are eliminated with it. Under certain conditions, however, the oil is found not to have a deleterious effect. In a study with adults one ounce of mineral oil taken at bed time was found not to impair the absorption of either vitamin A or carotene.

Storage and Excretion. In the human being about 95 per cent of the vitamin A stored in the body is found in the liver with small amounts in the lungs, body fat, and kidneys. The amount is least at birth and increases with age. The store of vitamin A in the liver depends on the quantity in the diet and the amount absorbed. It has been estimated that a normal liver may contain as much as 600,000 IU of vitamin A, enough to supply vitamin needs for one year.

Vitamin A is excreted by way of the intestine. It has been observed that the fat-soluble vitamins are excreted by the fecal pathway, the water-soluble vitamins by the urinary pathway.

The Assessment

One of the early symptoms of vitamin A deficiency is night blindness or subnormal dark adaptation. Various instruments similar in principle are used for the quantitative measurement. Among them is the photometer (adaptometer) (Fig.), which measures the amount of light (intensity) required for the subject to perceive objects before and after his exposure to bright light. The intensity of light and the time needed to regain vision after bleaching of the visual purple are indicative of vitamin A status. Symptoms of deficiency are recognized by the need for a greater intensity of light and a longer period of time for the regeneration of the visual purple.

Various sources of error are present in the photometer procedure, both in the instrument and in the subjective response of the person being tested. Conditions other than that of vitamin A deficiency may impair dark adaptation. If rapid and distinct improvement follows the ingestion of vitamin A, it is evident that a shortage of the vitamin was a causative factor.

The blood level of vitamin A and carotene bear a relationship to intake and are indicative of vitamin A status within certain limits. Adults placed on a vitamin-A-free diet are found to show no change in the level for several weeks, followed by an ultimate lowering in the concentration. A carotene-free diet results in prompt reduction of the blood carotene content. The body has stores of vitamin A that can be mobilized for use, whereas carotene is not stored.

Vitamin A deficiency in the population of this country appears to be relatively rare. The capacity of the body to store vitamin A provides for an effective emergency supply.

Hypervitaminosis A : Excessive consumption of vitamin A may lead to toxic symptoms in both man and animal. Hypervitaminosis A is being reported increasingly in both children and adults who have taken excessive dosages of concentrates of the vitamin for different lenghts of time. A case report of a 39-year-old woman who was taking approximately 150,000 IU of vitamin A daily in hopes of improving her vision has been described. The symptoms were swelling of the feet and ankles, fatigue, lassitude, weight loss, and intermittent pains in the shoulders, wrists, and knees. Discrete superficial hemorrhages in the retinae of both eyes in a 17-year-old boy who had been taking 200,000 IU of vitamin A daily for 18 months for acne was reported. Arctic travellers who have consumed polar bear liver, which is extremely high in vitamin A content, have manifested symptoms of acute hypervitaminosis A.

The symptoms of acute toxicity are not identical in all cases; most commonly observed are skin lesions, an itching rash in some instances, in others a coarsening of the skin, thinning of the hair, and frequently pain in the joints, with hemorrhage between the bone and its covering. These same symptoms may arise from other causes. A history of consumption of excessive amounts of the vitamin and improvement in physical condition with the discontinuance of its use led to conclusive evidence that hypervitaminois A was the responsible for the trouble in recognized cases. *Acute* toxicity may occur is an end stage of prolonged chronic toxicity resulting from overdosage over a period of time. The toxic dose is highly variable among individual especially infants and young children, but is belived to be over 100,000 IU daily. III effects from lesser amounts administered over a period of time have been reported. The Council on Foods and Nutrition of the American Medical Association state that "there is definite possibility of harm from the prolonged ingestion of vitamin A in excess of 50,000 USP units daily." The use of vitamin concentrates that need only a few drops for a dose afford an easy means of giving more than is desirable. For example, around 50 drops of several of the common concentrates were found to yield one-half teaspoon, providing in that small measure 15,000 IU of vitamin A. Concentrates of the vitamin are incorporated in multiple vitamin capsules in varying amounts, some contain 5000 IU per capsule, some 15,000 IU and others 25,000 IU and 30,000 IU.

Recommended Allowances : The Food and Nutrition Board Recommend Daily Allowances of vitamin A at levels believed to be "adequate to meet the needs of health. They are estimated to be about twice the minimal need. The recommendation for young men and women 18 to 35 years of age is 5000 IU, the same daily allowance is recommended for older adults. The allowance is made on the basis that 4000 IU are produced in the average diet by carotene and 1000 IU by vitamin A. Since carotene is only about one-half as well utilized by the body as is performed vitamin A, if all the vitamin A activity in the diet were from carotene the allowance would be 6000 IU instead of 5000 IU. On the other hand, if all the vitamin were provided by the performed vitamin the allowance would be 3000 IU instead of 5000 IU.

Food Sources

Performed vitamin A occurs only in foods of animal origin; carotene is found in both plant and animal products. The most potent sources of vitamin A are fish-liver oils, liver, butterfat, fortified margarine, and egg yolk. The concentration of vitamin A in the liver increases with age and varies with the species. The potency of butterfat and egg yolk are influenced by the diet of the cow and the hen, respectively. In the spring and summer when green grass becomes more available, the vitamin A content of the milk and eggs produced by the animals is higher. However, the improved and enriched livestock feed used today makes possible the provision of an ample source of vitamin A in animal rations the year round.

The most important sources of the provitamin are the yellow, yellowishred, and green vegetables and fruits. The yellow carotene in the green products is masked by the presence of chlorophyll. Among the excellent sources of carotene are carrots, sweet potatoes, apricots, spinach, tomatoes, and kale. Some vegetables and fruits deviate from the general colour rule; even though yellow in colour, oranges, rutabagas, and yellow wax beans are not good sources of carotene. Head lettuce and avocados, two green-coloured foods, provide small amounts of carotene. Foods containing little vitamin A are nuts, grains, vegetable oils, muscle meats, and light-coloured fruits and vegetables. In most foods one-half or less of the carotene is converted to vitamin A. As a consequence, the carotene content of a food is divided by two in most instances to obtain the biological activity in terms of vitamin A.

The vitamin A values of some common foods are given in Table. The term vitamin A *value* of foods is used instead of vitamin A *content*. It would be incorrect to indicate that carrots, for example, contain 12,000 units of vitamin A per 100 gm; actually they contain no vitamin A at all; they are merely rich in crotene content.

In compiling the tables of food composition published as Agriculture Handbook 8, Wait and Merrill converted ug of carotene in foods to IU of vitamin A activity on the basis that 0.6 ug of beta-cartone, and 1.2 ug of the sum of the other carotenes with vitamin A activity or of cryptoxanthin are equivalent to 1 IU of vitamin A value.

The Stability

Since both vitamin A and carotene are insoluble in water and stable to ordinary cooking temperatures, there is little loss in food preparation. Canning and freezing cause little loss, but drying and dehydration result in considerable destruction through oxidation.

The development of rancidity in fats causes destruction of both vitamin A and carotene. Storage of fats in a cool, dry place delays the onset of rancidity and thus helps to preserve the vitamin. Magarines were found to lose than 25 per cent of the original vitamin A content when stored at 14° F for a period as long as 2 years, but when stored at 64°F, 75 per cent of the original vitamin A content was lost in only 17 weeks

Daily Need

Dietary records of college students often show an intake of vitamin A less than the Recommended Allowance. An evaluation of these dietaries shows a low intake of dark green and yellow vegetables. It is estimated that vegetables and fruits provide about 60 per cent of the available supply of the vitamin A value in the United States. The inclusion daily or three to four times a week of a serving of a dark-green or yellow vegetable is an easy way to lay the foundation for an adequate intake of vitamin A. Table shows groups of foods for meeting the daily vitamin A need.

Individuals vary in their ability to absorb carotene and in its conversion to vitamin A. Therefore, there should be sufficient intake of carotene to allow for any discrepancy. Foods providing vitamin A, such as eggs, liver, and cheese, should be included.

The Functions

Growth. Failure of young rats to grow when they are deprived of certain foods was the initial symptom that led to the discovery of vitamin A. In the early studies before vitamin A had been isolated and chemical methods found for its determination, growth rate was the criterion of measurement of the vitamin A potency in a food. Animals maintained on a vitamin-A-free diet will cease growing when the body is depleted of its vitamin stores. Materials to be assessed for vitamin content were then fed in known amounts for a prescribed period of time and the growth response noted. Better growth was associated with higher vitamin potency in the food. If adequate vitamin A is not provided for normal growth, the bony structure will suffer a growth failure before the soft tissues manifest the stunting effect. The cessation of bone growth may result in overcrowding of the brain and central nervous system. It is now clear that the paralysis sometimes accompanying vitamin A deficiency in very young animals is due to failure of bone growth, that causes cranial pressure with consequent brain and nerve injury. In some instances there is pinching of the optic nerve, resulting in blindness.

Health of the Epithelial Tissues of the Body. The outer surface of the body is covered with epithelial tissue, the internal cavities of the body lined with it. The tissues are differntiated to better serve their individual functions; the external covering is the resistant, protective epidermis; the internal tissue is a secretory mucous membrane. Deficiency of vitamin A causes an abnormal condition of the epithelial tissues. The specialized functions of the tissue are suppressed, and it is transformed to a keratinized (dry, horny) type of epithelium. Excessive dryness of the skin may result from an insufficient intake; the mucous membrane may fail to secrete normally and becomes less resistant to bacterial invasion.

***Xerophthalmia* (also known as keratomalacia).** It is a condition of the eye resulting from severe deprivation of vitamin A (Fig.). The tear glands cease to function, the eye is dry, the cells of the *cornea* (the transparent outer covering of the iris and pupil) become dry and in serious cases ulcreated. If the condition is not arrested by provision of vitamin A, blindness ensues. The cornea, the part of the eye suffering the severe damage in xerophthalmia, is an epithelial tissue. The condition occurs in both man and animal. Xerophthalmia rarely occurs in the United States. It is, however, not an uncommon occurrence among the young in China. It is a cause of blindness among the rice and cassavaeating people of the southern part of India and Ceylon and is found in Indonesia, Malaya, the Philippines, and in parts of Central and South America. The majority of the cases occurs in children two to six years of age. In 1914-1918 children in Denmark were afflicted; butterfat was shipped from the country, and skimmed milk containing little vitamin A used. The condition can be developed in experimental animals deprived of vitamin A.

Composition and Functioning of Visual Pigments. The retina of the eye contains visual pigments that are a combination of vitamin A and a protein. The retina is the light-sensitive layer at the back of the eye; it may be likened to the sensitive film of a camera on which the image appears. Two kinds of receptor cells are in the retina—the rods that function in vision in dim light, and the cones that function in bright light and colour vision. The terms "rods" and "cones" arise from the shape of the cells. Rhodopsin is a vitamin A-containing pigment in rods, iodopsin in cones.

Relation with Vision

From the viewpoint of the chemistry of nutrition, the outstanding relationships of vitamin A to vision are: that vitamin A is essential to the regeneration of visual purple, the bleaching of which in the retina is one of the chemical reactions involved in vision; and (2) as a consequence it follows that the ability to see in a dim light, or to adapt to a changed intensity of light, is very directly related to the abundance and immediate availability of the supply of vitamin A. The mechanism of vision is also two-fold: that of the cones of the (diagrammatic) physical structure of the retina.

Jeans and coworkers (1936), using visual adaptation tests, found seriously frequent evidences of shortage of vitamin A among school children, especially in low-income families. Subsequently much attention has been given to the improvement of instruments for use in such tests, to the critical study of details of technique and interpretation, and to the extension of this general method both in surveys of populations as to their nutritional status and in laboratory research upon the quantitative problems of vitamin A requirement in human nutrition.

Whether measurement of visual adaptation or of the concentration of vitamin A in the blood will prove the more valuable criterion is a research question, the answer to which will doubtless depend both upon the development of techniques and upon fuller and more critical knowledge of the nutritional significance of vitamin A. It is probable, that a real shortage of vitamin A in the body will inevitably impair the efficiency of visual adaptation; but also that such impairment may sometimes result from other causes. And because bodily storage may be such a large and imperfectly understood factor, the recent dietary and therapeutic history may not always allow a sound judgement as to whether or in what sense a real shortage exists.

Vitamin A and Metaplasia : The relation of vitamin A to mucous membranes is a fact to very far-reaching significance. It has been found to be important in the eyes, in certain secreting glands, in the respiratory system, the alimentary canal, and the urinary tract; and it is readily conceivable that still other such relationships may remain yet to be discovered.

Mori, Bloch, and Blegvad independently described diseased conditions of children which in the light of all evidence now available seem to have been due primarily to shortage of vitamin A. While xerophthalmia was the symptom which called attention to these cases, it was noted that there was also a high incidence of respiratory disease which was probably due to the weakening of the respiratory system by the same dietary deficiency. Blegvad argues that the deaths which occured among the children who showed the eye trouble were due in most cases to respiratory disease.

Experimental Deficiency in Laboratory Animals : In experiments with laboratory animals the easily observable sign of vitamin A deficiency is the development of an ophtalmia (xerophthalmia, keratomalacia, conjunctivitis). This was first recorded by Osborne and Mendle in 1913 and has occurred so regularly (at least in the experience of most observers) as to be considered as characteristic result of a lack of vitamin A. But it is only one of several results, and not the most important. Further studies have shown that vitamin A deficiency results also in an increased incidence: of respiratory disease; of skin, car, and sinus infections; of inflammations and infections of the alimentary tract; and even of renal calculi (Osborne and Mendle, 1917). The underlying cause of the weakness that may show itself in so many ways is the histological change that Wolbach, Howe, and Church have described as follows: "The specific effect of the absence of fat-soluble vitamin A in albino rats, guineapigs, and humans is found in epithelial tissues. This effect is the substitution of strarified keratinizing epithelium for the normal epithelium in different parts of the respiratory tract, alimentary tract, eyes and paraocular glands, and the gennito-urinary tract."

The higher incidence of infection observed among individuals on diet of low vitamin A value may be due to the displacement of normal by squamous epithelial tissue implies not only a loss of local secretion but also a loss of the *cilia* that act with the secretion in cleansing the surfaces of the respiratory tract by expulsion of foregin particles. Plainly the normal surface thus tends to protect the respiratory system both from bacteria and from miscellaneous particles that might cause mechnical injury and so increase the danger of the establishment of infection by such bacteria as may remain. It has also been suggested that the local roughness and stickiness which accompany the breaking-up of the normal tissue (when it is first being replaced by the squamous) may also increase the danger of lodgment of bacteria which would normally have been expelled.

Thus even our present incomplete knowledge shows more than one way in which shortage of vitamin A may increase the dangers of infectious disease. Whether it specifically increases susceptibility, in the technical sense in which bacteriologists and patholohists are apt to use this term, we do not know. Emphasis tends to be laid upon the characteristic metaplasia of epithelial tissue which many investigators have observed to result from deficiency of vitamin A and which structurally impairs what the *journal of the American Medical Association* has called "the bodys first line of defense." This metaplasia of the epithelium has been observed in many parts of the body and in every species which has been thoroughly studied. It probably underlies the external eye symptoms as well as the other ills above mentioned. In Tildern and Miller's experiments upon monkeys, a shortage of vitamin A resulted in injuries to the digestive tract which were often fatal before the development of a conspicuous degree of ophthalmia.

The influence of vitamin A, unquestionably important at the first line of defense" against infections extends also beyond this, has been shown by several investigations (among others those of Boynton and Bradford, 1931; of Lassen, in 1932; and of McClung and Winters, 1932), in which infective agents have been *injected* under experimental control, and it has been found that the animals more liberally supplied with vitamin A survived better.

To avoid confusion, we should keep in mind that there are wide differnces in attitude among those presumably competent to discuss the relation of vitamin A to infection; and that these differences are to a considerable extent expained by the fact that anthorities vary in their modes of approach to the problem those who have sought to connect vitamin A with the immunological mechanism as commonly conceived have reported negatively. Those whose approach is somewhat broader but who still concentrate upon the search for clearly direct primary relations, tend to report either negatively, or inconclusively. But on the more comprehensive (if technically less precise) question, there is much positive evidence that the level of intake of vitamin A does (whether directly or indirectly, primarily or secondarily) influence the frequency, or severity, or duration, of infectious disease—not specifically of any one infection alone, and also not equally of all infections.

Both age and previous feeding may influence the results of a vitamin-A-poor diet.

In a typical series of observations, three-fourths of the rats placed upon vitamin-A-free food at the age of about one month (end of infancy) developed the characteristic ophthalmia before death; but developed in only about one fourth of the rats that were from two to nine months old when subjected to the same dietary deficiency (Sherman and Storms, 1925).

The older rats, however, proved more susceptible than the younger ones to lung infection, the frequency of which, in Columbia experience, like that of Steenbock and other investigators, is increased by diet deficient in vitamin A.

Requirements and Adequacy

The National Research Council's Recommended Daily Allowances for vitamin A values of dietaries, expressed in International Units (I.U.) are as follows: Men (70 kg.), 5000 I.U.; Women (5b kg.), 5000 I.U.; pregnancy, latter half, 6000 I.U.; lactation, 8000 I.U.; Children under 1 year, 1500 I.U.; 1-3 years, 2000 I.U.; 4-6 years, 2500 I.U.; 7-9 years, 3500 I.U.; 10-12 years, 4500 I.U.; Girls, 13-15 years, 5000 I.U.; 16-20 years, 5000 I.U.; Boys, 13-15 years, 5000 I.U., 16-20 years, 6000 I.U.

These allowances are materially higher than previous estimates of the amounts required for the prevention of such signs of deficiency as have been described above, and at the time of adoption early in 1941 were considered as carrying liberal margins above actual needs. Soon afterward, however, Kruse (1941) by use of a more delicate method, found evidence of a much greater prevalence of shortage of vitamin A than had previously been known or suspected.

Bitot had described the occurrence of elevated spots on the conjunctival membrane of the eye, and these had become known as Bitot spots, but were not generally as significant until Kruse observed repeatedly that they disappeared-although sometimes very slowly-under vitamin À therapy. He therefore considered Bitot spots a sign of vitamin A deficiency (*avitaminosis* A), and proceeded to a biomicrosopic study of the parts of the eye where they are most often seen. This investigation showed that the typical Bitot spot, visible to "gross" examination, is a relatively advanced sign, the earlier or lesser stages of which can be seen only with the aid of biomicroscopy but are then found to be of more frequent occurrence than the grossly visible spots.

If the evidence of this more delicate method be taken as conclusive, it follows that a large proportion of the people previouly considered as amply nourished had actually received too little vitamin A, and that the true human requirement for vitamin A is considerably higher than hitherto supposed. But not all nutritionally minded physicians yet regards these minute and hitherto unknown or disregarded signs as conclusive evidence of vitamin A deficiency.

At present, therefore, according to one's judgement of these biomicroscopic observations, one may regard the Recommended Allowances of vitamin A value as liberal or as scanty from the point of view of preventing deficiency signs.

Correspondingly one may feel that the average American dietary does or does not carry a fully satisfactory margin of vitamin A value. There seems to be no doubt that the actual vitamin A value of the American food supply has been fully maintained, if not increased, during recent years, due to better conservation of food (including fishery) resources and increased production of green and yellow vegetables and tomatoes in home gardens. These improvements easily can, and wisely may, be carried further, for medical experience and animal experimentation have clearly shown that very liberal margins of vitamin A value in the dietary give added benefit in the long run.

Mellanby has reported, from his combined hospital experience and animal experimentation, that increased allowances of vitamin A (or precursor) continue to diminish the dangers of infection up to levels of intake four times that required for normal nutrition in absence of exposure.

For people living under everyday conditions and not encountering special exposure to infection, perhaps the best evidence as to long-run desirability of a liberal intake of vitamin A value is that of the above cited full-life and successive-generation experiments with laboratory animals. These experiments showed that successively increased levels of vitamin A intake resulted in successively improved life histories and vigor of offspring, up to diets of about four times the vitamin A value which ordinary criteria would accept as adequate.

By more liberal use of green vegetables the vitamin A value of the American diet can be largely increased at little cost. Thus Gillett and Rice found that families of the same economic grouping in New York City, spending practically the same amount per capita for food in the same market, but according to their own habits and prefrences within their economic limitations were actually consuming four or five times as much vitamin A value per person in some families as in others.

20

Vitamin B

Ever since 1926, when Minot and Murphy observed that pernicious anemia (Addison's anemia) could be controlled by the ingestion of whole liver, many laboratories have been searching for the factor responsible for this therapeutic action. But it was not until 1948 that scientists isolated a red crystalline material that in very small amounts was beneficial in the relief of this form of anemia. The substance, which is the most recently discovered vitamin, was named vitamin B_{12}. It is a red crystalline compound with a complex formula ($C_{63}H_{88}N_{14}O_{14}PCo$) containing two minerals, cobalt and phosphorus. Vitamin B_{12} is not a single substance but may exist in several different forms, which have been given the generic name cobalamin. The two active members of this group are vitamin B_{12} (cyanocobalamin) and vitamin B_{12a} (hydroxocobalamin).

Vitamin B_{12}

This is the most recent discovery in this group, and is one of the most highly active substances known. The first vitamin B_{12} to be obtained in the pure state was extracted from liver.

It is now prepared on a large scale by a different process.

Its structural formula has not yet (1954) been fully elucidated. A rather unusual feature is that it contains cobalt. It is deep red in colour.

The Discovery

In 1926 Minot and Murphy found that pernicious anemia responded to treatment with whole liver or its extract, and during 1927 to 1980 West and Dakin worked actively on the fractionation of liver extracts, testing the fractions in actual therapeutic use with pernicious anemia patients whenever such patients were available. Other research teams also had worked actively in this same field for several years when, in 1948, Smith and Parker in England, and Rickers, Brink, Koniuszy, Wood, and Folkers in the United States were semultaneously successful in the isolation of a crystalline red substance containing cobalt and phosphorus, which was fully effective (in right conditions) against pernicious anemia. The pure substance, designated vitamin B_{12} in 1948, was found to be so active therapeutically that a single dose of 3 to 6 micrograms was clinically effective when injected intramuscularly in patients with pernicious anemia. Clinical tests have shown that vitamin B_{12} is fully effective in the treatment of pernicious anemia, arresting central nervous system degeneration as well as restoring normal blood values.

Extrinsic and Intrinsic Factor

The complete effectiveness of vitamin B_{12} in pernicious anemia led to intensive study of its place in the postulated theory: extrinsic factor + intrinsic factor—erythrocyte maturation factor; and the role of extrinsic factor has been assigned to vitamin B_{12}.

Berk, et al. (1948) found that vitamin B_{12} administered orally in amounts effective when injected was ineffective unless accompanied by a source of intrinsic factor, e.g., normal gastric juice, but that if very large amounts were fed, it was effective alone. This suggested the identification of vitamin B_{12} with the extrinsic factor, a theory which has been sustantiated by the work of various groups approaching the problem from different angles. Other sources of extrinsic factor, such as beef muscle, have been shown to be effective in proportion to their vitamin B_{12} content.

The role and identification of intrinsic factor is still, however, an unsolved problem. It has been suggested that the function of intrinsic factor is to facilitate the absorption of vitamin B_{12} in some manner. The finding of a substance in normal gastric juice that combines with vitamin B_{12}, making it unavailable for the growth of micro-organisms, has been reported by a number of investigators. Ternberg and Eakin have isolated from gastric juice and from hog mucosa a non-dialyzable, heat-labile material, apoerythein, which combines stoichoiometrically with vitamin B_{12}. The resulting complex, erythein, is inactive for micro-organisms, releases vitamin B_{12} on heating, and is highly selective in its binding action. Vitamin B_{12} is bound by other substances also, such as the lysozyme of eggwhite, but there is a distinction in the complex formed with apoerythein in that vitamin B_{12} cannot be separated from it by dialysis, a ready means of separating it from other protein substances.

A study of the reactions between normal gastric juice and vitamin B_{12} by Wolf, et al. indicates that although it enters into some form of combination, there is no great change in the vitamin itself, and thus strengthens the theory that the function of the intrinsic factor is to promote absorption of vitamin B_{12}, the extrinsic factor.

***Vitamin* B_{12} *and Growth* :** Vitamin B_{12} has been found to have growth-promoting activity for some micro-organisms, for animals, and in some cases for children.

Wetzel and co-workers (1949) have shown strikingly favourable results from vitamin B_{12} in 5 of group of 11 children who had been making less than normal growth. Although there were no characteristic or suggestive physical signs in the hair, skin, eyes, mouth, or nervous system prior to giving vitamin B_{12} the growth response was accompanied by improved vigor and alertness, demonstrating that this vitamin has a far more general nutritional significance than its anti-anemic effect alone.

The growth promoting property of the so-called "animal protein factor" has been reported to be due in part to vitamin B_{12}, but other effects of this factor, which appears to be of a multiple nature, are due to substances which await the results of further research for identification.

Apparently vitamin B_{12} occurs in at least two active reversible forms that vary in their effect on the growth of micro-organisms, although they are both available in analytical feeding tests. The vitamin is known to be formed by many of the lower plants, including certain bacteria, yeasts, and molds. In view of this, it now becomes quite gratuitously confusing to link it with the misnomer of "animal protein factor." Rather may this latter term be dropped in the interest of clarity, both here and in comparisons of proteins.

***Vitamin* B_{12}*folic acid, and ascorbic acid* :** Both vitamin B_{12} and folic acid have been shown to have their hematological effects enhanced in certain cases of refractory megaloblastic anemia by the administration of ascorbic acid. In a deficiency of ascorbic acid in infants, larger amounts of folic acid are required for the remission of anemia. A lack of response to vitamin B_{12} in megaloblastic anemias has been observed by various workers, success being achieved with folic acid alone. Holly has reported the successful treatment of such anemias by the combined use of ascorbic acid and vitamin B_{12}. Studies of Sealock, Day, Darby, and their associates, have demonstrated an interdependence between folic acid, vitamin B_{12}, and vitamin C in regulating the metabolism of phenylalanine and tyrosine. An inter-relationship between these three vitamins has also been studied in chicks by Elvehjem and his co-workers who suggest a function of vitamins B_{12} and C in stimulating folic acid synthesis.

Darby and co-workers have shown that either folic acid or vitamin C will correct the abnormal tyrosine metabolism of the scorbutic gunieapig.

Studies by Rodney, et al. with liver tissue of folic acid-deficient rats indicate that only folic acid will correct the metabolic defect *in vitro* but that *in vivo* either folic acid or vitamin C is effective.

Vitamin B_{12} seems to be both the much-sought anti-pernicious anemia factor and an essential nutrient for several species, probably including man. If it shows its indispensability to man less clearly than to some other species, this is thought to be due to the fact that under favourable normal conditions a part of the human adult's need is provided by the activities of the "friendly" bacteria in the intestinal tract; and also that infants may be born with prenatal stores from their mothers.

The Distribution

Methods of assay for vitamin B_{12} are based on its growth-promoting effect for certain bacteria, for the rat, or the chick. Liberation of the vitamin and its separation from interfering substances, as well as the formulation of suitable medium, have led to some uncertainties in the results of microbial assay, but with improvements in technique and with corroborative results by the growth of rats treated with iodinated case in it appears certain that vitamin B_{12} is limited in it distribution in common foods. Scheid and Schweigert report values of 40.9-62.3 mcg. per 100 gm. fresh liver, 14.8-23.1 mcg. per 100 gm. fresh kidney, and from 0.9 to 3.2 mcg. for 100 gm. fresh muscle meat. Recent work indicates that apparent vitamin B_{12} values reported for alfalfa and other plant products may in part be due to some unidentified factor or factors, and that the richest food sources of vitamin B_{12} thus far tested are liver and kidney, with eggs, milk, and muscle meats contributing substantial amounts. Robbins, et al. have suggested that the vitamin B_{12} values they find for root vegetables are due to the absorption of vitamin B_{12} formed by soil bacteria.

The Function

Vitamin B_{12} functions as a co-enzyme in the metabolism of the body cells, particularly for cells of the bone marrow, the nervous system, and the gastrointestinal tract. Its role in metabolism is not yet completely understood. In the past the function of vitamin B_{12} had been related to the effectiveness in relieving the abnormal blood pattern and neurologic symptoms of pernicious anemia, because of its initial use in the relief of this disorder. From the hematologic viewpoint, vitamin B_{12} seems to be a factor without which the reduced folic acid co-enzymes cannot be formed or function effectively to assist in the production of normal red blood cells in the bone marrow.

The existence of the B_{12} co-enzymes was first reported by Barker in 1958. Since then data have accumulated on the biological role of these factors, and at the present time the B_{12} co-enzymes have been related to seven different enzymatic reactions or processes.

Soon after the discovery of vitamin B_{12}, it was suggested that this nutrient might function to promote growth in animals and man. Although a few isolated studies have been reported on the growth-promoting effect of vitamin B_{12} in small numbers of underweight children, further work is necessary before the role of the vitamin as a human growth stimulant can be established.

Pernicious Anemia

Pernicious anemia is a chronic disease of genetic origin characterized by a macrocytic anemia, glossitis, and neurological disturbance. Stiffness of the limbs, irritability, drowsiness, and expression are some of the typical neurological symptoms. When liver was effective in treatment of the disease, Castle postulated that a substance in food (extrinsic factor) combined with a substance in the gastric juice (intrinsic factor) to form the antipernicious anemia factor. Vitamin B_{12} is now recognized as the extrinsic factor. The intrinsic factor is a mucoprotein formed in the gastric juice. Pernicious anemia patients lack the intrinsic factor that functions in the body to remove vitamin B_{12} from the material to which it is bound and with the help of calcium to attach vitamin B_{12} to the intestinal wall. Thereafter, by enzymatic action, the intrinsic factor facilitates the passage of vitamin B_{12} through the mucosal cells into the blood stream. Very minute doses of the vitamin, 10 to 15 ug per day given by injection (intramuscularly), are effective in relieving the disorder and larger amounts given every three or four weeks, prevent the recurrence of pernicious anemia in patients.

It is of interest to mention the observations of Herbert and Sullivan (10) on the relative effectiveness of vitamin B_{12} (cyanocobalamin) and coenzyme B_{12} in the treatment of pernicious anemia in a limited number of subjects. In the case of the deficiency with only hematologic damage, small doses of coenzyme B_{12} appear to be more effective in relieving the blood symptoms than do equal portions of cyanocobalamin. However, when both hematologic and neurologic symptoms were present, cyanocobalamin rather than coenzyme B_{12} was more effective in relieving the blood symptoms. From these limited data, these workers speculate that "... although coenzyme B_{12} seems more potent than cyanocobalamin when only the hematopoetic system is damaged, a very small dose of coenzyme B_{12} may go preferentially to the nervous system when the system is damaged, and thus be less hematopoietically potent in patients with both hematologic and neurologic damage."

Deficiency in Man

The first documented case of pernicious anemia in man caused by an actual lack of dietary vitamin B_{12} was reported in 1956 (16). The characteristic blood pattern was found in this patient who had existed on an animal-food-free diet for about eight years. The administration of vitamin B_{12}, however, brought about complete relief from the anemia. Deficiency manifestations may occur in persons infested by the fish tape worm/ sprue, and other malabsorption syndromes.

Recommended Allowances

The human allowance for vitamin B_{12}, under normal conditions of health, is unknown, but it has been suggested that a diet containing 3 to 5 ug per day should satisfy man's need. A diet containing adequate amounts of meat, eggs, and milk will more than satisfy this need.

In a study by Chung and associates of high-cost, low-cost, and poor diets, the daily vitamin B_{12} contents were 31.6, 16.0, and 2.7 ug respectively.

Food Sources

Animal proteins are the chief source of vitamin B_{12} in the food intake. A summary of many assay reports of the vitamin B_{12} value of foods shows liver, kidney, oysters, and clams to be excellent sources of the nutrient. Lean beef, milk, veal, lamp, poultry meat, and salt water fish are considered good vitamin B_{12} foods where as cereal grains, vegetables, and legumes are poor carriers of the vitamin.

Folic Acid, and Choline

Liver extract has been found effective in promoting growth and preventing the formation of fatty liver in rats on diets devoid of known methyl donors. Bennett (1945) reported that folic acid was not the active principle, nor could the methionine or choline content of the extracts account for the effect. Following the isolation of vitamin B_{12}, a sparing effect of this vitamin on the choline requirement of chicks and rats was observed by Schaefer, Bennett and others. Jukes, investigating the effect of vitamin B_{12} on the response of chicks to homocystino, concluded that vitamin B_{12} is involved in the synthesis of methionine from homocystine. This was confirmed by Bennett and co-workers in 1951.

Nutritional Function

The amounts of folic acid and of vitamin B_{12} required in normal nutrition show rather large and as yet little-studied individual variation.

Folic acid occurs in both free and bound forms, and as tri- and hepta- (as well as mono-) glutamate. Additional forms of folic acid contain an extra carbon atom linked as a formaldehyde or alcohol group. Apparently all of these forms are about equally available in animal and human feeding tests, but they vary widely in quantitative test with micro-organisms.

Both in the case of folic acid and of vitamin B_{12}, the output is more largely through the feces than through the urine, and little is yet known as to what part is due to the food and what part to bio-synthesis in the alimentary tract.

R.J. Williams and coworkers state that the folic acid intake level of 25 mcg. per 100 grams of food suffices for survival in the chick; while 35 mcg. support normal hemoglobin formation; 45 mcg. support normal growth; 55 mcg. support normal feathering and 50 to 110 meg. support optimal growth.

It is interesting and very significant to compare the graded responses to stepwise enrichment of diet by folic acid, with the results of graded intake levels of ascorbic acid as found by CG. King.

Register and Sarett studied ,the urinary excretion of vitamin B_{12} folic acid, and citrovorum factor in healthy young men on various diets, and found little effect of diet. In fasting the vitamin B_{12} excretion, but not that of the folic acid or citrovorum factor, was about double the normal value.

Pyridoxin

For a substance to be called a vitamin at least two species of animal must need it for the maintenance of normal body function. Although there are other factors of the B-complex family in addition to thiamine, riboflavin, and niacin that qualify as vitamins according to this definition, not all of them have been shown to be essential to man. It has been more difficult to induce experimental human deficiencies with these factors than it was with the vitamins discussed in the preceding chapters. Because of this fact, investigators have not only used deficiency diets in their human studies of these factors but also have found it essential to use chemical substances known as *antagonists.* An antagonist, which may or may not be related chemically to a specific vitamin, will counter act the beneficial effects of the vitamin and for this reason is sometimes known an *antivitamin.* For example, in the induction of experimental pantothenic acid deficiency in man the subjects were fed omega-methyl pantothenate, which is an antagonist of pantothenic acid.

Since most of these factors act as coenzymes in the enzyme systems of the body, they may be thought of as quite similar in their function. Because these vitamins are needed in minute amounts and are found in the commonly used plants and animal foods of the diet it is not likely that human beings will ever have a deficiency. In general the ordinary practices of food processing and food preparation do not significantly affect the content of any of these vitamins in the food intake.

Vitamin B_6, folacin, pantothenic acid, and vitamin B_{12} are essential in human nutrition and must be included in the daily diet. A human dietary need of biotin, choline, and inositol is questionable at this time. Choline can be synthesized in the body from several substances, and biotin is synthesized by bacteria tract.

Vitamin B_6

In 1934 another factor in the B_2 complex was recognized as being distinct from riboflavin. This was B_6. It may be wondered why it was not B_3, B_4 or B_5 but, as stated before, these symbols had already been used for other substances thought at the time to be new vitamins. The terms were dropped when it was shown that they were not separate entities.

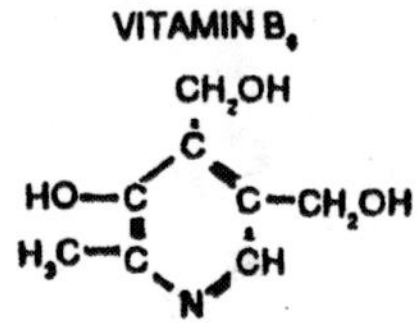

Vitamin B6
Pyridoxine

B_6 deficiency caused a form of dermatitis in rats (the so called rat pellagra). Several chemically related substances have B_6 activity but the commonest form is pyridoxin, present with vitamin B_1 in relatively high concentration in wheat germ. There has never been a unit of B_6 activity because by the time that enough was known about it to require measurement in foods, the pure substance was available for comparison.

Vitamin B_6 was the first designation of a then unidentified factor required for growth and for prevention of dermatitis in the rat. Because the dermatitis caused by its lack is, in the case of the rat, especially striking, vitamin B_6 was sometimes also called "the antidermatitis vitamin" and confused with the pellagra-preventive factor. Soon after the latter was identified as nicotinic acid, vitamin B_6 was given the name *pyridoxine* (adopted during the spring of 1940 by the American Institute of Nutrition, the American Society of Biological Chemists, and the Council of Pharmacy and Chemistry of the American Medical Association).

Spies, as we saw, reproted indications of a vitamin B_6 deficiency disease occurring as a complication of pellagra; but Elvehjem has designated it as a doubtful factor in human nutrition. Its role in nutrition and its therapeutic possibilities are problems for further research.

The conspicuous relationship of pyridoxine as observed in laboratory experimentation is to the prevention of the acrodynia-like dermatitis in rats. There is some evidence that both vitamin B_6 and a fatty acid or fat-soluble factor are concerned in the prevention of this dermatitis; and the latter factor shows similarities to the essential fatty acid of Burr and Burr and to the fat-soluble antidermatitis factor of Hogan and Richardson.

In another aspect of its functioning, vitamin B_6 occurs in several body enzyme systems involved in the metabolism of proteins and amino acids.

A number of recent studies have suggested that pyridoxine is required by man and that blood levels and urinary exertion rise and fall with intake. A naturally occurring deficiency has not been clearly demonstrated but the therapeutic usefulness of pyridoxine in the nausea of early pregnancy has been associated with a possible insufficiency of this factor in such cases.

Pyridoxamine and *pyridoxal* are distinct but chemically related substances which also have vitamin B_6 activity. There is now a tendency to prefer the term "vitamin B_6" where the group of active substances is met, and reserve chemical designation for a single specific substance. This factor was named pyridoxinc by Gyorgy in 1934. However, there are three closely related chemical compounds with potential vitamin B_6 activity — pyridoxine, pyridoxal, and pyridoxamine (Fig.). The pyridoxal and pyridoxamine forms occur mainly in animal products whereas pyridoxine is found largely in products of vegetable origin.

The Functions

Vitamin B_6 functions as part of coenzyme systems in the body that help in the metabolism of amino acids, fatty acids, and in the release of energy. It is found in tissues predominantly as pyridoxal or pyridoxamine phosphate.

Pyridoxal phsophate (vitamin B_6 coenzyme) functions in a number of reactions that are essential in the metabolism of amino acids: decarboxylation (removal of CO_2 group), transamination (transfer of NH_2 group), and desulfuration (removal of H_2S group). The conversion of the amino acid, tryptophan, to the vitamin, niacin, depends on the action of this coenzyme. There seems to be a relationship between vitamin B_6 and essential fatty acid metabolism. Witten and Holman postulate that vitamin B_6 is responsible for the conversion of linoleic acid to arachidonic.

A pyridoxine deficiency has been found to affect antibody formation in man and animals. Hodges and associates reported that an induced pyridoxine or pantothenic acid deficiency resulted in impairment of the human antigenic response to tetanus and typhoid immunizations.

The Effects

In Animals : All species of animals appear to need vitamin B_6 for normal growth, yet different kinds of symptoms result when differents species of animals are deprived of it. Although the rat loses weight very quickly on a vitamin B_6-deficient ration, the characteristic lesion in this species is a skin disorder called *acrodynia,* which may develop on the tail, paw,nose, mouth, or ears (Fig.). As the deficiency advances, however, the rat may suffer from muscular weakness and convulsions.

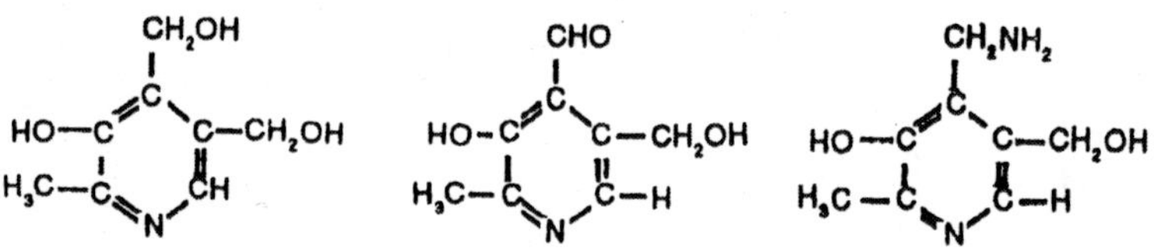

Pvridoxine Pyridoxal Pyridoxainine

Chemical formulas of the vitamin B_6 group.

Vitamin-B_6-deficient dogs, pigs, and monkeys do not show the acrodynia shown in rats but instead develop an anemia in which the red blood cells are small and low in hemoglobin (microcytic hypochromic anemia). Nevertheless, in the later stages of the deficiency, these animals too may suffer from convulsions. Neither the acrodynia nor the anemia is a characteristic symptom of calves deprived of vitamin B_6. These deficient animals show anorexia, listlessness, a poor coat of hair, and sometimes convulsions.

In Man : The human need for vitamin B_6 has been established from deprivation studies of both adults and infants. It was essential to resort to experimentally induced vitamin B_6 deficiency to establish the need for the vitamin. So widespread is the occurrence of vitamin B_6 in foods that a deficiency does not occur naturally.

The three symptoms observed in these human experiments where skin lesions, anemia, and convulsive seizures.

To study the effect of vitamin B_6 deprivation in adults, eight subjects were fed a low vitamin B_6 diet as well as desoxy-pyridoxine, the antagonist of the vitamin. After two to three weeks on this diet seven of the subjects developed a greasy scaliness on the skin about the nose, mouth, and eyes (Fig.). These lesions remained unchanged when thiamine, riboflavin, and niacin were ingested but disappeared in several days after the administration of vitamin B_1.

The effect of a vitamin B_6 deprivation is more dramatic in the infant than in the adult. The two infants that were studied experimentally by Snyderman and her group stopped gaining weight some weeks after being fed a diet deficient in the vitamin. Although one of the infants developed convulsive seizures after 76 days on the diet, the other one showed no convulsive activity pattern but did develop anemia after about 130 days of vitamin deprivation. Both of these conditions, however, were quickly relieved by the administration of pyridoxine. Between 1951 and 1953, reports came from various parts of the United States describing symptoms of a nervous irritability and convulsive seizures in infants less than 6 months old. It was found that all the babies affected had been fed the same commercial, canned, liquid-milk formula. A history of the symptoms and treatments of this disorder in a 2½ months went into convulsive seizures. Three minutes after the baby had received an intramuscular injection of 100 mg of pyridoxine hydrochloride, the convulsions had stopped and the child's pattern of activity was normal. It is now known that the cause of these convulsive seizures was a lack of vitamin B_6 in the diets of the infants fed that particular preparation because the natural vitamin B_6 of the milk formula had been destroyed by the heat processing during its manufacture.

The sole use of non-fat milk (skim) powder in infant diets might also produce a vitamin B_6 deficiency because of the destruction of the vitamin during processing. Because of this experience, processed milk formulas are now analyzed and those on the market contain a sufficient amount of the vitamin. A daily allowance of 400 ug of vitamin B_6 has been suggested for the artificially fed infant. Although vitamin B_6 relieves the convulsive seizures resulting from a deficiency of the nuttrient, it has been shown to be of no value in the control of convulsive seizures in the treatment of epilepsy in children.

There are other abnormalities that may be related to a deficiency of vitamin B_6 in animals. Several studies support the finding that a deficiency of the vitamin favors the deposition of kidney stones (calcium oxalate) in animals and man. Certain types of human anemias have been observed that respond to pyridoxine therapy that the vitamin may offer some protection against dental caries during human pregnancy has been suggested also by the study of Hillman and his associates.

Recommended Allowances

The human allowance for this vitamin has not been definitely established but at this time a daily intake of 1.5 to 2.0 mg per day has been suggested as a reasonable allowance for the adult. Because vitamin B_6 is widespread in both plant and animal foods, this amount will be provided in the average mixed diet. It has been reported that in high and low-cost adequate diets and even in poor diets in the United States, the vitamin B_6 content is about 2 mg., 2.7 mg, and 1 mg, respectively, per day.

Food Sources

Vitamin B_6 is widely distributed among the plant and animal foods of the diet. Although the best sources of the nutrient are found in muscle meats, liver, vegetables, and whole-grain cereals, few foods can be classified as poor sources of vitamin B_6. For example, a slice of whole-wheat bread (0.096 mg) contains about four times as much of the vitamin as does a slice of white bread (0.023 mg). Milk, a widely used food in the American diet, contributes about 0.175 mg of vitamin B_6 per pint.

Niacin

The pellagra-preventive (P-P) factor was recognized as part of the B_2 complex in 1935 and found to be nicotinic acid. A related compound nicotinamide has similar properties. In the United States the term niacin is used, to avoid confusion with nicotine, which has no vitamin activity.

The B_2 complex is now known to include at least seven other factors, i.e. biotin, inositol, pantothenic acid, para-aminobenzoic acid, choline, folic acid and vitamin B_{12}.

Some of these have been identified as constituents of physiologically active food factors discovered early in the century as essential for the growth of yeasts and known collectively as 'bios.' As research progressed, it became clear that bios consisted of several factors all associated with other B vitamins and at present time the term is no longer used.

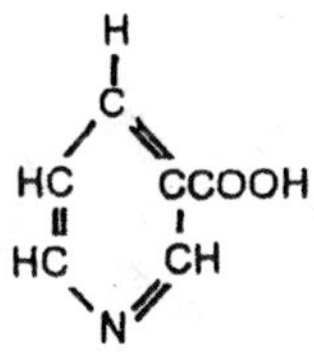

NICOTINIC ACID

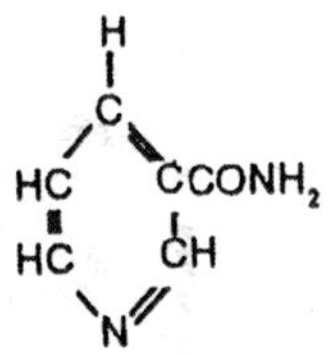

NICOTIN-AMIDE

The disease, *pellagra,* which is now known to be prevented by an adequate body supply of niacin, has been known for centuries. Although it was once an endemic disease caused by the intake of a diet referred to as the "Three M's" (meat— salt pork, whichi was almost all fat and no lean—maize, and molasses) and was found to occur in the southern part of the United States. Pellagra is still prevalent in certain sections of Italy, Spain, and Rumania where corn products are a staple part of the diet. The characteristic symptoms of pellagra involve the skin, the gastrointestinal tract, and the nervous system. Due to its specific lesions, pellagra is known as the disease of the "Three D's" (dermatitis, diarrohea, and dementia). The characteristic dermatitis of pellagra, which is aggravated by sunlight and heat, is bilaterally symmetrical and occurs on the hands, feet, face, forearms, and other parts of the body (Fig.) The skin becomes red and the lesions itch and burn. When the mucous membranes of the gastro-intestinal tract become affected, the early symptoms include *glossitis,* anorexia, and abdominal pains. As the disorder becomes more severe, diarrohea occurs. Insomnia, irritablility, fear, depression, and forgetfulness are the first neurological changes observed but as the disorder advances, dementia and paralysis may occur. If pellagra remains untreated, it may prove fatal.

Active work to determine the cause of pellagra began in 1914 when the disease was very prevalent in southern states of U.S.A. Although it had been suggested that pellagra was related to the intake of a poor diet, there were many people who believed it was an infectious disease. Goldberger and his coworkers, under the auspices of the U.S. Public Health Service, were the first to induce and then cure pellagra in healthy men by dietary means. Soon after this classic experiment it was found that brewer's yeast eaten in sufficient amounts would protect a human against developing pellagra.

Approximate Contribution of Some Foods to the Day's Recommended Allowances of Riboflavin for College Men and Women over 18 Years

Food	*Measure*	*College Men, per cent/day*	*College Women, per cent/day*
Apple	1 med	1	2
Beans, green, cooked	2/3 c	5	6
Beef, ground, cooked	3 oz	12	15
Bread, white, enriched	12 sl	5	6
Broccoli, cooked	2/3 c	11	15
Butter	1 tbsp	0	0
Carbonated beverage	8 oz	0	0
Carrot	1 med	1	2
Cheese, American	1 oz	5	6
Chocolate bar	1 oz	5	7
Egg 1 med	9	12	–
Frankfurter	1	6	8
Ice cream	1/8 qt	8	10
Liver, beef, cooked	2 oz	139	182

Contd...

Food	*Measure*	*College Men, per cent/day*	*College Women, per cent/day*
Milk, whole	1 pt	49	65
Orange juice, frozen	½ c	1	1
Pizza, cheese	5*Vi* in.	sector 7	9
Potato chips	10 med	1	
Potatoes mashed	1 c	6	8
Shrimp, canned	3 oz	2	2
Spinach, cooked	2/3 c	10	13
Sugar, white	1 tsp	0	0
Tuna fish, canned	3 oz	6	8

In the meantime, the discovery of a disorder in dogs known as *blacktongue,* which is analogous to human pellagra, was the symptom in the test animal that led to the final identification of this dietary factor. At the University of Wisconsin in 1937, Elvehjem and his group isolated a compound called nicotinamide from a liver concentrate and then demonstrated its effect and that of a related compound, nicotinic acid, in curing and preventing blacktongue in dogs. Although the chemical nature of nicotinamide had been known since 1870, its biological value was not recognized until this time. The first successful treatment of human pellagra with either nicotinic acid or nicotinamide was reported the same year. To avoid confusion with the name of the tobacco extract, nicotine, the name of the vitamin, nicotinic acid, was changed to niacin soon after its discovery. It was Cowgill at Yale who coined the name, niacin, from the letters of the three words—nicotinic acid, and vitamin.

The Properties

Niacin is found as colourless needles (Fig.) having a very bitter taste. It is soluble in water. It is not destroyed by acids, alkalies, light, air, or heat.

The word pellagra literally signifies a rough red skin. The victim, known a pellagrin, has in addition to such a dermatitis, a sore mouth. *Black-tongue* in dogs may be considered an approximate analogue of human pellagra. While the dermatitis and accompanying inflammation of the tongue are the most outwardly characteristic symptoms of the disease, there are generally digestive and nervous or psychotic disturbances.

In 1937 Elvehjem showed that blacktongue in dogs can be prevented or (if not too advanced) cured by administration of nicotinic acid, much as beriberi is prevented and cured by thiamine. And within a few months at lest four medical research

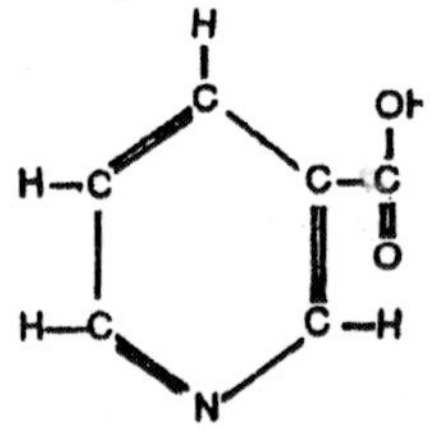

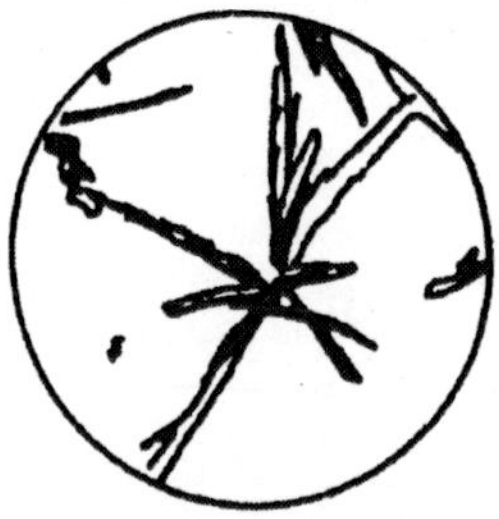

groups had shown the antipellagric value of nicotinic acid and of foods like yeasts which are rich in it. The inapprppriateness of the name nicotinic acid for a normal nutrient was recognized and *niacin* was coined as a name which in speaking or writing nutritionally may stand for the nutrient *factor* whether actually present and functioning as nicotinic acid itself or its amide, or the chloride or hydrochloride, or some conjugate.

The Functions

The principal role of niacin, as with thiamine and riboflavin, is as a constituent in coenzymes. Niacin is a constituent of two coenzymes that are essential for certain oxidation reactions to take place in the body. One of the chief functions of the coenzymes is the removal of hydrogen (dehydrogenation) from certain compounds and the transfer of it to others in a series of complex reactions.

The Tryptophan-Niacin Relationship : It has been demonstrated that one of the essential amino acids, tryptophan, can be converted into niacin in the animal body. This relationship between tryptophan and niacin was first reported in 1945 by Krehl and others at the University of Wisconsin, and since that time Reports have shown that human beings as well as most animals are able to change the amino acid to the vitamin. Because of this biological phenomenon, workers until 1945 had been unable to produce a human niacin deficiency by the deprivation of niacin alone in an otherwise adequate diet.

The conversion ratio of tryptophan to niacin in the human has been investigated. Horwitt and his group at Elgin State Hospital found the conversion ration of tryptophan to niacin in adult men to be about 60 to 1. Using adult women as subjects, Goldsmith and coworkers at Tulane University have reported similar values to confirm the validity of this conversion ratio. Studies with pregnant and non-pregnant women indicate that the conversion of tryptophan to niacin is more efficient in the pregnant than in the non-pregnant state.

Nutritional Chemistry

The normal nutritional functioning of niacin is largely, if not mainly, through the formation of compounds which act as coenzymes.

Like thiamine and riboflavin, niacin serves in the system of tissue enzymes involved in the intermediary and oxidation-reduction metabolism. The complex inter-relationships are still in process of being worked out. However niacin alone does not enable the typical pellagrin to become a healthy person, for the food supplies which have been responsible for the illness of most pellagrins are generally also deficient in one or more other nutrients. The metabolism of nicotinic acid as indicated by the urinary exertion of it and its derivatives has been studied by Perlzweig. It has been stated that: "the major urinary excretory products in human subjects are 'N'-methylnicotinamide and N-methyl-6-pyridone-3-carboxylamide." There is cited a presumably typical experiment where in a subject on a "normal" diet excreted daily 6.3 mg. of N'-methylnicotinamide and 14.2 mg. of pyridone; after taking 100 mg. of niacinamide (by mouth), an excess of 26 mg. of N'-methylnicotinamide and 62 mg. of pyridone appeared in the urine; while after taking 100 mg. of niacin (i.e., nicotinic acid) less of the end-products were recovered, 10 mg. of N'-methyl-nicotinamide and 26 mg. of pyridone. When 3 gm. of tryptophane were taken, 2.8 mg. of "extra" N′-methylnicotinamide and 16.6 mg. of "extra" pyridone were excreted— indicating a conversion of this amino acid into niacin or its products.

From such studies it appears that, due to its more complicated intermediary metabolism, niacin yields less clearcut results than do vitamin C and thiamine in studies of the relation of nutritional intake, concentration level in the body, and urinary output. Nor does there seem to be conclusive evidence as to whether or not it is advantageous to maintain bodily saturation with niacin habitually.

Niacin need not be supplied totally performed of certain precursor substances are available in sufficient quantity, for the human body which has considerable ability to form niacin from precursors. Krehl and his coworkers have demonstrated the transformation of tryptophane to niacin in the rat. It has now been established that man, several species of animals, micro-organisms, and the growing corn plant, can accomplish this chemical change. The mechanism of this conversion, the site where it occurs to what extent the body tissues accomplish it and to what extent it is due to the activity of intestinal micro-organisms-still are not clear, but isotopic studies indicate current progress. Methylation may take place and account for some of the recoverable end products.

As explained below, it is still not clear to what extent the human requirement for niacin is met by the niacin formed by intestinal bacteria. It may be largely influenced by the diet.

Kruse (1942, 1942 b) studied the influence of *aniacinosis* (niacin deficiency) upon the condition of the tissues, especially of the surface of the tongue. He devised a system of appraising the condition of the tongue in aniacinosis which takes account of the form, intensity, and stage of the departure from normal. the two main categories of form are acute and chronic; then the condition is further classified as to intensity or degree; and finally, as to the stage which the pathological process has reached. Kruse states that in practice it is possible to gauge intensity in three degrees; and stage in four steps if acute, or five if chronic. Also, that "if acute and chronic processes are both present, — as most frequently they are — they can be appraised separately" and then combined in a dual rating of status.

In niacin deficiency the tongue and its papillae show vascularity and hypertrophy or proliferation, followed by atrophy of the papillae. Usually the fungiform precede the filiform papillae in undergoing change. Different sites on the tongue may show different stages, the usual order being: tip, anterior and antero-lateral edges, anterior and antero-lateral margins of the dorsum, anterior border of predominantly filiform zone, mid-dorsum. Late in the process, ulcers and erosions may occur on the tongue, usually beginning on the edges. Those desiring a fuller account should read Kruse's original papers.

Niacin or its amide cured all these symptoms, the cures being quicker in the acute than in the chronic cases. When both acute and chronic pathology were present in the same person, the symptoms of the acute process disappeared first. Kruse emphasizes the same principles here as in analogous cases of other avitaminoses, the chronic cases of long standing can be fully cured only by long-continued treatment. He thinks many patients may have been discharged with only their acute and not their chronic injuries cured.

The specific pathologic changes on the tongue are considered by Kruse (1942b) to be the earliest and most universal and reliable symptoms of niacin deficiency and so of pellagra when thus defined. Kruse also speaks of this *glossitis* as the earliest discernible tissue manifestation of niacin deficiency. He holds that, as the tongue symptoms usually appear earlier than the dermaitis in the development of pellagra, so correspondingly the dermatitis disappears earlier than the glossitis in a cure. And he suggests that too often the pellagra patient is discharged as cured when the skin symptoms are relieved whereas therapy ought to continue until the tongue symptoms also have entirely disappeared which may require a year or more.

Prominence of maize in the diet has been charged, especially in European medical literature, with responsibility for pellagra; but this is possibly only true in the same sense that rice eating is related to beriberi, that is, too great dependence upon cheap sources of food calories which are but poor sources of the needed vitamin. Possibly contributing to the effect also is the fact that maize proteins are notably low in tryptophane, which, as mentioned earlier, may serve as a precursor of niacin. That cornmela is not actively pellagra-producing when the rest of the diet is right, is reaffirmed by Kooser and Blankenhorn (1941) who found that one of two otherwise comparable communities studied has freed itself of pellagra by increased consumption of milk and eggs, and in lesser degree of lean pork and chickens, with little if any decrease in the consumption of cornmeal.

Other Factors

Ball (1939), Elvehjem (1939), and S.G. Smith (1940) and some others have emphasized, at least three vitamins of the B group (thiamine, riboflavin, and niacin) are collectively concerned in biological oxidations and any investigation of the role of one of these demands of consideration of the other two.

Independently, Sydenstricker, McLester, Spies, and others emphasize the general experience in pellagrous regions that while nicotinic acid cures the dermatitis and inflamed tongue which are the conditions to which the name pellagra specifically applies, and usually also alleviates the digestive and nervous disturbances, yet the clinical pellagrin is not thereby rendered a fully healthy person. In fact, Sydenstricker (1941) construed pellagra as a combined deficiency of niacin and riboflavin. Multiple nutritional deficiencies seem, in the actual experience of American people, to be more frequent than deficiencies in niacin alone.

The typical diet of the poor pellagrin contains so little of anything else than milled cereals, sweets, and fats or fat meat that it is all too apt to induce other nutritional deficiencies at the same time with that of niacin.

Physicians dealing clinically with pellagra have repeatedly reported that while niacin cures the pellagra itself, many if not most clinical pellagrins need also thiamine or riboflavin or both. And Spies has reported a group of cases who after receiving all three of these vitamins were still not entirely well, but showed further improvement after receiving pyridoxine (vitamin B_6) in addition.

What is most needed in the pellagrous regions is a better food supply. Sebrell gives as the most important foods to add, "milk, liver, lean meats, fish, eggs, tomatoes, green peas, and a variety of green and leafy vegetables such as kale, mustard greens, turnip greens, and collards."

The deficient diet of the pellagrous regions is, "caused by the cultivation of a money crop instead of food and forage crops." When even a small fraction of the time and land hitherto devoted to cotton is used for raising vegetables for the family and forage for a family cow, the health, and with it the earning power, of the people should be greatly improved.

The field studies of Sandels and Grady (1932) and of Stiebeling and Munsell (1932) both showed a very clear connection between food habits and pellagra incidence.

It is gratifying to know that the educational work based on such observations as the above, together with opportunity for fuller employment and consequent increased purchasing power have recently very greatly reduced the prevalence of pellagra. Fuller reporting of cases that still occur may tend to obscure the fact in the official statistics, but personal communications from Dr. V.P. Sydenstricker are conclusive that at least in a considerable part of the South there has been a very real reduction in the prevalence of pellagra, attributable to improved food habits of the people generally and increased purchasing power of previously low-income families.

Practical Problem

Largely before the discovery of the relation of niacin to pellagra, the United States Department of Agriculture had made extensive quantitative field studies of the relations of food consumption to pellagra incidence and the Public Health Service had experimented intensively with human subjects to ascertain the amounts of different foods that suffice to prevent pellagra. Bringing together the findings of these Government researches, it may be concluded that any one of the items of Table contains a full day's quota of pellagra preventive value:

Food Items Preventive of Pellagra

A quart of milk or buttermilk;
A pint of evaporated milk;
One third to one half pound
of dried skimmed milk,
of lean meat,
of canned salmon,
of peanut meal, or
of wheat germ;
One pound of fresh or canned collards,
of kale,
of green peas, or
of turnip greens; or
Two to three pounds of tomatoes, fresh or canned,
or of tomato juice.

Knowing the extreme carefulness of these governmental investigations, particularly in the direct experimentation with human subjects. It is believed that the findings constitute important evidence. That for some foods it does not run parallel with the blacktongue preventive values derived from experiments with dogs may be partly due to pellagra as ordinarily met being less purely a niacin deficiency than is blacktongue and partly to the fact that some foods are better utilized by human beings than by dogs.

With some foods there is, and with some there is not, a good quantitative correlation between the above direct evidence of pellagra preventive value and the current analytical estimates of niacin content. This corresponds with other evidence that present niacin assays, both "microbio" and *in vitro,* have more validity for some types of food than for others. These methods are fairly trustworthy for comparisons among meats, or among cereal products, or for control of enrichment in breadstuffs and cereals; but much less so for evaluations of widely different types of food. Another complication lies in the fact that foods may have both direct and indirect niacin values. In addition to the niacin that they actually contain, some foods to a large extent— others to only a small extent, if at all—increase the body's niacin intake by favouring the activity of niacin forming bacteria in the digestive tract. If this is a considerable factor in the niacin supply of human nutrition, it must also be subject to large individual variations, which would explain the fact that some patients do not derive the usual amount of pellagra-preventive value from certain foods.

The National Research Council's Recommended Allowances of niacin were set at ten times the corresponding allowances of thiamine for both sexes and all ages. Obviously this was a tentative "first approximation." In 1951-52, however, it is still in force.

As yet, we do not know: (1) what part of the niacin requirement need be contained in the food as eaten, and to what extent a well-balanced diet will be "biologically enriched" in niacin by the bacteria of the digestive tract, or (2) the degree of trustworthiness of present estimates either of the niacin requirements of human nutrition or the niacin contents of some types of food.

The Effects

In Animals : Only the dog develops characteristic deficiency symptoms due to a lack of niacin. Although other animals show symptoms such as reduced growth, loss of appetite, and weight loss, these conditions are only the standard symptoms observed in most vitamin deficiencies. Blacktongue, the term used to describe a niacin deficiency in a dog, manifests itself in serve lesions of the mouth, a characteristic halitosis, diarrhoea, emaciation, and a lack of appetite for both food and water.

In Man : The first report of experimentally induced pellagra in man by dietary means was made by Goldberger and Wheeler in 1915. However, after the identification in 1937 of niacin as the antipellagra factor, numerous workers tried without success to produce experimental pellagra in man by niacin deprivation. It was not until 1952 Goldsmith and her associates at Tulane University were successful in inducing human pellagra by the use of corn diets low in niacin and tryptophan. Now it is known that these other workers who had failed to produce a human niacin deficiency had used experimental diets adequate in all known nutrients except niacin, and that the stumbling block was the presence of adequate tryptophan, which indirectly supplied niacin to the body.

The subjects of Goldsmith's study developed pellagra during the fifty days they were ingesting a corn diet that was low in niacin and low in tryptophan. At the end of the study they were cured of the disorder with tryptophan, which further proved the importance of this amino acid in niacin metabolism. Prior to this work the intake of vitamins of the B-complex group other than niacin was recommended to improve the multiple lesions of pellagra. In this study, however, even though added amounts of these vitamins were present in the diet, the typical lesions of the disorder were produced. This observation indicates that pellagra is caused by a deficiency of one specific factor, niacin, and not by a group of factors.

In 1955 the same group of workers conducted a similar human experiment using a wheat diet instead of the typical corn diet that is associated with pellagra. It was found that the niacin deficiency took longer to develop and was less severe on the wheat diet than when the corn diet was used, even though the diets were comparable in niacin and tryptophan content. The authors point out that the low tryptophan content of the corn may not be the only explanation for the pellagragenic effect of this cereal. It may be that the relatively low digesibility of corn and the low availability of niacin in corn contribute to the earlier development of pellagra on corn diets.

Recommended Allowances

The Food and Nutrition Board (NRC) has expressed the allowances for niacin in terms of *niacin equivalents* (10). Equivalents of niacin include the dietary sources of the performed vitamin as well as the precursor, tryptophan; conversion ratio is 60 mg tryptophan equals 1 mg niacin.

The daily food intake of a college or young adult man (18 to 35 years of age) should provide 19 mg equivalents of niacin; a middle-aged man (35 to 55 years) should have 17 mg equivalents; and 15 mg equivalents have been proposed for older man (55 years and older). College and young adult women (18 to 35 years) need about 14 mg equivalents of niacin; those over 35 years have an allowance of 13 mg equivalents of the vitamin each day.

Food Sources

Data on the pellagra-preventing potency of foods was reported by Goldberger and others long before the recognition of niacin as a vitamin. This information was based on studies of humans suffering rom pellagra and of blacktongue in dogs. Foods were rated as *good, fair,* and *little or none,* depending upon their curative effect of pellagra and blacktongue. Today, with the extensive information on the occurrence of niacin in foods, it has been shown that this early rating system corresponds well to the niacin values measured by present-day procedures.

Although niacin occurs both in plant and animal products, brewer's yeast and peanut butter are the best food sources of the vitamin. Glandular meats, muscle meat, poultry, and fish are rated as excellent niacin foods, whereas eggs are rated as niacin-poor. Milk, although low in actual niacin content, must be classed as a good anti-pelalagra food because of its high tryptophan content. Fruits and vegetables vary in their content of niacin. Whole-grain cereals contain more of the vitamin than do highly refined cereals even though cereals are low in niacin. The enrichment of corn and wheat flours in this country has done much to increase the niacin content of our diets. A report from the U.S. Department of Agriculture has shown that the national food supply in 1962 contributed about 28 per cent more niacin than it did for the period (1935-1939) prior to the adoption of cereal enrichment.

Coffee is a new item to be added to the list of niacin-contributors to the human diet. It has been reported that ordinary retail coffee contains about 10 mg niacin per 100 gm and instant powdered coffee about 30.6 mg per 100 gm. Apparently the niacin level is dependent on the degree of roasting of the coffee beans: light roast equals about 1 mg niacin per cup, and dark roast equals 2.4 to 3.4 mg niacin per cup. In certain areas of the world where the diet is low in niacin the tryptophan, the high consumption of coffee may explain the low incidence of pellagra among these groups of people.

Equivalents of Some Typical Foods : At this time the niacin values in food tables are expressed in terms of the performed vitamin, milligrams of niacin, not in niacin equivalents. It is not yet realistic to express food values in equivalents of niacin because of the lack of sufficient data for the tryptophan content of various foods under different conditions of processing, storage, and preparation. In the 1963 edition of *Composition of Foods —Raw, Processed, Prepared* niacin values of foods are given in terms of the performed vitamin.

To illustrate the contribution of the tryptophan precursor in foods, the niacin equivalents of some commonly used foods have been calculated (Table). The niacin assay of a cup of milk is listed as 0.1 mg. This same portion of milk contains about 120 mg of tryptophan. Using the conversion factor, 60 mg tryptophan equals 1 mg niacin, a cup of milk contributes 2.0 mg of the vitamin from its tryptophan source. Thus a cup of milk contributes 2.1 mg equivalents of niacin to the diet, not 0.1 mg.

Losses during Food Processing and Preparation : Niacin, which is the most stable vitamin of the B-complex group, is not affected by the common factors encountered in food processing or storage. Because of its water-soluble nature, however, a food may lose some of its niacin during food processing or preparation. The use of a small amount of water in cooking niacin-containing foods will minimize the loss of the vitamin during the cooking process. The liquid fraction from the processing or preparation of a food should be used in order to reclaim any of the original niacin that was lost from the food by leaching.

A Guide to the Selection of Foods to Meet the Niacin Need of College Men and Women : The approximate contribution of some typical foods to the day's recommended allowances of niacin equivalents for college students over 18 years is presented in Table 9.3. The importance of high-protein foods in the college diet as carriers of the precursor of the vitamin is evident from the contribution listed for beef, chicken, milk, and shrimp. A 3-0z hamburger in a roll (equivalent to about 2 slices of bread) will provide over 50 per cent of the mg niacin equivalents suggested for college-age students. Peanut butter, a favorite snack food of students, is another rich source of niacin equivalents.

Approximate Niacin Equivalents of Some Typical Foods

Food	Measure[1]		Tryptophan Content[2]	Niacin Equivalents from Tryptophan[3]	Niacin Content[1]	Niacin Equivalents
	Serving	gm (A)	mg/100 gm (B)	mg/Serving (C)	mg/Serving (D)	mg/Serving (C+D)
Beans, green, canned	1 c	239	14	0.6	0.7	1.3
Beef, ground	3 oz	85	187	2.7	5.1	7.8
Bread, white, enriched	1 sl	23	91	0.4	0.5	0.9
Carrot	1 med	50	10	0.1	0.3	0.4
Cheese, Cheddar	1 oz	28	341	1.6	trace	1.6
Chicken, broiler	3 oz	85	250	3.5	7.4	10.9
Corn, Canned	1 c	256	12	0.5	2.3	2.8
Egg	1 med	50	211	1.8	trace	1.8
Lettuce	2 lg lv	50	12	0.1	0.2	0.3
Milk, whole	1 c	244	49	2.0	0.1	2.1
Orange juice	1 c	249	3	0.1	0.8	0.9
Peanut butter	1 tbsp	16	330	0.9	2.4	3.3
Peas, green, canned	1 c	249	28	1.2	2.2	3.4
Potatoes, white	1 med	136	18	0.4	2.0	2.4
Salmon, canned	3 oz	85	200	2.8	6.8	9.6
Tomato	1 med	150	9	0.2	1.0	1.2
Yeast, brewer's	1 tbsp	8	710	0.9	3.0	3.9

Approximate Contribution of Some Foods to the Day's Recommended Allowances of Niacin Equivalents for College Men and Women over 18 Years[1]

Food	*Measure*	*College Men, per cent/day*	*College Women, per cent/day*
Banana	1 med	6	8
Beans, green,canned	2/3 c	5	6
Beef, ground	3 oz	41	56
Bread, white, enriched	2 sl	9	13
Butter	1 tbsp	0	0
Carrot	1 med	2	3
Cheese, Cheddar	1 oz	8	11
Chicken, Broiler	3 oz	57	78
Corn, canned	2/3 c	10	14
Cornflakes	1 oz	5	6
Egg 1 med	9	13	
Frankfurter	1	12	16
Lettuce	2 lg lv	1	2
Milk, whole"	1 pt	22	30
Orange juice	½ c	3	4
Peanut butter	1 tbsp	18	24
Peas, green, canned	2/3 c	12	16
Potatoes, white	1 med	13	17
Shrimp, canned	3 oz	22	30
Sugar, white	1 tsp	0	0
Tomato	1 med	.6	8

Enrichment of Cereal Grains : Nearly 72 per cent of the Calories consumed by the people of the world come from rice, wheat, corn, and other grains. Wheat and rice, however, because of consumer preference and a better keeping quality, are milled to a highly refined form causing a loss of the nutrients that are important in regulating the body's use of this food energy. Moreover, diets rich in corn products may be pellagra-producing because of the insufficient amounts of niacin and tryptophan in this cereal food. In order to combat the clinical and subclinical deficiency diseases which may result from inadequate amounts of the B-complex vitamins in cereal-rich diets, the enrichment of wheat, corn, and rice products has been adopted.

The committee on Food and Nutrition of the National Research Council, now known as the Food and Nutrition Board, deserves much credit for the adoption of cereal enrichment in this country. Although the enrichment of white flour and bread began in 1941 under a War Food Order of the Federal Government, South Carolina in 1942 became the first state to enact legislation which demanded enrichment of all wheat products sold within its boundaries. After the Second World War, when wartime directives ceased to exist, the idea of enrichment was well established as nineteen states had passed legislation requiring it. In 1962 thirty states and Puerto Rico had passed enrichment laws, and in the other states approximately 92 per cent of the white bread sold was enriched. In five southern states, where cornmeal and grits are used in large amounts, legislation has been extended to include the enrichment of these products. Only one state, South Carolina, has enacted legislation to include the enrichment of rice.

Thiamine, riboflavin, niacin, and iron are the four nutrients that must be added to white flour or bread under the enrichment law. The addition of vitamin D and calcium are optional but not mandatory. In order for one to see the value of the enrichment of wheat flours, a comparison of the vitamin and iron content of whole wheat, enriched white, and unenrichment white flours in presented in Fig. Enriched white flour contains about twice as much riboflavin but only four-fifths as much thiamine, niacin, and iron as is found in whole wheat flour. On the other hand, enriched flour contains nine times as much thiamine, six times as much riboflavin, and about four times as much niacin iron as does unenriched flour. The addition of these nutrients to white flour causes no change in taste, color, or cooking quality. The present cost to enrich wheat flour is about 3.6 cents per hundred pounds.

The value of white flour enrichment to the nutritional well-being of a population has been shown in the famous New-foundland surveys (1,2). Prior to the introduction of white enriched flour into Newfoundland in 1944, medical surveys were conducted in two areas of the country. Four years later a resurvey was made in the same areas to determine the benefits derived from such a program. The clinical signs of malnutrition, following the period of the addition of thiamine, riboflavin, and niacin to the diet through the enriched flour, had decreased. By 1948 the general health of the people had improved, but there was no improvement in the lesions related to ascorbic acid malnutrition where no dietary supplement of that nutrient had been provided.

The diet of the people of the United States were improved notably by enrichment of bread and flour. As shown in Table lower income families benefited relatively more than higher income families because of a greater consumption of flour and bread.

Rice can be enriched by coating it with a mixture of thiamine, niacin and iron, and this mixture is called *Premix*. When Premix is used in amount of 1 part to 200 parts of rice, the enriched product has a nutritive value comparable to brown rice. In the past riboflavin was not used in rice fortification because it discolored the grains. Now a new process has been developed that overcomes this yellow spotting of the rice grains.

Effect of Flour and Bread Enrichment on the Thiamine, Riboflavin, Niacin, and Iron

Content of Urban Diets, Spring 1948

Annual Income Class (dollars)	*Per Cent of total Nutrient in Diet Added by Enrichment With:*			
	Thiamine	*Riboflavin*	*Niacin*	*Iron*
All income classes	16	3	13	12
1000 to 2000	20	5	15	14
5000 to 7500	13	2	9	10

The peoples of Asia, who constitute almost one-half of the world's population, eat about 80 to 90 per cent of their Calories in the form of rice. The nutritional shortcomings of rice as a staple food are evident in the high incidence of beriberi among the peoples of this continent. The solution for the control of beriberi in the Far East appears to be in the enrichment of rice — demonstrated by studies conducted in the Bataan Province of the Philippine Islands. Nine months after the introduction of enriched rice to the seven experimental towns it was found that nearly 90 per cent of the people who had previously shown mild to definite beriberi symptoms were greatly improved. The incidence of death due to beriberi was reduced by 67 per cent during the first year of rice enrichment, and not one death due to this disease was reported during the latter quarter of the second year. Through the efforts of Dr. R. R. Williams of the Williams-Waterman Fund (8,9) and other scientific and civic leaders, both in the United States and in the Philippines, a plan was adopted in 1955 that brought the entire Philippines under rice enrichment by law late in 1956.

Counters of areas other than the United States, Newfoundland, and the Philippines have flour enrichment programs. Flour enrichment is mandatory in the State of Sao Paulo (Brazil), Chili, Denmark, El Salvador, Guatemala, Honduras, Israel, Panama, and the United Kingdom. In Australia, Belgium, Canada, Germany, Hong Kong, Italy, Japan, Portugal, Singapore, Sweden, and Switzerland, the enrichment of wheat flour is a voluntary practice.

21

B-Complex

The B-Complex vitamins are a group of related nutrients that are important for the well-being of every cell in the body. Because these factors are a part of the human enzyme systems, they are essential for the normal metabolism of food. When any of these vitamins are missing, in an otherwise adequate diet, characteristic deficiency symptoms are produced in both man and animals. At this time, the ten vitamins included in the B-complex family are thiamine, riboflavin, niacin, vitamin B_6, pantothenic acid, biotin, folacin, vitamin B_{12}, choline, and inositol.

B Vitamins

The first step was the recognition that vitamin B was not a simple substance but consisted of at least two factors. These were distinguished first as one fraction unstable to heat and another stable. The heat-labile fraction became vitamin B_1 and was identified as the anti-beriberi or antineuritic factor; the heat stable fraction was B_2, known in America as vitamin G.

Vitamin B_1: In 1926 Jansen and Donath (in Denmark) extracted a highly potent vitamin B_1 fraction from rice polishings and were able to obtain this in crystalline form.

Their work led to other efforts at concentration and it was

found that from an aqueous extract the vitamin could be concentrated and dried on finely divided powders such as acid clay. This process was used in the manufacture of the first vitamin B_1 concentrates and the resulting powder (Pulv. vitamin B_1) was given a place in the British Pharmacopoeia. It was such a powder as this which was used as the first international standard for vitamin $B_{1.}$

From this stage to the synthesis of vitamin B_1 involved much chemical research, first not the discovery of the formula of the vitamin and then into methods of building this up in the laboratory and it was nearly ten years before vitamin B_1, known chemically as aneurin, or thiamin, was synthesized in 1936. Its formula is

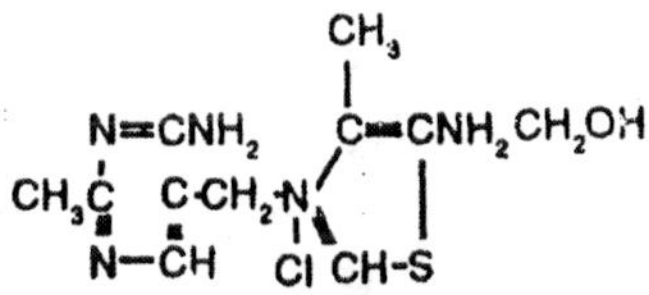

It contains nitrogen and sulphur in addition to carbon, hydrogen and oxygen, and its commonest salt is the hydrochloride.

The discovery of thiamine resulted from work along two paths of research: laboratory experimentation in normal nutrition, and medical studies of the disease beriberi.

In 1906 Professor (later Sir) Frederick Gowland Hopkins of Cambridge University reported feeding experiments with rats which showed the need in normal nutrition of some unidentified organic substance or substances soluble in water and in alcohol. Seeking identification of this factor, Hopkins postponed publication of the details of his experiments until 1912. Meanwhile, Dr. Casimir Funk, a Polish chemist working in the Lister Institute (London), had postulated the existence of a factor, the lack of which causes beriberi, and coined for it the name *beriberi vitamin* (e). Also he predicted (as had Hopkins) that other diseases might be found to be due to shortages of analogous substances.

Thiamine

Beriberi, a nervous system disease that is caused due to lack of thiamine in the diet, has occurred in the Far East for centuries and has resulted in a high morbidity and mortality rate among the peoples of these countries. Not until 1882 was it shown by Takaki, at that time Director General of the Medical Department of the Japanese Navy, that beriberi could be prevented in man by the intake of a good diet. A decade afterwards the observations of Christian Eijkman, a Dutch medical scientist in Java, initiated the research that finally led to the identification of thiamine as the anti-beriberi vitamin. Eijkman observed that the chickens housed in the hospital yard near his laboratory developed a disorder, that was similar to human beriberi, when fed a ration solely of polished rice. Because of the general paralysis and head retractions which occurred in the birds, he called this disease *polyneuritis.* He found that the symptoms of the polyneuritis disappeared when rice polishings, the outer layers and embryo of the rice kernel, were fed to the birds. In 1911 the Polish biochemist, Funk, prepared a small amount of a crystalline material from a rice bran extract which he found cured experimental beriberi in birds. Since he called this substance *beriberi vitamin,* he is credited with coining the word vitamin. Four years later McCollum confirmed the existence of a single substance which not only cured polyneuritis but also promoted growth and he named this factor *water-soluble* B. Jansen and Donath, at Eijkman's laboratory in Java, demonstrated the vitamin's curative effect in human beriberi in 1926, after they had prepared sufficient crystals of the vitamin from rice bran. Ten years later R.R. Williams and his group in this country identified the chemical structure of thiamine and then developed the first method for its synthesis.

The Properties

It is white crystalline /substance soluble in water. It has nutlike, salty taste and a characteristic odour which in recognizable in the smell of yeast.

It is stable in acid but is rapidly destroyed in neutral or alkaline solution. It is not readily oxidized dry and can be heated in acid solution for half an hour at 120°C without loss but is destroyed by sulphite — a fact which would have some slight importance in fruit bottling but for the additional fact that fruit contains very little vitamin B_1.

It is also rapidly destroyed in the dehydration of meats and vegetables.

Commercially the vitamin is sold in the form of thiamine hydrochloride.

The Function

It is important to the body as one of the keys in the release of energy during the metabolism of carbohydrates where, as thiamine pyrophosphate, it functions in the enzyme system. Carbohydrate metabolism may proceed to the formation of pyruvic acid, an intermediate compound; thiamine is essential for the breakdown of pyruvic acid. It is believed that the enzyme systems containing the vitamins of the B-complex family, along with other vitamins and certain minerals, work cooperatively in the utilization and storage of energy and in the normal functioning of body processes that involve growth and body maintenance.

Experiments suggested that thiamine also may have a specific role in the functioning of peripheral nerves.

Thiamine seems to have severd indirect functions in the body because of its role in energy metabolism. The maintenance of the appetite, the tone of the muscles, and a healthy mental attitude are all related to thiamine nutrition.

One of the earliest symptoms of a thiamine deficiency in man is the loss of appetite or *anorexia.* In several human studies involving thiamine deprivation anorexia was observed as early as the fourth day and became so severe as the deficiency progressed that the subjects had to force themselves to eat food. With the restoration of the vitamin to the diet, however, the loss of appetite disappeared very quickly. Indirectly thiamine will promote growth because the stimulation of the appetite results in an increased consumption of food.

Constipation and tenderness of the muscles of the leg calf are two symptoms frequently found in experimental thiamine deficiencies. The sluggishness of the intestinal muscles associated with constipation may be a because of insufficient thiamine in the diet. Soon after the vitamin is restored to the food intake after periods of thiamine deprivation, subjects showed relief from both of these symptoms.

Thiamine is often called the *morale* vitamin because of its ability to restore a healthy mental state in man following periods of experimental deprivation. When human subject are fed restricted quantities of thiamine in their diets, changes in the emotional stability of normal subjects have been observed. Williams observed that his subjects on restricted intakes of thiamine became irritable, quarrelsome, and moody; they failed to cooperate; and at times they suffered periods of mental depression. Although thiamine is not beneficial in the treatment of mental disorders, it has been observed that dietary restriction of the vitamin in mental patients appears to intensify the psychiatric symptoms of their mental state. The interest, ambition, attention, and sociability of these patients have been observed to be definitely affected by sub-optimal intakes of the vitamin as shown by a decrease in talking and a slowing of body movement.

The Effect

In Animals : All species of animals require thiamine for their normal energy metabolism. Although the ruminants (cattle and sheep) can synthesize adequate quantities of the vitamin in their rumens, other animals must obtain a supply of thiamine in their daily food ration so as to meet their requirements.

Irrespective of species, the symptoms of a thiamine deficiency in animals are quite similar. It is believed that the symptoms are related to the disturbances in carbohydrate metabolism accompanying thiamine deficiency, although the precise relationship is not known as yet. The albino rat, which may be used in the classroom to demonstrate the effects of this vitamin deficiency, develops anorexia and soon begins to lose weight when fed on a ration low on thiamine. As the deficiency progress, the animal develops a spastic gait because of a partial paralysis caused by affectation of the nerves; shows signs of a slowing of the heartbeat; regains its equilibrium with much difficulty after being rotated; and suffers from convulsions. The condition is called polyneuritis. The recovery of these thiamine-deficient rats within eight hours after receiving supplements of the vitamin (Fig.) has been observed.

In Man : Beriberi results from a severe deficiency of thiamine. Although beriberi occurs very rarely in the Western countries, it is a common disease in the countries of the world where polished rice is the main item of the diet. During the period from 1954 to 1958, beriberi ranked fourth as the cause of death among the people of the Philippine Islands.

The three general symptoms of beriberi are polyneuritis, ether edema or emaciation of the body tissues, and disturbances of the heart. The polyneuritis begins with a soreness in the muscles of the legs and is soon followed by a decrease in the reflex of the knee and a drop in the muscles that support the toes and foot. This latter symptom of the disorder accounts for the high-stepping gait observed in beriberi patients (Fig.). As the disorder advances, however, the arms and other parts of the body become affected due to a degeneration and lack of coordination of the body muscles.

When edema occurs in beriberi, the body tissues become swollen due to of the accumulation of fluid in the body cells(Fig.). This condition is called *wet beriberi* and is noticed first in the legs and thighs of the patient. Occasionally a reverse condition wherein edema does not occur but rather emaciation exists; this is diagnosed as *dry beriberi.* Inasmuch as the beriberi sufferer breathes with great difficulty and his heart beats rapidly and becomes enlarged (Fig.), death in beriberi is because of these cardiac impairments. Beriberi heart disease has been found among some elderly hospitalized persons.

Cases of mild thiamine deficiency are quite common throughout the world. R.R. Williams has written "... there are scores of millions, perhaps hundreds of millions, in Asia, who are suffering from mild beriberi and have done so for years and they still do not know they have it..." In U.S.A., cases of extreme deprivation of the vitamin are rare, but sub-optirnal intakes of thiamine have been reported to exist among people of all age groups as shown by nutrition surveys from various regions of the United States. From human studies of the effects of mild thiamine deprivation in the diet, one can recognize some of the typical symptoms that accompany subclinical cases. Fatigue, anorexia, constipation, labored breathing, lassitude (weariness), nausea, tenderness of the leg calf, depression, changes in the electro-cardiogram measurement, low excretion of thiamine in the urine, and an increase of pyruvic acid in the blood are the symptoms which accompany a thiamine deficiency. Although some of these symptoms are specific for thiamine, others may result from other nutrient deficiencies.

Nutritional Nature of Beriberi

The disease beriberi was long familiar to the Orient before it became well known to the Western world, so long and so widely, in fact, that the origin of the name is uncertain. Some trace it to one language according to which it means a great weakness; others to the word for sheep (in a different language) because the first outward sign is usually a stiffness of the ankles giving the victim a "sheep-like" gait.

In formal terminology it has been called a *multiple peripheral neuritis,* a disease of the nerves of both feet. In severe cases it may result in paralysis of the legs, and sometimes ultimately in heart failure. Sometimes the heart shows injury first.

In 1878-1883, the total strength of the Japanese navy was about 5000 men, and of these 1000 to 2000 each year were at some time on sick-list with beriberi. The enormous morbidity of 20 to 40 per cent from one disease led to an investigation by Takaki, a high medical officer of the Japanese navy, who worked upon the theory that the fault was most likely to be found in the food since climate appeared to have no influence and the sanitary conditions on the Japanese ships were as good as those in the European navies which were not troubled with the disease. A Japanese naval vessel with 276 men on a 9 months' cruise from Japan to New Zealand, Valparaiso, and Honolulu had 169 cases of beriberi, with 25 deaths. On another vessel with a similar crew, sent over the same route, but with a ration in which the rice was decreased, the barley increased, and vegetables, meat, and condensed milk added, only 14 men had beriberi and each of these had failed to eat his full allowance of the new foods. With this experiment, Takaki secured the adoption of a new ration for the entire Japanese navy with the result that the number of cases of beriberi soon became practically negligible.

From his government, Takaki received generous recognition for this really great achievement; and yet it failed to accomplish what might reasonably have been expected of it. Opinion continued to be divided, as it had been before, as to whether beriberi were a nutritional or an infectious disease, with the majority of the medical men of the Orient (where the disease continued to be common) inclining to the theory of an infection. Takaki, as we now know, was right in his belief that the disease was essentially nutritional and he was quite explicit in saying so, but his work did not carry conviction because he offered no adequate theoretical explanation of it. He discussed his improvement of the ration only in terms of protein, which others rightly considered as unconvincing. It was a time of great activity in the sanitary applications of the then new science of bacteriology. Sanitary improvements had naturally been made in the Japanese navy during the same years in which Takaki was succeeding in getting the ration changed, and the sanitary improvements were generally given credit for the eradication of the disease because the germ theory appeared adequate even with the germ unidentified, whereas Takaki offered no satisfactory scientific explanation of what we he had accomplished.

In 1897, Eijkman, a Dutch physician working in the East Indies, published observations upon "an illness of fowls similar to beriberi." He had noticed that fowls living in the bare yard of a prison hospital and subsisting almost exclusively upon the left-over rice from the prisoners' table, often showed stiffness and weakness of the feet and legs very suggetive of that of his beriberi patients. By systematic experimentation he found that this disease could regularly be induced in fowls by confining them strictly to a polished-rice diet. This is now generally considered as the first clear-cut case of experimental production of a dietary deficiency disease. At the time, however, Eijkman's presentation and discussion of his findings was so strongly coloured by what English investigators afterward called "the pharmacological bias" that the bearing of his work upon normal nutrition was not made clear. Reading Eijkman's early work in the light of later knowledge, we may consider it as marking the beginning of our experimental knowledge of dietary deficiency diseases; but it was not until some years later, and until the experiments had been extended by Grijns (1901), that the results of this work were stated in terms of a clearly nutritional concept.

Meanwhile British and American workers in the Orient were also beginning to give systematic attention to the nutritional aspects of beriberi.

For example, when American Army officers took over the administration of the Bilibid prison in Manila they sought to improve the treatment of the prisoners by furnishing them a better commercial grade of rice-plump, pure-white rice of highest market grade instead of the "low-grade" irregularly shaped brownis rice previously purchased. Shortly thereafter, the proportion of cases of beriberi in this prison population began to increase; but for some time this was considere as an epidemic to be combated by sanitary measures. After several months during which everything possible in the way of sanitation had been done, but without conquering the epidemic of beriberi, attention was turned to the "nutrition hypothesis." Dietary changes analogous to those which Takaki had made in the Japanese navy were tried; and the beriberi was then brought under control.

At about the same time Fletcher, by experimenting with the diet in a hospital for the insane, showed that when 28 ounces of rice were fed daily with only small amounts of other food, the use of white polished, or of brown unpolished, rice was alone sufficient to determine the occurrence or non-occurrence of beriberi.

In 1909, convinced that beriberi was related to diet, the United States Army Medical Commission in the Philippines initiated changes in the rations of the Philippine Scouts and in 1911 Champberlain was able to announce the eradication of beriberi from these troops by the improvement of their ration. In 1908-1909, when beriberi was at its worst among the scouts, the ration consisted essentially of 12 ounces of beef, 8 ounces of white flour, 8 ounces of potatoes and 20 ounces of rice (ordinarily polished). The change in the ration, as finally decided upon after some months of experimentation, consisted in giving, in place of the 20 ounces of polished rice, 16 ounces of unpolished rice and 1.6 ounces of dried beans. Experiments, made largely upon fowls, had shown that while beef has some effect in preventing beriberi, an equal weight of beans, peas, or peanuts in much more efficacious. Chamberlain stated, that the disease had practically disappeared because of adding the beans to the ration, before the substitution of unpolished for polished rice had been completed. He held, "that the consumption of beans to the daily amount of 6 ounces would, unaided, have prevented a recurrence of beriberi but it would obviously be difficult to make sure that all the men ate their share of this article over long periods, and it is therefore much safer that the largest component of the diet, the rice, should be of the unpolished variety and by itself sufficient to prevent neuritis." Several other investigations gave similar results.

Repeated demonstrations of a close connection between diet and the occurrence or prevention of beriberi naturally gave a great impetus to experiments designed to find, and to identify chemically, the precise substance which has the property of preventing and curing the neuritis which is the outstanding feature of this disease.

Antineuritic Substance

As early as 1902, Hulshoff-Pol showed that the antineuritic property could be demonstrated not only with the effective foods themselves, but also with extracts made from them, and even after clarification of such extracts with basic lead acetate and subsequent removal of the lead.

Fraser and Stanton, working in the Malay States, and Chamberlain, Vedder, and Williams in the Philippines, together showed the antineuritic constituent of foods to be an organic substance more stable in acid than in alkaline solution, probably a nitrogen compound, but not identical with any known amino acid or alkaloid. Funk studied the substance further and named it *vitamin* (e). Here the introduction of the name of the substance followed by about 10 years its clear apprehension as a definite chemical concept, and preceded by about 15 years its isolation by Jansen and Donath in 1926, while nearly another decade of research was invested by Williams in working out a method of separating the pure substance in sufficient quantity to permit of the establishment of its molecular structure and full chemical nature.

Thus Williams, at great personal sacrifice of his "spare" time for a quarter century, succeeded in identifying the substance and making it available at phenomenally low cost; and now has the satisfaction of seeing the saving of thousands of human lives through the use of the results of his work.

Chemical Nature

Larger scale production was made possible by the method of Williams, Waterman, and Keresztesy (1934), and ample evidence soon supported, and synthesis confirmed, the structural formula proposed by Williams (Fig.).

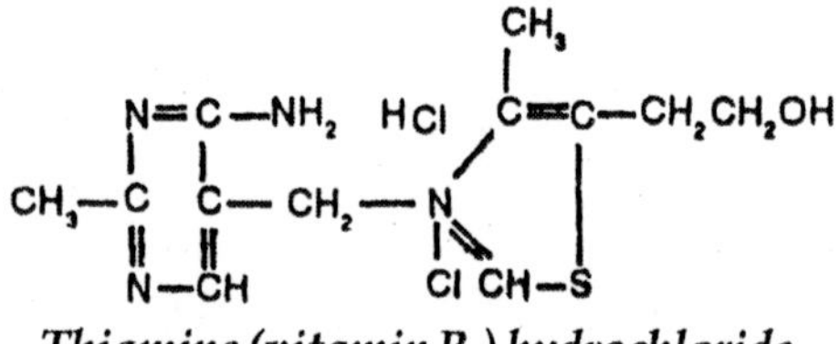

Thiamine (vitamin B_1) hydrochloride.

The molecule contains a substituted pyrimidine nucleus and a substituted thiazole nucleus linked by a $-CH_2-$ group. This was the first time that the thiazole nucleus had been found in a naturally-occurring substance.

The study of the chemical nature of this vitamin was greatly advanced by Williams' discovery that sulfite at pH 5.0 causes a quantitative cleavage of the molecule not two man fractions. The one which proved to be a thiazole derivative was identified and synthesized by Clarke and Gurin (1935); and Williams and Cline (1936) completed the synthesis and structural identification of the vitamin whose chemical nature Williams had been consistently investigating for about twenty-five years. Williams and Spies (1938) give full evidence for the structure, including an account of the synthetic work.

Thiamine as the Modern Name : After the molecular structure had been established, the name thiamine (thiamine chloride, thiamine chloride hydrochloride) was chosen as the official name of the substance by the Federal Food and Drug Administration, the American Medical Association, and the United States Pharmacopeia. It has also been adopted or endorsed by American scientific societies generally. The term vitamin B_1 continues to be used frequently synonymously with thiamine. The additional synonym, aneurin, sometimes encountered in European literature; but its use in some countries is discouraged.

Recommended Allowances

The human requirement for thiamine is based on the energy need of an individual. The 1958 revision of the Recommended Dietary Allowances proposed a thiamine value of 0.5 mg per 1000 Cal. In the 1963 revision, however, the suggested allowance for thiamine has been lowered to 0.4 mg per 1000 Cal.

College men and young male adults, 18 to 35 years of age, need 1.2 mg of thiamine each day. A slightly lower allowance, 0.8 mg per day, has been suggested for the college and young adult woman of the same age group. In the stress periods of life (growth, pregnancy, and lactation) the thiamine need is increased. As one grows older, the need for the vitamin decreases with the decrease in caloric requirement. However, new experimental findings indicate that the minimum need for thiamine may be greater for the older person than for the younger one.

Food Sources

All the foods of the diet with the exception of fats, oils, and refined sugars contain some thiamine. However, there are no foods particularly high in thiamine content as there are for ascorbic acid and vitamin A.

Although brewer's yeast and wheat germ head the list of the thiamine containing foods (Table), they are not very important in the typical food intake because relatively few people eat either one of these in large amounts. The best sources of the vitamin in the average diet are pork products, legumes, and whole-grain and enriched cereals. Nuts vary in their thiamine content but when they are used in large amounts, prove to be rich contributors of the vitamin. The thiamine in the egg is confined to the yolk. A quart of milk each day will contribute a good share of the thiamine need.

Thiamine Values of Some Typical Foods

Food mg/100 gm	*Portion*	*mg/Serving*[2]	
Brewer's yeast	15.61	1 tbsp	1.25
Wheat germ	2.01	1 c	1.36
Pork chop, cooked	0.96	3½ oz	0.63
Brazil nuts, shelled	0.96	1 c	1.34
Pecan nuts, shelled	0.86	1 c	0.93
Ham, smoked, cooked	0.47	3 oz	0.40
Bran flakes, 40 per cent	0.40	1 oz	0.11
Green peas, fresh, cooked	0.28	2/3 c	0.30
Bread, whole-wheat	0.26	1 sl	0.06
Bread, white, enriched	0.25	1 sl	0.06
Lima beans, fresh, cooked	0.18	2/3 c	0.20
Peanut butter	0.13	1 tbsp	0.02
Rice, white, parboiled, cooked	0.11	1 c	0.19
Beef, ground, cooked	0.09	3 oz	0.08
Egg	0.09	1 med	0.05
Orange juice, frozen	0.09	1 c	0.21
Milk, whole	0.03	1 c	0.08
Sugar, white	0	1 tsp	0
Butter	0	1 tbsp	0

The use of enriched cereal grains has bolstered the thiamine value of the American diet. Data from the U.S. Department of Agriculture show that the national food supply contributed about 12 per cent more of the vitamin in 1962 than it did for the period prior to the adoption of cereal enrichment (1935-1939).

The Losses

Vegetables : The amount of thiamine retained in a vegetable during freezing or canning is apparently related to the nature of the food used. Peas have shown no significant loss of the vitamin during the freezing procedures, but spinach and asparagus have been reported to lose about 51 per cent and 28 per cent, respectively. In a study to determine the effect of canning proceures on the vitamin retention of vegetables, it was found that the thiamine retention was good in tomato juice, only far in asparagus, and relatively poor in snapbeans, Lima beans, corn, peas, and spinach.

Under ordinary conditions of home cooking, most of the thiamine lost in a food is because of the leaching of the vitamin into the cooking water rather than to heat destruction. This fact was observed when twelve different vegetables were processed in family-size amounts by four different cooking methods. The retention of thiamine in the vegetables averaged 91.0 per cent (waterless cooker), 84.1 per cent (tightly covered saucepan with ½ cup water), 78.9 per cent (pressure cooker), and 62.6 per cent (tightly covered saucepan with water to cover). In view of this report, it appears that the thiamine content of a vegetable can be conserved if it is cooked in the least amount of water compatible with good palatability.

Because thiamine is highly soluble in water, the liquid fraction of a canned vegetable as well as the soaking or cooking water from a vegetable will contain an appreciable amount of the vitamin lost from the vegetable solid during processing or food preparation. To utilize all the thiamine contained in the foods of the diet, this liquid fraction should be used and not discarded.

Meats : Because pork and its products are excellent sources of thiamine in the American diets, it is important to know the effect that cooking methods have on the retention of the vitamin in these foods.

Studies have shown the retention of the vitamin in fresh hams and pork loins, fresh-ground pork, as well as in cured hams before and after cooking by standardized procedures. Although the thiamine content of different cuts from the same pork loin or ham may vary considerably, the retention of the nutrient after cooking is quite uniform. Pork loins and fresh hams have been shown to retain about 70 per cent of their thiamine after roasting and broiling and about 50 per cent after braising. When the vitamin content of the droppings from these pork products is added to the thiamine value of the cooked meat, there is about 70 per cent retention for all three methods of preparation. As much as 85 per cent of the vitamin was retained in ground pork patties when they were cooked by pan-broiling, oven-broiling, or electronic cookery. In the curing of ham, almost three-fourths of the original thiamine of the fresh meat has been reported to the retained. Fried cured ham has a higher thiamine content than cured ham that has been roasted, braised, or broiled. Regardless of the processing procedure or the method of cookery, it appears that cooked pork products are worthwhile contributors of thiamine to the diets.

A Guide to the Selection of Foods to Meet the Thiamine Need of College Men and Women : Serving of commonly used foods, some rich in thiamine, have been used to show their contribution to a college student's need for this nutrient. Two sets of values are presented in Table because of the difference in the need for thiamine between college men (1.2 mg) and college women (0.8 mg).

That pork products are good food sources of the vitamin is evident from the data presented. However, common food items that are less rich in thiamine than pork products are the main carriers of the vitamin on a daily basis. The bread, breakfast cereals, milk, etc., eaten by college students all carry thiamine. Here gain, the importance of milk in the college-age diet is evident.

Approximate Contribution of Some Foods to the Day's Recommended Allowances of Thiamine for College Men and Women over 18 years

Food	*Measure*	*College Men, per cent/day*	*College Women, per cent/day*
Apple	1 med	3	5
Beans, Lima, cooked	2/3 c	17	25
Beef, ground, cooked	3 oz	7	10
Bran flakes, 40 per cent	1 oz	9	14
Bread, white, enriched	2 sl	10	15
Butter	1 tbsp	0	0
Carbonated beverage	8 oz	0	0

Contd..

Food	Measure	College Men, per cent/day	College Women, per cent/day
Chicken, leg, fried	2 oz	3	4
Chocolate bar	1 oz	2	3
Egg	1 med	4	6
Frankfurter	1	7	10
Ham, smoked, cooked	3 oz	33	50
Ice cream	1/8 qt	3	4
Milk, whole	1 pt	13	20
Oatmeal, cooked	1 c	16	24
Orange juice, frozen	½ c	9	13
Peanuts, shelled	Vic	19	29
Peanuts butter	1 tbsp	2	3
Peas, green, cooked	2/3 c	25	37
Pizza, cheese	5½ in.sector 3	5	
Pork chop, cooked	3½ oz	53	79
Potato chips	10 med	3	5
Potatoes, mashed	1 c	14	20
Shrimp, canned	3 oz	1	1
Sugar, white	1 tsp	0	0
Tomto juice	½ c	5	8

Normal Nutrition

While earlier experiments of Lunin and of Pekelharing may now be considered to have dealt with what have since come to be known as vitamins, it was the work of Hopkins which first made the concept effectively clear to most students of normal nutrition. His alcohol extracts of dried milk and of certain dried vegetables were clearly shown to contain some substance or substances essential to growth. We now know that these extracts contained both fat-soluble and water-soluble nutritional factors, and that one of the latter was the substance now named thiamine.

Thus the longest-known functions of thiamine are the promotion of growth and the prevention of polyneuritis.

Thiamine also has an important function in relation to appetite. While all or most nutritional deficiencies tend to result in gradual decline of appetite, thiamine deficiency causes (after a short period in which the body is presumably using up its reserve) a characterically more abrupt loss of appetite than does any other known vitamin deficiency. Also, thiamine promotes the recovery of appetite in a more prompt and specific way than does any other known nutrient.

This vitamin makes the food more appetizing and has an effect upon the appetite as a physiological function. For even the feeding of the thiamine *separately* causes the animal to return with appetite to the food which it had refused.

Such observations supported the unqualified statement that this vitamin stimulates appetite. More recently there has been a tendency from two points of view to qualify this statement. From the viewpoint of what is "permissible to claim" it has been proposed to limit the claim for thiamine's action on appetite to a statement that it promotes recovery of an appetite which has been lost because of its lack. And in answer to the question whether stimulation of appetite is normally desirable it has been said that thiamine promotes recovery from a condition of subnormal appetite; and stabilizes the appetite at the normal level.

It may perhaps be questioned whether the evidence fully supports these somewhat drastic limitations upon the hitherto accepted simple statement that one of the functions of this vitamin is to promote the appetite. Certainly some scientifically-trained people have thought they observed increases in already-normal appetites after taking thiamine. And it appears reasonable to suppose that a positive promotion of appetite is involved in the greater growth which results from the higher levels of thiamine feeding among animals normal throughout.

Thiamine appears to function in the maintenance of the normal motility of the digestive tract. Both L.J. Harris and M.S. Rose wrote of a general lack of tone of the gastrointestinal tract due to a shortage of this vitamin. Cowgill, experimenting with dogs, observed a diminution of vigor in the normal contractions of the stomach as an earlier result of thiamine shortage than the loss of appetite. As the loss of normal motility may extend to the intestine, the frequently reported constipating effect of dietaries too largely composed of artificially refined food may perhaps be partly due to paucity of thiamine.

Thiamine Functions Very Importantly in the Intermediary Metabolism : An excellent series of experiments by peters and his coworkers of Oxford has shown that thiamine, presumably acting mainly in the form of its pyrophosphate (cocarboxylase), catalyzes the transformation of pyruvic acid and thus helps the carbohydrate metabolism through its intermediary stages. Thus thiamine carries the intermediary metabolism of carbohydrate through an essential step, whether on the way to combustion as fuel or to conversion into fat. This shows thiamine to be a substance of great importance to the working of the active cells in all the bodily systems.

Williams and Spies (1938) emphasized this repeatedly and held that it affords a scientifically satisfactory explanation of the clinical reports of helpful results from thiamine therapy in a very wide variety of ills.

They also suggested that the abnormal accumulation of pyruvic acid (and perhaps also other intermediates) in the body when thiamine intake is too low may be due to failure of appetite and of decline in gastro-intestinal motility and general bodily tone.

To what extent the apparent "toning-up" property of thiamine is limited to "curing the consequences of a previous lack of the same substance," and to what extent it stimulates an already normal appetite and bodily tone, is not yet clear.

Doubtless this is largely because "B vitamin" has been found to include other things as well as thiamine.

Probably ten or more fairly independent substances may now (1951-2) be considered as generally accepted members of the B group of vitamins: (1) thiamine; (2) riboflavin; (3) niacin (niacinamide); (4) folic acid (pteroylglutamic acid or glutamates); (5) pantothenic acid; (6) pyridoxine (with pyridoxal and pyrdoxamine); (7) vitamin B_{12}; (8) choline; (9) biotin; (10) lyxoflavin; and from some viewpoints: inositol, para-aminobenzoic acid, and perhaps other substances.

Thiamine in Reproduction and Lactation : There is abundant evidence that "the vitamin-B complex" is importantly concerned in the nutritional support of successful pregnancy and lactation; but just how this responsibility is shared between thiamine, riboflavin and other factors is still a problem for further research.

The Demonstration

Methods Utilizing the Influence of Thiamine upon Growth of Mammalian Species : Rats served as subjects in the work of Hopkins, of Osborne and Mendel, and of McCollum which established the existence of this "new" factor as essential to normal nutrition, and the growth methods for its determination have often made use of this species.

Such quntitative feeding tests can also readily made to furnish striking demonstrations of both the prevention and cure of nutritional polyneuritis by thiamine and visible evidence of its effect upon growth.

While the effect on the weight curve is usually the best basis for *measurement*, the most *characteristic effects* of shortage of thiamine are failure of appetite and development of polyneuritis. As the appetite fails there is also a development of much the same condition as is produced by starvation. *Complete* deprivation may even result in death without the appearance of the characteristic nerve symptoms.

Other Methods of Thiamine Assay : Methods based on cure of thiamine-deficient rats or pigeons have been largely used, but more recently thiamine assays are usually made either by measuring its effect upon the activities of microorganisms (microbio-assays) or by *in vitro* methods.

Quantitative Expression

Whatever the method used in measuring thiamine values, good practice now requires that all quantitative testing include experiments with known amounts of the vitamin side-by-side with those upon the food or other material under investigation.

When properly conducted and interpreted such quantitative experiments permit statement of the value found in terms of the weight of the actual substance (thiamine, thiamine chloride, or thiamine chloride hydrocholoride).

When vitamin B_1 itself was available as a standard of comparison, units of activity were no longer needed and it was possible to express the potency of any food or pharmaceutical preparation in terms of the actual weight of vitamin naturally occurring in a given weight of material, such as, for example :

Food	*Vitamin* B_1 *in mg. oz.*
Wholemeals bread	0.06
Ox Liver	0.11
Fresh egg (whole)	0.04 (Ministry of Food, 1953)

Human Requirements

We have seen that thiamine helps carbohydrate (and doubtless also the glycerol radical of fat and the deaminized radicals of some of the amino acids) through the pyruvic acid stage of metabolism. Hence thiamine deficiency causes in a relative accumulation of pyruvic acid in the body; and so, determinations of pyruvic acid in the body's tissues or fluids may be used as evidence as to whether there is a shortage of thiamine in the body at any given time.

Several investigators have made use of this as well as other criteria, including that of "thiamine clearance" (Melnick and Field, 1942) which may be considered as analogous to the determination of percentage recovery in urine of test doses of ascorbic acid in the studies of vitamin C requirement.

In 1942 Melnick emphasized the objective nature of his thiamine balance studies as evidence of requirements; and considered that from this evidence the thiamine requirements of the human adult could be estimated to be 350 micrograms (millionths of a gram) per 1000 Calories of food. To provide a margin of safety and to cover individual differences he recommended 500 micrograms per 1000 Calories.

Williams, Mason, and Wilder(1943), making use both of objective chemical methods such as Melnick used and the more subjective neurological tests and observations of preclinical symptoms in subjects living for long periods on thiamine-low diets, concluded that the daily allowance of 0.6 mg. (600 micrograms) per 1000 Calories, as then recommended by the Food and Nutrition Board of the National Research Council, is "none too high."

In a later paper, Melnick (1944) concludes, from a consideration of the influences of various factors as well as individual differences, that the margin of insurance provided by the allowance of 0.6 mg. of thiamine per 1000 Calories of food is to be considered as desirable.

In the 1948 revised Recommended Allowances, the thiamine is generally set at 0.5 mg. per 1000 calories; but in pregnancy is increased in greater proportion than the energy allowance, while for men and boys with very high energy needs the thiamine allowance is not increased quite in proportion to the caloric intake.

It is to be remembered that the Recommended Allowances of the National Research Council are intended to represent intakes capable of supporting optimal nutritional well-being rather than as minimal requirements. Thus we have, in use by the Federal Government at the same time, but for different purposes, an estimated minimum requirement of 1000 mcg., and a recommended allowance of 1500 mcg., per day for in average-sized man of such activity as to metabolize 3000 Calories per day.

The Recommended Allowances are : Men (70Kg.), sedentary, 1.2 mg.; physically active, 1.5 mg.; doing heavy work, 1.8 mg.; women (56 Kg.), sedentary, 1.0 mg.; moderately active, 1.2 mg.; very active, 1.5 mg.; pregnancy, latter half, 1.5 mg.; lactation, 1.5 mg.; children under 1 year, 0.4 mg.; 1-3 years, 0.6 mg.; 4-6 years 0.8 mg.; 7-9 years, 1.0 mg.; 10-12 years, 1.2 mg.; Girls, 13-15 years, 1.3 mg.; 16-20 years, 1-2 mg.; Boys, 13-15 years, 1.5 mg.; 16-20 years, 1.7 mg. of thiamine per day.

Thiamine in Typical Foods and in the American Diet : Thiamine is widely distributed in significant amounts in foods of both plant and animal origin, except such as have been artificially refined or otherwise denatured. Table shows typical foods as illustrations.

Thiamine in Edible Portion of Some Typical Foods: Milligrams per 100 grams

Food *Range*	*Number of Studies*		*Number of cases within which the normal average may be expected to be found*
Apples	14	32	02-04
Asparagus	7	32	11-20
Bananas	16	20	04-10
Barley (entire)	9	61	50-65
Beans, mature	13	37	40-70
Beef muscle	20	64	08-16
Cabbage	15	38	04-08
Oats (Oatmeal)	19	56	54-82
Orange or juice	12	19	07-09
Tomato	16	17	06-10
Wheat (entire)	24	400	55-60

Plants synthesize thiamine and while it occurs in the various plant organs in their natural state, there is a general tendency toward a higher concentration of thiamine in the mature dry seed than in other parts of the plant.

Unmilled cereal grains are richer in thiamine than the diet as a whole needs to be. Thus we saw above that the liberal recommended dietary allowance of the National Research Council is 0.5 mg. of thiamine per 1000 Calories of food; but whole wheat, whole barley, and oats average about 1.6 mg. per 1000 Calories, and brown rice contains about 0.8 mg. per 1000 Calories. The greater part of this thiamine is in or near the embryo (germ) and outer layer of the grain. Hence such *milled products* as white flour, white rice, and degerminated corn (maize) meal or hominy, are relatively poor in this vitamin, as well as in others and in mineral elements. While the quantitative relationships differ with the grains and with the milling processes used, one in general can expect that the highly refined grain product will contain only about one to two tenths as much thiamine as the original whole grain, — or not nearly enough thiamine to provide adequately for the metabolism of the carbohydrate carried by these foods. Thus diets made up largely of refined grain products together with refined sugars and fats thus enriched and a large proportion of breakfast tend to be undesirably low in thiamine. *Fortification* of such products (legalistically designated as *enrichment* in the case of breadstuffs, and *restoration* in breakfast cereals) by additions of thiamine with or without other nutrients came to be advocated by many food chemists and to receive governmental approval and encouragement about 1940. By 1944 much the larger part of the flour and bread sold in the United States was *enriched*, and some of the States had made such enrichment compulsory. With breadstuffs cereal thus enriched and a large proportion of breakfast *restored* to approximately whole grain levels, the thiamine content of the American diet was (and is) greatly improved. The average thiamine intake of the people of the United States is thus brought to a level about 30 per cent above the liberal recommendation of the National Research Council, and the distribution of our national consumption of thiamine is probably more equitable than of most other nutrients because in general more bread is consumed by low-income families. But, of course, to ensure this benefit to those who need it most, it is important that enrichment of white flour and bread be made as nearly universal as possible.

Mature dry legumes rank with the whole-grain cereals in thiamine content. In this respect as in some others, there is overlapping between these two groups of foods, while within each group there is considerable variation among species and even among cultural varieties.

The other vegetables and the fruits vary still more widely, partly because of specific differences among plants, partly because vegetables include different organs of plants, and partly because both fruits and vegetables are marketed sometimes fresh with very high natural water content and sometimes dried.

Among foods of animal origin the thiamine content is relatively low in milk, higher in eggs and lean meats, and among the meats it is particularly high in pork muscle, while liver and kidney show an amount intermediate between lean pork muscle and the other muscle meats.

Inositol

Like choline, there has been some questions as to whether inositol should be considered a member of the B-complex family of vitamins. It appears that inositol may be a component of the body tissue rather than a specific nutrient because it is found in foodstuffs and required by animals and micro-organisms in far greater amounts than are any of the other members of the vitamin group. For the time being, however, inositol is included here because it does occur with the B-complex group in foods.

This is another of the group of bios factors, needed for the growth of yeast.

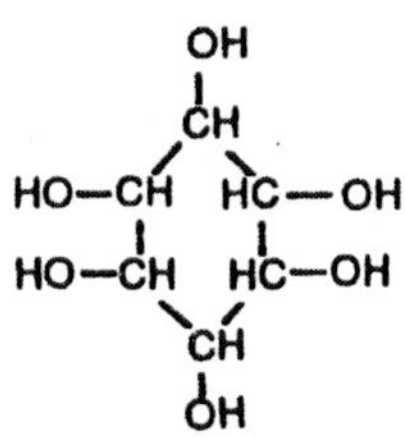

INOSITOL

It is, like biotin, present in all plant and animal tissues. The highest concentration in animal tissues is in muscle, brain, red blood cells, heart and kidneys. In plants it is found in yeast fresh fruits, vegetables and cereals, where, combined with phosphoric acid it exists as phytin.

The Deficiency

In Animals and Birds : Although the function of inositol in the animal organism is not known, it has been observed that there is a retardation of growth when the factor is missing in the ration of the mouse, rat, guinea pig, and chick. Rats also show a loss of hair that is characterized by a "spectacle eye condition" suggestive of the lesion observed in a biotin deficiency of this species.

Food Sources

Inasmuch as there is no evidence of the effects of inositol deprivation in man, the amount of this factor that one needs each day is not known.

Heart, liver, wheat germ, yeast, and whole grain cereals have been found to be good food sources of inositol.

Para-aminobenzoic Acid (Paba) : This has been known as a constituent of plant and animal cells for about 100 years, but it was only in 1941 that it was recognized as a vitamin in the B-complex group, being needed for growth by chicks and for normal lactation and hair pigmentation of rats. It is also needed by bacteria.

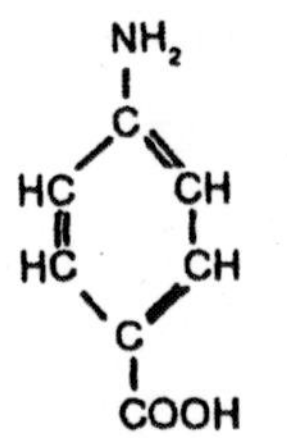

PARA - AMINOBEZOIC ACID

It is present in most tissues but the best sources of supply are yeast, liver, wheat germ and rice polishings.

Citrovorum Factor

The existence of this substance was reported by Sauberlich and Baumann in 1948, who found that *Leuconostoc citrovorum* — a non-pathogenic bacterium commonly found in milk and dairy products—failed to grow in media containing all previously known B vitamins and amino acids unless certain crude materials, such as liver concentrates used in the treatment of pernicious anemia, were included. This unidentified substance could be partially replaced by high levels of folic acid and by high levels of thymidine. Because of its occurrence in liver extracts and the low activity of folic acid, it was at first thought that there must be a metabolic relationship of the substance to vitamin B_{12} but it was soon established by Sauberlich and Baumann (1949) and other investigators that the citrovorum factor (CF) differed from vitamin B_{12} and that the latter had no growth-promoting action for

L. citrovorum. Sauberlich reported further that concentrates of CF overcame the toxicity of aminopterin (an inhibitory analogue of folic acid) for *L. citrovorum*, and that whereas rats on a folic acid-deficient diet excreted very little CF in the urine, administration of folic acid either with thc diet or by injection resulted in large increases in the CF content of the urine of both rats and man. These results pointed to conversion of folic acid *in vivo*. Such a conversion in rat liver slices was reported by Nichol and Welch (1950), who also found that addition of ascorbic acid with folic acid greatly favoured the conversion.

Investigations of the activity of liver extracts in counteracting the inhibitory action of anti-vitamins related to folic acid (e.g., methylfolic acid) by Bardos, et al., and Bond, et al. suggested a relationship of CF to formylfolic acid, a synthetic compound, which had been reported by Gordon, et al. to be more effective than folic acid in counteracting the inhibitory action of methylfolic acid for the organism.

Streptococcus Faecalis : This was confirmed by Shive, et al., who reported the production from formylfolic acid of a mixture of compounds which was highly active in promoting the growth of both *Lactobacillus casei* and *Leuconostoc citrovorum* 8081.

Next came the isolation of a crystalline compound possessing the high activity of this reaction mixture as a barium salt by Brockman, et al., and as the free acid by Flynn and coworkers. Later work by Pohland, Flynn, Jones and Shive (1951) on the synthesis of this acid led them to propose for it the tentative structure of 5-formy 1-5, 6, 7, 8-tetrahydropteroylglutamic acid. It is commonly spoken of as folinic acid—SF (synthetic factor). That this synthetic factor is not identical with the natural factor has been established by Keresztesy and Silverman who reported the isolation from horse liver of the crystalline barium salt of the citrovorum factor. Assayed with *L. citrovorum*, this barium salt was found to be approximately twice as active as the synthetic factor.

Studies in higher animals with the synthetic factor have given results quite similar to those obtained in micro-organisms. For example, Brockman, et al. (1950) and Broquist, et al. (1950) have reported that in mice 15 to 30 micrograms of SF overcame the toxic effects of 10 mcg. of aminopterin; in contrast, 20 mcg. of folic acid had no effect. Very much higher doses of folic acid-up to 5 mg. — were, however, found to be effective by Erschoff and coworkers.

Both citrovorum factor (as a concentrate) and folinic acid (as the crystalline synthetic material) have been shown to have anti-anemic properties for man. The results following intramuscular injections in pernicious anemia patients are less striking than those obtained with vitamin B_{12} or folic acid, but improvement in the blood picture, return to normal of the bone marrow and improvement in clinical condition, but no change in neurological status, have been reported by several independent groups.

Since CF and folic acid have similar activity for all organisms thus far used for the assay of folic acid, it follows that assays for the folk acid content of natural materials must represent the CF plus folic acid content. What proportion of the total activity is due to each form is not as yet known. Schwarz (1951) has reported that in liver 95 per cent or more appears to be in the form of CF.

All studies thus far indicate that folic acid must be transformed to CF within the living organism before folic acid can carry out the catalytic functions for which it is required. In other words, the citrovorum factor either represents the catalytically active form of folic acid or is more closely related to this form that the folic acid itself.

Much has yet to be done to determine the relationship between synthetic and natural CF, the form in which these factors occur in the enzymes of which they are undoubtedly a part, and the nature of the reactions which these enzymes catalyze. The relationship between the function of these factors and that of vitamin B_{12} also needs to be determined.

Choline

There is some doubt as to whether choline should be included in the list of the vitamins because it occurs in the body in very large amounts and has never been related to a deficiency in man. Beside these two reasons, it is known that choline may be synthesized in the body when sufficient amounts of the amino acid, methionine, are available. Nevertheless, choline has tentatively been included with the vitamins of the B-complex family because it does occur with them in foods. Choline is a methylated nitrogen-containing base which was first isolated from bile in 1862 (Fig.).

This is another substance which has been known for over a century but which has only recently been classed as a vitamin.

It was isolated from body fluids in 1849 and is known to be present in plasma, being closely related to another highly active physiological compound acetyl-choline. Choline is, in fact, present in many animal and plant tissues, the best food sources being eggs, meat, cereals and some vegetables.

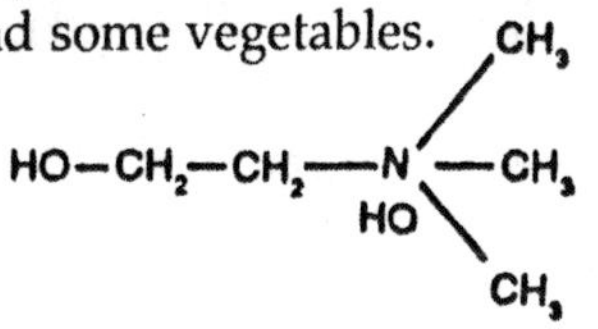

CHOLINE

While differing from most of the substances which we call vitamins in that its chemical nature and wide natural occurrence were well known many years before its nutritional importance became evident, *choline* is now often considered as a member of the "B-vitamin group."

It was noted by Best that choline prevented the abnormal accumulation of fat otherwise occurring in the liver of depancreatized dogs or certain types of diet; and later this same so-called *lipotropic* activity of choline was demonstrated in normal rats fed purified diets of high fat or cholesterol content.

The diets in some of the early experiments were so distorted as to leave considerable doubt whether choline need be considered a factor in normal nutrition; but with increasingly successful efforts to eliminate choline from the experimental rations, it became possible to produce symptoms of choline deficiency without resort to grossly unbalanced diets. One of these symptoms is the disturbance of fat metabolism already indicated, which manifests itself in the development of so-called fatty livers. But, as Griffith has shown, other tissues may be affected as well. Thus he found that on diets still more rigorously freed of choline than is necessary to induce fatty livers, young rats within a very short time showed "hemorrhagic enlargement and degeneration of the kidneys, a regression of the thymus, and an enlargement of the spleen"; and he suggested that choline is essential for the maintenance of the normal structure of tissues as well as for its lipotropic action.

These observations become very suggestive when it is recalled that choline is a component of some of the phospholipids which, as we have already seen, may have both structural and metabolic roles. The effect of choline in preventing the accumulation of fat in the liver was early interpreted as indicating that it facilitates lecithin synthesis, which, according to one theory, is an important step in fat metabolism. Direct evidence in support of this view was obtained by Perlman and Chaik off who, using radioactive phosphorus as a tracer, showed that the rate of turnover of liver phospholipids is markedly accelerated by the administration of choline.

In ways not yet altogether clear, choline seems to be interrelated in its action with the sulfur-containing amino acids.

It seems also that choline may serve as a methylating agent in body processes. Thus it was found by du Vigneaud and his associates that homocystine (the next higher homolog of cystine) failed to replace methionine in a diet rigorously freed of choline; but that it became a nutritionally satisfactory substitute for methionine when choline was also supplied. Their suggested explanation was that a methyl group may be transferred from the nitrogen of choline to the sulfur of homocystine, converting the latter to methionine. The reverse transfer was also demonstrated by the finding, after deuteromethionine (methionine containing deuterium in the methyl groups) had been fed, of a higher proportion of deuterium in the body of rats maintained on a methionine-choline-free basal diet. This observation may afford at least a partial explanation of the suggested interrelationship of choline and the sulfur-containing amino acids.

Experimentally, choline has been found nutritionally essential to all species of animals with which it has been fully tested. The requirement appears to be relatively greater in the young than in the adult, and to be increased in lactation.

Lyxoflavin **:** G. Emerson and K. Folkers have recently reported identification of what appears to be a new member of the vitamin B complex, *L-lyxoflavin.* This new nutrient, a growth factor for micro-organisms and rats, is reported to be identical with riboflavin except that the sugar L-lyxose is present instead of D-ribose.

Functions of Choline : Choline functions in protein and fat metabolism. It is a source of labile methyl (CH_3) groups which are used in the synthesis of vital protein substances such as epinephrine, the hormone of the adrenal medulla. Choline is used in the formation of phospholipids such as lecithin, an important emulsifying substance. Choline apparently functions in another way be serving as the precursor of a related compound (acetyl choline) which is necessary for the transmission of nerve impulses in the nervous system.

The Effects of a Choline Deficiency : *In Animals and Birds.* Not all species of animals and birds need dietary choline, but it has been shown to be an essential factor in the ration of the rat, chick, and dog. A lack of choline in the diet of the rat results in a marked accumulation of fat in the liver and an enlargement and hemorrhaging of the kidney. The fatty infiltration of the liver tissue and the kidney damage can be reversed with the administration of either cholinet or methionine to the animal. Vitamin B_{12}, however, has also been shown to be effective in relieving the symptoms of this deficiency. Although the characteristic effect of a choline deficiency in chicks is perosis or "slipped tendon disease," it is believed that both choline and manganese are necessary for the prevention of this crippling effect in the chicken.

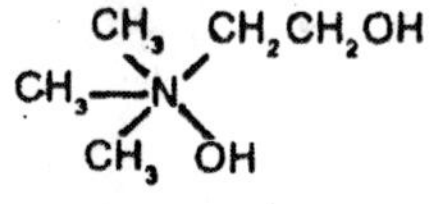

Chemical formula of choline.

Recommended Allowances and Food Sources of Choline : Since the effects of a choline deficiency have not as yet been observed in man, no human allowances for the factor has been suggested. The average American diet has been estimated to contain 250 to 600 mg of choline.

The richest animal sources of this factor are found in egg yolk, glandular meats, and brain; wheat germ and the legumes are the richest plant sources of choline.

Riboflavin

After the recognition of the multiple nature of vitamin B, its separate factors have been one by one identified. These tend to occur together in nature and to interact in the body, so, although they differ from one another chemically, they are all included in the B-complex.

This was the first of the heat-stable factors to be distinguished. It is yellow, fluorescent pigment, which had been discovered in milk in the 19th century, but this was before we had any knowledge of vitamins and it was only when, in 1933, its property of promoting growth in rats was discovered that it became included among the other vitamins. It occurs in eggs, liver, yeast and cereals. It was synthesized in 1933 and is still sometimes referred to as vitamin B_2 though this is obviously misleading now that we are aware of the multiple nature of what was originally called, vitamin B_2.

As early as 1920 some workers started questioning the belief that *water-soluble* B was a single nutritive factor. Studies indicated that after a food containing this dietary essential had been autoclaved, if still had the ability to promote growth.

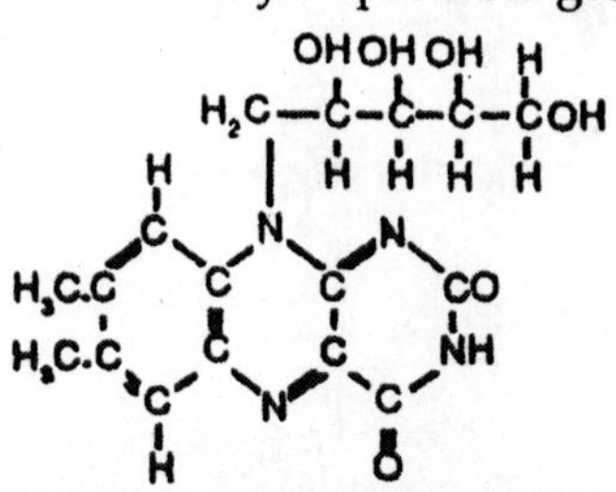

Riboflavin : In animals even though its antineutritic powers were destroyed. This observation indicated that the water-soluble, vitamins B was not a single factor but rather a mixture of factors, one that was destroyed by heat and the other that heat did not affect. About 1928 the fact had definitely been established that there were at least two different vitamins contained in this complex, an antineuritic factor (Vitamin B_1, now called thiamine)

and a growth-promoting factor (Vitamin G or B_2, now called riboflavin).

The first steps in the isolation of riboflavin began in Germany about 1930 when a yellowish-green fluorescent material, which stimulated growth in animals, was isolated from egg white. Soon it was found that similar fluorescent substances were present in milk, liver, yeast, muscle, and certain plant materials. The fluorescent materials all proved to belong to a chemical family called the *flavins* and so were named after the foodstuff from which they were isolated, such as *ovoflavin* from egg and *lactoflavin* from milk. Later chemical studies showed that all of these fluorescent materials were identical with one another and the name, riboflavin, was the general term adopted to identify them. Riboflavin has been produced synthetically since 1935.

Occurrence of Riboflavin: Riboflavin (sometimes free, sometimes in combination with phosphoric acid, often in combination with both phosphoric acid and protein) is very widely distributed in both plant and animals tissues. The so-called respiratory enzyme (Warburg's yellow enzyme) is a combination of riboflavin phosphate with protein. Several such combinations are now known. The working-out of their inter-relationships as catalysts of the oxidation processes in the tissues is an active field of research.

The riboflavin-phosphoric acid-protein combination of typical yellow respiratory tissue enzyme was described by Kekwick and Pedersen (1936), on the basis of sedimentation, diffusion, and electrophoresis measurements, as a chemical entity with a molecular weight of the order of about 80,000 and with one flavin group in such a molecule.

In milk, the riboflavin exists partly as an analogous compound, but also very largely in readily diffusible form, evidently free from combination with any large protein molecule.

Whether riboflavin in food is in the free state, or in combination with phosphoric acid, or with phosphoric acid and protein, probably makes little difference to its nutritive value. According to one view, riboflavin, if free when it reaches the intestine, is combined with phosphoric acid in the course of its absorption through the intestinal wall.

Hence statements of riboflavin values of foods usually do not distinguish as to whether or in what proportions, it exists in a free or a combined form.

Properties of Riboflavin : Riboflavin occurs in yellow-orange crystals (Fig.), which are only slightly soluble in water; solutions of the vitamin have a yellowish-green flourescence. The presence of acids, air, or heat have no destructive effect upon the vitamin, but riboflavin is easily destroyed by alkalies and light.

Function of Riboflavin : Like thiamine, riboflavin serves as a coenzyme in several of the enzyme systems of the body that are involved in the metabolism of the energy nutrients.

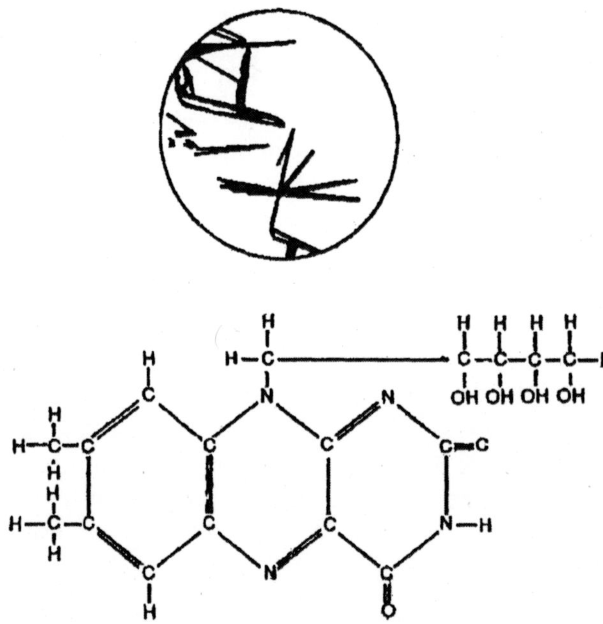

Crystalline riboflavin and its chemical formula.

Effects of a Riboflavin Deficiency

In Animals : Because the rat was used as the experimental animal in the early studies of this vitamin there is available a wealth of information on the effect of a riboflavin deficiency in this species. Growth failure occurs even though the animal is ingesting a sufficient quantity of food, is one of the *early* symptoms of

Riboflavin deprivation in the rat. Soon changes in the condition of the skin and eyes become apparenty then a *dermatitis* develops in these areas (Fig.). The eye lesions in a riboflavin deficient rat vary in severity and may range from *conjunctivitis* to the formation of a permanent cataract. Digestive and nervous disturbances are also observed as the deficiency becomes more pronounced, and death will occur if the vitamin is not restored to the animal's ration. All animals need riboflavin for normal growth and body maintenance.

In Man : Riboflavin deficiency on man may be complicated by a deficiency of other vitamins of the B-complex family, and because of this reason the clinical symptoms of human riboflavin deprivation are not as clear-cut as are those of thiamine. The first experimentally induced riboflavin deficiency in man was reported in 1938 by Sebrell and Butler. They found that the women subjects of their study developed *cheilosis,* a reddening of the mucous membranes of the lips, and *seborrhea.* All of these lesions disappeared when riboflavin was restored to the diet. In a more recent study of induced human *ariboflavinosis,* the same kind of skin lesions were found to develop in the male subjects maintained on sub-optimal intakes of the vitamin for a period of 9 to 17 months. Another lesions reported by these workers was scrotal dermatitis. The changes in the tissues of the eyes due to riboflavin deprivation in humans are not as specific as the lesions of the skin. It is believed that *conjunctivitis* as well as *lacrimation* (a watering of the eye) are the ocular lesions which can be attributed to human ariboflavinosis. The condition involving the *vascularization* of the *cornea* (where the capillaries invade the cornea tissue) has not been proven as yet to be a specific lesion of riboflavin insufficiency in man.

Factors that May Affect Riboflavin Metabolism : The excretion of riboflavin in the urine is a useful index to reflect the state of the metabolism of the vitamin in the animal organism. It is assumed that a marked increase in urinary riboflavin indicates a loss of the nutrient from the tissue whereas a decrease indicates a conservation of tissue of riboflavin. Recent studies with animals and man suggest that environmental, nutritional, and/or physiological factors may affect the normal metabolism of riboflavin.

The effects of exposure to sunlight (ultraviolet irradiation) for different periods of time on the riboflavin metabolism of the rat has been investigated at the Institute of Nutrition, USSR Academy of Medical Sciences. Moderate exposure of the animals to these rays apparently increased the capacity of the tissues to retain the vitamin, but there was a marked loss of tissue riboflavin during prolonged exposure. An adequate supply of dietary riboflavin, however, increased the animal's tolerance to ultraviolet light.

The influence of sleep, bed rest, work, diuresis, heat, acute starvation, and thiamine intake on the riboflavin exertion in man has been studied. Sleep retarded the exertion of the nutrient whereas the rate of exertion was increased in the subjects during enforced bed rest. There was a decreased excertion of riboflavin during short periods of hard physical work. Diuresis, induced by water, did not effect the excretion rate of the vitamin. Both heat stress and acute starvation increased the excretion of riboflavin in the subjects. There was an immediate increase in riboflavin excretion when thiamine was eliminated from the diet but with a combined deficiency of thiamine and riboflavin, the increase in the loss of riboflavin was delayed.

Recommended Allowances for Riboflavin : The need for riboflavin is related to the allowance for energy, as are those for thiamine and niacin.

The daily riboflavin need of college and young adult men, 18 to 35 years of age, has been suggested to be 1.7 mg whereas college and young adult women of the same age group have been assigned 1.3 mg of the vitamin each day. Middle-aged men, 35 to 55 years, need 1.6 mg and older men, over 55 years, need 1.3 mg of riboflavin each day. After the age of 35, the daily need for women has been proposed to be 1.2 mg per day. Here again, the need for this vitamin is increased during growth, pregnancy, or lactation.

Food Sources of Riboflavin : Since the best food sources of riboflavin, brewer's yeast and glandular meats, are not used to any great extent in an average diet, their value as dietary contributors is not very significant. However, because it is so widely used as a dietary food, milk is probably the chief source of vitamin in most diets. The ingestion of 3 cups of whole milk per day contributes approximately 1.26 mg of the nutrient, which is about 74 per cent and 97 per cent, respectively, of the daily riboflavin requirement of college men and women over 18 years of age. Cheese, eggs, veal, beef, leafy green vegetables, and salmon are other foods that are good sources of the vitamin. In addition, the use of whole grain or enriched cereal products rather than those that are unenriched will bolster the amount of the vitamin ingested. According to statistics from the U.S. Department of Agriculture, the national food supply in 1962 contributed about 12 per cent more riboflavin than it did in the period (1935-1939) prior to the adoption of cereal enrichment. Students often wonder why a 100 gm portion of whole wheat bread (0.12 mg) has less riboflavin than an equal amount of white enriched bread (0.21 mg). The answer to this is that more of the nutrient is added to the wheat flour duringthe enrichment process than was originally present in the whole wheat grain.

Riboflavin Losses during Food Processing, Storage, and Cooking: The two factors in the processing, storage, or cooking of a food that causes a loss of riboflavin are the leaching of the vitamin into the processing or cooking water and the destruction of the vitamin by light.

The loss of riboflavin during food preservation and preparation takes place generally during the blanching (boiling for a short period prior to preservation by freezing or canning) or cooking process. It has been reported that peas lost 20 per cent of their riboflavin content while being blanched, but relatively no further loss of the vitamin occurred in the subsequent stages of freezing or canning. Losses of the vitamin because of varying quantities of cooking water have been observed. There appears to be about a 20 per cent loss of riboflavin from vegetables cooked in a small amount of water (one-half cup), and when the water is increased the retention of the vitamin is decreased. However, very little riboflavin is destroyed in the range of temperatures used in common cooking practices because of the heat-stable nature of the vitamin. Although the solubility of this vitamin is not as great as was found either for ascorbic acid or for thiamine, the liquid fraction from a processed food should be used in order to reclaim that part of the vitamin that has leached out into the processing water.

The destruction of riboflavin by light is a serious problem because of the fact that one of the important dietary sources of the vitamin, milk, is often exposed to it. When a bottle of milk is allowed to stand on the doorstep on a bright sunny day for a period of two hours, as much as 85 per cent of its riboflavin value may be destroyed. In contrast to this, the exposure of a bottle of milk for one hour on a rainy, dark day may result in only about 10 per cent of the vitamin being lost. Most dairies recognize this problem and have made a constructive effort to preserve the riboflavin quantity of this food product. Both paper cartons and amber-coloured bottles have been introduced to eliminate light destruction of milk riboflavin.

Selection of Foods to Meet the Riboflavin Need of College Men and Women : Like ascorbic acid, riboflavin is another of the vitamins that is often low in the food intake of college men and women.

The daily college-age need for the vitamin will be more than satisfied with as little as 2 oz of cooked beef liver. A pint of milk supplies approximately one-half of the riboflavin needed by college students. The college student who does not drink milk and rejects leafy-green vegetables and liver cannot hope to meet his need for this important nutrient by dietary means.

Nutritional Chemistry of Riboflavin and its Significance for Well-being : *The journal of the American Medical Association* has spoken of riboflavin as "essential to the defense powers of the organism"; and Pinkerton and Bessey (1939) working in the Harvard Department of Pathology have shown that the level of nutritional intake of riboflavin greatly affects the susceptibility of experimental animals to rickettsial disease.

There is enough evidence to indicate that the amount of riboflavin in the food has a large influence upon the level of health and efficiency of people living under ordinary environmental conditions. This is a fact of fundamental importance and should be held clearly in mind when one meets differences of view in regard to the significance of diagnostic signs.

For, while there may be doubt as to the relations of riboflavin to disease, there is no doubt as to its great importance to health, efficiency, vigour, and resistance.

Relation of Level of Riboflavin Intake to Growth, Development, Aging, and Longevity : The relation of riboflavin to growth is such that through an important period of the development of our knowledge of riboflavin values of foods, these were determined by means of growth experiments. That the growth rate (of animals receiving all that the need of other nutrients) is so directly dependent upon the riboflavin intake is readily intelligible when we remember that riboflavin is an essential constituent of all the active body tissues that have been examined for it, and that the riboflavin enzyme is known to be a fundamentally important part of the muscle tissue of which the body is so largely composed.

Because these riboflavin enzymes are so definitely built into the essential substances of muscle tissue, that tissues retain their riboflavin so tenaciously under dietary deprivation. And riboflavin deficiency does not cause so early and rapid a loss of health and body weight as does an equal shortage of thiamine. And when, there is not a complete deprivation but rather a suboptimal intake of riboflavin, there may be nothing which ordinary medical examination would detect as a deficiency disease. In the young, there may be only a slowness of growth and development which the ordinary examiner would consider within the normal range and would either ignore or attribute to inherent physiological variability.

The slowing of growth which results from suboptimal intake of riboflavin is accompanied by a corresponding retardation of general development. And if shortage of riboflavin on the food supply continues throughout the period of growth we must expect that the individual's development will remain below the level of his inherited or inborn potentiality.

It may be expected that such chronic paucity of riboflavin intake will shorten life, and that to a still greater extent the period of enjoyment of the individual's full adult capacity will be curtailed by the early onset of senility.

Whether or not one considers riboflavin deficiency as a frequently occurring disease, it is clear that riboflavin is an important factor in health, and in the building of already-normal health to higher levels.

Hence riboflavin is essential to growth, and to normal nutrition at all ages. When the food is distinctly poor in riboflavin for any considerable length of time, digestive disturbances, nervous depression (different from the symmetrical polyneuritis of thiamine deficiency though both may occur in the same person), general weakness and deterioration of tone, and poor conditions of the eyes and skin are not to develop; the incidence of infectious disease is likely to be increased, vitality diminished, life shortened, and the prime of life curtailed by the unduly early development of the physical manifestations of old age.

In comparison to adequate or minimal adequate intake of riboflavin, a diet of higher riboflavin value tends to result in better development, higher adult vitality, greater freedom from disease at all ages, somewhat longer life, and (more significantly) a longer "prime of life", i.e., a longer segment of the life cycle between the attainment of adult capacity and the onset of old age.

This extension of the "period of the prime" — a property not of riboflavin alone but of the superior nutritional status or internal environment in which riboflavin is apparently one of the major factors — has been especially emphasized by Simms as important both to the individual and the community because in terms of human affairs it means longer life with a *smaller percentage of years of dependence.*

In controlled experiments extending through two generations of laboratory animals, it has appeared: (1) that riboflavin ranked with calcium and vitamin A as the major factors in the improvement of nutritional welbeing which resulted from an increase of the proportion of milk in already-adequate diet; and (2) that in experiments in which riboflavin was the sole significant variable, successive increments of intake resulted in successively increased benefits to nutritional wellbeing, up to levels more than twice that of minimal adequacy.

Relation to the Health of the Mouth, Skin, and Eyes : Sebrell and Butler in their first description of riboflavin deficiency as a human disease for which they coined the name *ariboflavinosis,* emphasized chiefly the symptom which consists in an inflammation of the lips with cracking at the angles *(cheilosis).*

Frequently accompanying symptoms are a specific type of glossitis (sore mouth with an abnormally smooth purplish-red tongue) and a dermatitis around the folds of the nostrils. Similarly, dermatitis was a frequent symptom in the human subjects maintained by Horwitt, et al. (1949) for long periods on a low intake of riboflavin. In riboflavin-deficiency experiments with rats, the dermatitis frequently manifests itself in any of three ways: as "sour nose," or by loss of hair around the eyes, or on the shoulders, the latter two giving rise respectively to the familiar names "spectack-like" and "saddle-like".

Day and his coworkers showed that riboflavin deficiency deterioration not only of the skin but also of the eye (which by origin is a specially developed skin-spot). The correlation of the skin and eye conditions with each other and with cessation of growth in the young, and the experimental conditions under which these three signs of deficiency are governed by riboflavin, were clearly developed in the independent series of papers by Day and by Bessey with their respective coworkers. Their discovery of the sensitiveness of the eye to shortage of riboflavin, made by means of controlled experiments with laboratory ahimals, has since been strikingly confirmed in the clinical experience of Sydenstricker, Sebrell, Kruse, and Others:

Directy clinical evidence that riboflavin-deficiency is not only a possibility but and actually occurring disease in the United States, is furnished by the work of Sebrell and Butler (1938), Sydenstriker, et al. (1939), Oden, Oden and Sebrel (1939), Kruse, Sydenstriker, and Sebrell (1940), Sydenstriker, Sebrell, Cleckley, and Kruse (1940), Kruse et al (1943), and by several other medical observers. Thus Spies and others have emphasized strongly the frequency of its occurrence in our South; and Aykroyd in a lecture at the U.S. Department of Agriculture in 1943 stated that riboflavin deficiency is the most frequent of the vitamin deficiencies in India, white Braun, et al. (1945) found riboflavin deficiency in 190 of 900 pregnant women examined in Palestine. It has also been reported as of frequent occurrence in several other countries. Medical reports have emphasized its occurrence both independently and in conjunction with pellagra. In the latter cases, the much more dramatic symptoms of the pellagra, along with the fact that the two diseases usually both yield promptly to good diet, make it likely that a large proportion of the riboflavin deficiencies are overlooked. For the same reasons it is probable that the pathology of ariboflavinosis as thus far known constitutes the beginnings rather than a complete picture of the consequences of riboflavin deficiency.

Slit-lamp examination showed an abnormal vascularity technically designated as "a superficial vascularizing keratitis" with "congestion and proliferation of the limbic plexus" as the earliest and universal symptom in a series of 47 cases, all of which were shown to be riboflavin deficiencies by the readiness with which they were cured by riboflavin alone. Some of these people were hospital patients who had been on "pellagra-producing" diets but were saved from the pellagra symptoms as now understood by means of nicotinic acid. Others of the people in this series of 47 were members of the hospital staff who considered themselves well nourished and who came under observation merely because of cracked lips, eyes unduly sensitive to light, dimness of vision, or "eyestrain." While the latter people had chosen their food at will from an apparently adequate food supply, investigation revealed that: "Bad dietary habits with inadequate intake of milk, eggs, and green vegetables were prevalent in the entire group." This investigation adds strong evidence for the view that a large proportion of people need a different emphasis in their choice of food so as to get enough riboflavin to support them at the best level of health of which they are capable.

However, two other facts are to be remembered in this connection. There is not yet an entire agreement of medical opinion as to how far the eye symptoms may be taken as evidence of riboflavin deficiency. And, while unwise choice of food may be responsible for much of the riboflavin deficiency that exists, poverty is also undoubtedly often a primary factor.

Relation to Incidence of Infectious Disease : It has now been definitely shown that the level of intake of riboflavin has a marked influence upon a certain type of infectious disease in experimental animals; and it is probable that this is likewise true for man, and for an as yet undetermined number of other infections. As noted above, the American Medical Association based its official "recognition" of riboflavin on the general ground of its function in promoting and maintaining the resistance of the tissues. And this view is, of course, strongly supported by the growing knowledge of the importance of riboflavin as a constituent of tissue enzymes and in the promotion of a high general level of vitality and vigor. Thus there is a broad and strong basis for the general recognition of liberal riboflavin intake as a factor in resistance.

Problems of Interrelation of Riboflavin with other Factors : The occurrence and significance of riboflavin in various combinations and its possible interrelations with analogous derivatives of thiamine and of nicotinic acid are subjects of active and complicated research.

Here it is sufficient to call attention to the fact that the findings of several investigators show that compounds (we do not yet know how many) of thiamine, of riboflavin, of nicotinic acid, and of other B-complex vitamins all function in the complex enzyme-coenzyme systems which catalyze the oxidation process in the body tissues, and that there may be interrelations in their action. It is partly through studies of the enzymic reactions *in vitro* and partly through very comprehensive feeding experiments under exceptionally accurate laboratory control that one may hope to obtain further light upon such questions as : to what extent do the thiamine, the riboflavin, and the nicotinic acid compounds function independently; to what extent are any of their functions interlinked, so that shortage of one impairs the functioning of another; or do any two of them act in parallel to a sufffcient extent that an abundance of one may in some degree ameliorate the effect of a shortage of one of the others?

The discovery by Ellis and Zmachinsky (1937 and subsequent experiments) that as high level of riboflavin intake increases the ability to withstand a deprivation of thiamine; may afford an affirmative answer to this last question; or may be interpreted upon the mass-action principle with the hypothesis of a common dissociation product inasmuch as thiamine and riboflavin both contain the pyrimidine nucleus.

Riboflavin has also been found to increase hemoglobin

production in dogs which had been rendered anemic by bleeding and then fed standard basal diet.

Interestingly also is the work of Stannus (1944) identifying riboflavin deficiency with *"pellagra sine pellagra"* and recording his findings as to the sequence in which different body tissues are attacked in riboflavin deficiency: which Stannus prefers to call *hyporiboflavinosis* rather than ariboflavinosis.

Quantitative Measurement and Expression of Riboflavin Values: At present the riboflavin values of foods are considered as due to the one substance and within the limits of formal nutrition at least, the nutritional availability, and efficacy of riboflavin is essentially the same whether taken in the free state or in the form or forms of combination in which it largely occurs in plant and animal tissues.

Riboflavin is now so readily available in pure form that all quantitative work on riboflavin values of foods should be controlled, standardized, and interpreted by paralled operations with the pure substance, and the results are best expressed in terms of weight of actual riboflavin. Milligrams micrograms (mcg) of riboflavin per 100 grams of food are two scales of expression now generally regarded as most. As in the case of thiamine, *in vitro* methods and microbioassays for the determination of riboflavin have been developed and are in general use.

Meanwhile quantitatively controlled feeding methods in which side-by-side experimental animals receive in some cases graded amounts of riboflavin and in other cases graded amounts of the food under investigation, may be used as a "court of ultimate resort."

In drawing the averages for the accompanying table and in compiling the riboflavin data in the Appendix, results obtained *in vitro,* by microbioassay, and by rat-feeding experiments have all been used.

Foods as Sources of Riboflavin : Riboflavin is formed in the growth of green plants whose leaves and stems seem to be richest in riboflavin when in their most succulent stage, notwithstanding the high water content.

Liver and kidneys, milk (whole or skimmed, fresh, canned, or dried), cheese, cream, eggs, lean meats, legumes, and green leaves and buds are typical foods of relatively high riboflavin content. We should not lean too heavily upon comparisons of reported riboflavin contents in foods of different types, because some of the methods of determination are less accurate for some types of food than for others. Hence comparisons *among* meats, for example, are presumably more accurate than comparisons of miscellaneous vegetables *with* meats.

According to current figures, dark meat is much richer in riboflavin than light meat of poultry. Unlike the case of thiamine, riboflavin values of lamb, beef, and pork muscle do not differ strikingly. On the other hand, liver and kidney contain at least ten times as much riboflavin as the. muscle meats, and are among the most concentrated food sources of this factor. It probably is significant that eggs average higher in riboflavin than muscle meats; and cheese (of Cheddar and similar types) higher than eggs. Among the vegetables and fruits, the deep-green leaves and green flower buds (broccoli) are outstanding, containing several times as much riboflavin as do typical fruits and other-than-green vegetables. The fact that sweet corn with its high water content nevertheless has about the same riboflavin content as does mature dry corn illustrates our general expectation of vegetables in the succulent stage as good sources.

Riboflavin in Edible Portion of Some Typical Foods: Milligrams Per 100 Grams

	Range within which the normal average may be expected to be found
Almonds	0.5-0.8
Apple, raw	0.02-0.05
Beans, snap or string	0.10-0.15
Beef muscle	0.15-0.3
Broccoli	0.15-0.3
Cabbage	0.05-0.3

Contd..

	Range within which the normal average may be expected to be found
Cheese (Cheddar type)	0.05-1.0
Eggs	0.3-0.5
Kale	0.25-0.4
Liver	2.0-4.0
Milk	0.17-0.25
Peas, fresh green	0.1-0.3
Pork muscle	0.2-0.4
Potatoes	0.03-0.05
Sugar	none or negligible
Sweetpotatoes	0.05-0.1
Tomatoes	0.03-0.05
Turnips	0.05-0.10
Wheat, entire	0.10-0.22
Wheat germ	0.6-0.8

The dry, mature, entire grains of barley, corn, oats, and wheat show very similar amounts of riboflavin, only about one fourth as much as their thiamine contents.

These whole grain cereals thus contain significant but not large amounts of riboflavin. Hence the addition of riboflavin in the enrichment of white flour and bread, while bringing the riboflavin content to about "whole-grain-levels" does not make nearly so large a contribution toward meeting the consumer's need in the case of riboflavin as in the case of thiamine.

Riboflavin Requirements in Human Nutrition and the Problem of Adequacy of the Food Supply : In its 1948 Allowances, the Food and Nutrition Board of the National Research Council recommended as follows:

Recmmended Dietary Allowances of Riboflavin

Man of 70 kg., 1.8 mg., *Woman* of 56 kg., 1.5 mg.,-in both cases regardless of activity, as in the parallel allowances of protein.

Woman in second half of pregnancy,	2.5 mg.
Woman in lactation	3.0 mg.
Child under 1 year	0.6 mg.
Child, 1-8 years (12 kg.)	0.9 mg.
Child, 4-6 years (19 kg.)	1.2 mg.
Child, 7-9 years (26 kg.)	1.5 mg.
Child, 10-12 years (35 kg.)	1.8 mg.
Girl, 13-15 years (49 kg.)	2.0 mg.
Girl, 16-20 years (55 kg.)	1.8 mg.
Boy, 13-15 years (49 kg.)	2.0 mg.
Boy 16-20 years (64 kg.)	2.5 mg.

The discussion accompanying the Recommended Allowances is, in part, "Evidence for assessing the desirable allowance of riboflavin is still rather incomplete....Keys *et al.* (1945) observed no physical or functional deterioration in a group of active young men who for nearly six months were supplied with only 0.29 mg. riboflavin per 1000 calories; whereas, Sebrell, Butler, and Wooley (1941) found that in 3 to 8 months a group of sedentary women receiving 0.21 mg. riboflavin per 1000 Calories developed distinctive physical signs of deficiency, which were responsive to riboflavin therapy.... the young men, because of their greater food consumption, received a total of nearly 1 mg. of riboflavin daily whereas the sedentary women ingested but o.5 mg per day....

"Williams, *et al.* (1943) found no apparent handicap in four female subjects maintained for 9 months on 13 mcg. of riboflavin per kg. per day. However, neither 13 nor 17 mcg. per kg. prevented a certain degree of depletion of tissue riboflavin...."

"It seems likely that very long experimental periods would be required before the consequences of milder deficiency states would become evident."

Thus the situation is that all the experiments with human beings have been for too short periods to show what the allowance would need to be to bring optimal well-being (so far as this factor is concerned) throughout the normal life cycle. The present Recommended Allowances are, in the judgment of the present writer, more than minimal adequate but less than fully optimal.

There are more reasons than the above noted for liberal intakes of riboflavin in human nutrition.

Spies, et al. (1940) have found that riboflavin deficiency "appears to be the commonest of the clinical deficiency diseases among children of the Alabama area studied."

Sydenstricker considered riboflavin deficiency to be perhaps the most frequent of all vitamin deficiencies in the United States; and Akroyd thought this also true of India.

It has been found that dogs on diet deficient only in riboflavin show no ill effects for a relatively long time (102 to 140 days), then undergo a sudden decline of collapse. Pure riboflavin given by injection or added to the diet prevents the breakdown or cures it if given in time.

It may be taken as strong evidence of the importance of riboflavin to the internal environment, that the body tends to hold its riboflavin content tenaciously, both in cases of shortage of intake of this vitamin, or of pellagra patients.

Urinary exertion thus becomes a means of judging the adequacy of the body's supply of riboflavin.

Feder, Lewis, and Alden (1944) found "a widespread incidence of riboflavin deficiency," as judged by measurements of urinary excretion of riboflavin, in the south-eastern area of the United States.

Stiebling and Phipard's dietary studies show also a marked prevalence of riboflavin-poor dietaries in the south-eastern states; and in low-cost dietaries of other regions as well.

The wide unexplained variation of riboflavin contents of normal meals found by Kerly (1944), for example, from a low of 0.28 to a high of 0.93 mg. in 14 "entire dinners" served in school canteens, gives added emphasis to the advice that dietary allowances of riboflavin may well be made more liberal than those which are usually accepted as standard.

The per capita consumption of riboflavin in the food of the people of the United States as officially estimated for 1943 and 1944 is practically the same as the Recommended Allowance figured to a per capita basis from the Census data of the age distribution of the population. Hence dietaries of the riboflavin-content indicated by the Recommended Allowances are not unduly difficult of attainment, assuming average purchasing power and an up-to-date knowledge of food values. But when decisions as to what foods to buy, and how much of each, are not guided by nutritional knowledge, the higher-income families with their larger per capita purchases of meats, eggs, milk, cheese, cream, and ice cream get more, and the lower-income families get less, than their *pro rata* shares of the riboflavin in the nation's food supply.

One objective of wise nutritional guidance is to raise the lower levels of riboflavin consumption without necessarily lowering the higher levels. In the United States this need not be difficult. It can most readily be bought about by increasing the production and consumption of milk and its products other than butter. This group of foods supplies at present levels about half the riboflavin of the typical American family dietary; and to increase the use of food of this group by perhaps one third in the average lower-income family will give such a family a better-balanced dietary, whatever be the shift in the food budget by which the purchase price is provided.

To some extent the needed shift of emphasis in food production can be made a matter of national policy, but probably the most potent incentive to increased milk production will be an informed consumer demand. An increased proportion of dairy cattle in the farm animal population of the United States will mean both a more efficient use of the Nation's, and the individual farmer's, food production resources and a better balanced food supply for the human population.

22

Vitamin C

Only a year after the production of crystals B1, vitamin C was extracted from both the adrenal gland and cabbage and soon afterwards from orange juice. In 1933 this vitamin was synthesized and given the name ascorbic acid.

Its formula is:

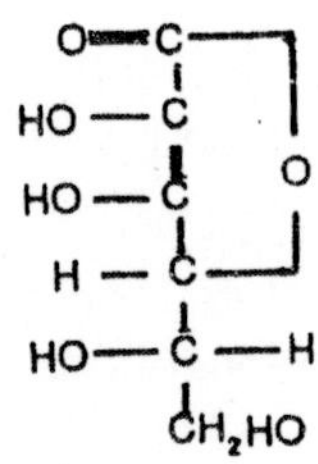

Ascorbic Acid

The crystals are colourless and readily soluble in water. The vitamin is unstable, being liable to destruction by oxidation. This destruction is more rapid at high temperatures and in alkaline solutions. Because of this, it is difficult to preserve vitamin C in foods and special precautions have to be taken in cookery.

The production of pure ascorbic acid crystals made it possible to discard lemon juice as a standard for the measurement of vitamin C potency and the amount in foods is now expressed in mg. per 100 g. or per ounce.

Discovery, and Chemical Identification : Five centuries or so ago scurvy was one of the most prevalent diseases in Europe. In fact one writer then suggested that all diseases might be considered as different modifications of outgrowths of scurvy. European medical literature does not show how long the people had suffered from this disease. It is known to have been one of the afflictions of the Crusaders; and when Vasco de Gama, near the end of the fifteenth century, made his historic voyage around the Cape of Good Hope, he reported the death by scurvy of nearly two thirds of his crew. In 1535 cartier, obliged to winter in Canada, lost a quarter of his men by scurvy, and nearly all the others were ill with it. From the natives they learned that decoctions of the twigs and leaves of certain trees served to cure and prevent the disease. Captain Cook and his men escaped scurvy by making a regular practice of stocking their ship with the fresh fruits of all the shores that they visited, giving the same careful attention to this as to the replenishment of their drinking water.

What the Europeans suffered from scurvy during their exploration and colonization of the New World was compensated for in two ways: they learned the antiscorbutic values of fresh vegetables and citrus fruits, so that as early as 1720 Kramer wrote of these as the sure preventive; and they brought the potato from America to Europe where, as potato culture became common, scurvy became relatively rare.

It continued, however, to afflict sailors on long voyages, and Lind as surgeon of the *Salisbury,* finding himself obliged to treat an epidemic of scurvy with only scanty supplies, decided to give the available oranges and lemons to two of his twelve patients while others received respectively, cider, cream of tartar, elixir of vitriol, or other drugs which had been recommended by different medical writers. The account of this experience which Lind published in 1757 had, therefore, much of the character of a report upon a controlled experiment. The men receiving the oranges and lemons were completely cured, those who had cider showed improvement, while the other patients gradually became worse.

In 1841, Budd explicitly ascribed the *antiscorbutic property* possessed by certain foods to a *definite substance* which, he wrote, "it is hardly too sanguine to state, will be discovered by organic chemistry or the experiments of physiologists in a not far distant future." Ninety years later, Budd's prediction was fulfilled by C. G. King, using the methods both of organic chemistry and of experimental physiology in his researches at the University of Pittsburgh.

King's chemical identification of vitamin C, quickly confirmed by Szent-Gyorgyi in 1932 and soon followed by the synthesis of the substance, made this the first of the vitamins to be conclusively established as a substance of known molecular structure; as it may also be considered to have been the first to be clearly postulated as a definite chemical entity.

In the meantime, however, the vitamin concept had been independently developed through studies of normal nutrition and of beriberi, so that when the anti-scorbutic substance was formally incorporated in the vitamin scheme of 1920 by Drummond, it was given only third place in the vitamin alphabet, McCollum having already designated a fat-soluble A and a water-soluble B.

Here both the two names, vitamin C and ascorbic acid will be used in text. The early explorers of both land and sea were menaced by outbreaks of a disorder known as *scurvy*, which is a nutritional disease caused by a severe deficiency of ascorbic acid. The exploration of the world during the fifteenth and sixteenth centuries was hindered by the outbreaks of this malady. On Vasco de Gama's famous voyage around the Cape of Good Hope in 1497, as many as two-thirds of his crew died of scurvy. All the members of Jacques Carrier's second voyage of Newfoundland in 1553 suffered from this disorder when they were forced to spend the winter in Quebec. Because fresh foods were not available, some of the crew died from scurvy before the Indians taught them to prepare a tea of bark and leaves of the spruce tree which acted both as a remedy and a preventative for the disorder.

James Lind, a ship's surgeon in the English Navy, had the genius to observe, record, and experiment with scurvy patients on board the H. M. S. Salisbury in 1747, and his classical work led to the elimination of scurvy among the crews of ships. He made the following report of his experiences in relieving securvy on this new famous voyage:

On the 20th May 1747, I took twelve patients in the scurvy.... Their cases were as similar as I could have them.... They lay together in one place and had one diet common to all.... Two of these were ordered each a quart of eyder a-day.... Two others took twenty-five gutts of elixir vitriol three times a-day.... Two others took two spoonfuls of vinegar three times a day. Two of the worse patients were put under a course of sea-water.... Two of the others had each two oranges and one lemon given them every day.... The two remaining patients took the bigness of a nutmeg three times a-day of a electuary recommended by a hospital surgeon.... The consequence was, that the most sudden and visible good effects were perceived from the use of the oranges and lemons; one of these who had taken them, being at the end of six days fit for duty.... The other was the best recovered of any in his condition; and being now pretty well, was appointed nurse to the rest of the sick....

Not only did Lind show that scurvy could be cured by the intake of orange and lemon juice, but also he developed a method for the concentration and preservation of the juices from citrus fruit for use at sea. In 1795 the Royal Navy began to provide a regular daily ration of lime or lemon juice to all its men, and the name "limey," which is still used to designate an English sailor, refers to this practice.

Throughout the nineteenth century there are records of scurvy epidemics in different parts of the world. The most famous epidemic was the potato famine of 1845-1848, which killed many people in Ireland where potatoes were the main food of the diet. Even the settlers of this country suffered from scurvy in their march westward. In 1846-1848 the Mormons, on their way to Utah, were forced to spend the winter near Omaha, Nebraska. Here they lived on a diet that consisted primarily of mush with the result that 300 children and adults died, most of them of scurvy.

Although man has known for years that the use of certain kinds of foods in his diet would prevent the occurrence of scurvy, it was not until 1933 that Waugh and King at University of Pittsburgh and Szent-Gyorgyi in Hungary identified the antiscurvy factor as ascorbic acid. Since that time, the vitamin has been manufactured in the synthetic form and today can be purchased at a very low cost.

Molecular Structure

The designation ascorbic acid has been generally accepted as being distinctive and historically appropriate. In molecular structure, ascorbic acid is closely related to the monosaccharide sugars, and to several of them. This fact makes possible the synthesis of ascorbic acid from different carbohydrate starting-points. Thus the substance, after King's identification, soon became relatively inexpensive, and so is widely available as research material, as well as for therapeutic use.

Even before the substance had been isolated it was known to be readily oxidized. Its oxidization is much more rapid in alkaline than in acid solutions, and is markedly catalyzed by even traces of copper. Although ascorbic acid is, in general, the most labile of the known vitamins, and its destruction under ordinary conditions is chiefly due to oxidation,. it is not autoxidizable within the normal pH range of plant and animal tissues. Rather, the rate of its oxidative destruction depends both upon its oxidation-reduction potential and a delicate interdependence of pH and the concentrations of other factors.

Due to the possible presence in urine of small quantities of some substance or substances of oxidation-reduction potential so similar to that of ascorbic acid as to involve the danger of being counted with it, Roe has developed a method for blood and urine analysis in which ascorbic acid is determined not by oxidation but by means of a carbonyl condensation reaction. This new method is of much value in clinical research.

The Destruction

Then conservation of vitamin C values of foods depends upon protecting the vitamin from oxidation not only by the exclusion of air in the ordinary sense, but also by freeing the material from dissolved air, by the avoidance of copper and other such catalysts as might gain entrance from even brief contact with utensils as well as the longer contacts with containers, and by the maintenance of a low temperature, and of an acid reaction. It is also to be kept in mind that the oxidation hazards are not all from without; for if cabbage juice is carefully acidified to the same pH as tomato juice, and the two are treated with exactly the same precautions in all the respects mentioned above, the vitamin C value will still fall more rapidly in the juice of the cabbage than in that of the tomato, because of the natural presence in the cabbage of substances of higher oxidation potential than in the tomato.

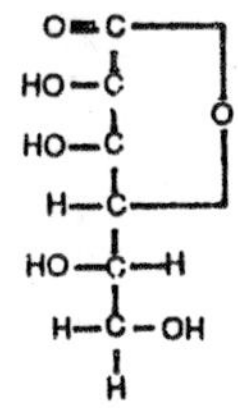

Numerical Expression

Vitamin C values and requirements are most satisfactorily expressed in terms of milligrams of the actual substance, and this plan is followed in the book. The now rarely used Internationl Unit of vitamin C is defined as 0.05 mg. of the actual substance.

The Properties

Ascorbic acid is the chemical name for vitamin C; it occurs in white crystals that are easily dissolved in water (Fig.). Althogh the crystals of the vitamin are quite stable in the dry form, a water solution of ascorbic acid will undergo destruction when exposed to air, heat, light, or metals like copper and iron. It is unstable in alkali but relatively stable in an acid medium.

The Functions

The exact bio-chemical function of ascorbic acid in man and animals is not yet established. It is known that the vitamin is important in the formation and maintenance of *collagen*, the cementing material that holds the cells of the body together. Also, it has been suggested that ascorbic acid plays a role in the normal metabolism of the amino acid, tyrosine, and in the function of adrenal gland.

The role of ascorbic acid in the formation and maintenance of this collagen was first observed in a study using guinea pigs that was conducted at Harvard University. When the tissues of these scorbutic animals were examined under the microscope, it was found that the collagen material between the cells had disappeared. Similar examinations of the teeth and bones of the animals revealed that the cells of the dentin, the interior part of the tooth, and those at the end of the long bones were affected.

When guinea pigs are placed on an ascorbic acid-free diet, they start to lose weight rapidly after 16 days, and between 25 and 28 days they will die if the vitamin is not restored to the ration. Loss of appetite as well as dull and rough hair are the characteristic gross symptoms of this disorder. There are also signs of marked hemorrhages throughout the tissues and organs of the body when the animals are autopsied.

In Man : Human scurvy is rarely found in the United States today. The few cases that do occur are found in infants between the ages of 6 and 12 months who have not received adequate quantities of the vitamin in their diets and in aged or older adults who have poor food habits.

Adult Scurvy : The first symptoms of scurvy that precede the clinical lesions of the disorder are weakness, loss of weight, and fleeting pains in the arms and legs. Usually the first noticeable clinical symptom is the formation of a horny growth in the hair follicles called *follicular keratosis,* which may occur on the legs, buttocks, arms, and back. This is followed by the appearance of redness (hemorrhages) around the hair follicle. Soon the scorbutic changes in the gums become apparent. It has been observed in cases of experimentally induced scurvy that the gum lesions begin to appear after 26 weeks of ascorbic acid deprivation. At first the gums are red and swollen, and as the disorder progresses, the swelling increases, hemorrhages occur, and the gum tissues take on a blue-red colour. If the condition persists, the teeth may become loosened.

The failure of new wounds to heal and the tendency of old wounds to become red and break open have been observed in scurvy patients for centuries. This incapacity for wound healing, however, was first demonstrated in 1940 when Crandon used himself as an experimental subject and ate an ascorbic acid-free diet for a period of six months (4). At end of this dietary restricted period a wound was made in the mid-area of his back to observe the effect of the vitamin deprivation on the formation of collagen tissue during healing. Ten days after the wound was made an examination showed that only the surface cells had knitted together and that the inside of the wound had not healed (Fig.). Yet ten days after ascorbic acid had been put back in his diet, normal healing was observed upon the examination of same wound.

An interesting study on the serum cholesterol levels before, during, and after recovery from scurvy among Bantu men has been reported. It was found that the scorbutic subjects had singnificantly lower mean serum cholesterol values than a group of healthy controls. Furthermore, the serum cholesterols of these men with scurvy failed to respond to the ingestion of fats known to elevate blood cholesterols in healthy subjects. When ascorbic acid was given to those afflicted with scurvy, there was an immediate rise in the serum cholesterol, more rapid when the vitamin was given orally than intramusecularly. Healthy subjects, however, showed no significant change in cholesterol levels when given either high-fat or low-fat diets supplemented with ascorbic acid. This relationship between ascorbic acid and cholesterol metabolism apparently is only evident in man during the stage of marked depletion of the vitamin.

Subclinical Scurvy : In this country some people who have low intakes of ascorbic acid in their diets undoubtedly suffer from subclinical or mild cases of scurvy. Nevertheless, it is impossible to describe lesions which might occur in these mild cases because of the non-specific nature of the symptoms of scurvy. Crandon, in his classic self-induced scurvy investigation, showed the symptoms of great weakness by his inability to perform work in addition to the inability to heal wounds. However, one of the characteristic symptoms of scurvy, the changes in the gums, did not occur even after six months of ascorbic acid deprivation. In contrast to this, hemorrhagic gingivitis that responded to ascorbic acid therapy has been observed among Navajo children in Utah. Arctic travellers, trappers, and Eskimos are known to live on a daily ascoubic acid-intake of less than 15 mg without showing any clinical symptoms of scurvy. A study from England on longtime deprivation of the vitamin in human beings demonstrated that an intake of as little as 10 mg of ascorbic acid each day was sufficient to relieve the clinical symptoms of scurvy. In view of these observations, it is easy to see why mild scurvy might be hard to detect.

Recommended Allowances

The Recommended Allowances for ascorbic acid are higher in this country than in either Canada or the United Kingdom. A daily allowance of 30 mg has been recommended for the English people in view of the report that the clinical symptoms of scurvy were relieved when 10 mg of the vitamin was ingested. In spite of this suggestion, there is adequate evidence that higher intakes of ascorbic acid are beneficial to individuals of all ages.

In the past, somewhat higher ascorbic acid allowances have been proposed for men than for women. The 1963 revision of the Recommended Daily Dietary Allowances suggests, however, that all individuals over the age of 18 years need about 70 mg of the vitamin. An additional 30 mg of ascorbic acid has been proposed for the periods of pregnancy and lacation. The recommended amount of the vitamin needed for the period of growth varies.

Food Sources

Fruits and vegetables are the main sources of ascorbic acid in diets. Citrus fruits (oranges, grapefruit, lemons, and limes), berries, melons, tropical fruits (pineapple, guava, and others), leafy green vegetables, broccoli, green pepper, cabbage, and tomatoes are all good sources of ascorbic acid (Table). The large amounts of potatoes used in low-income diets make this vegetable an important ascorbic acid food for many people.

The best natural sources of ascorbic acid have been found in the rose hip, the acerola fruit (West Indian cherry), and the *camu-camu* (a tropical fruit). Rose hips, which constitute the base of the rose bloom, are not eaten as such but in some countries are made into a sirup or extract for human consumption. During the wartime rationing of food in England, rose hip sirup was issued by the British Ministry of Food to help fortify the ascorbic acid intake of the English people. There is much interest today in the acerola fruit because of its high ascorbic acid potency. It has been reported that the ascorbic acid content of juice from ripe fruit averaged 1100 mg per 100 ml and 1250 mg per 100 ml from half-ripe fruit. Acerola juice, which has little effect upon the colour or flavour of other juices, is used commercially to fortify low-ascorbic-acid fruit juices. The *camu-camu*, a burgandy-red tropical fruit that grows on bush-like trees along the rivers of the Amazon River Basin, has been reported to contain 2415 mg of ascorbic acid per 100 gm pulp. This fruit is used in the preparation of a sweetened fruit drink and in jelly.

Many of the values that were once listed on food tables for the ascorbic acid content of foods were based only on reduced ascorbic acid measurements. Because of the newer knowledge of man's ability to use dehydroascorbic acid, food table values for ascorbic acid have been revised to include the total ascorbic acid contribution of a food.

Ascorbic Acid Losses during the Processing and Preparation of Foods : It is known that a food may lose much of its original ascorbic acid from the time it is harvested until it is eaten because of the many stages of processing and storage it may undergo. Because ascorbic acid is soluble in water and in solution is so easily destroyed by other factors, it has come to be known as the most *labile* or easily destroyed of the vitamins.

Ascrobic Acid Values of Some Typical Foods

Food	*mg/100 gm*[1]	*Portion*	*mg/Serving*[2]
Acerola, pulp + jc, raw	1300	—	—
Green Pepper, raw	128	1 med	79
Broccoli, cooked	90	2/3 c	89
Brussel sprouts, cooked	87	2/3 c	75
Turnip greens, cooked	69	2/3 c	66
Kale, cooked	62	2/3 c	45
Strawberries, raw	59	2/3 c	58
Orange	50	1 med	70

Contd..

Food	*mg/100 gm*[1]	*Portion*	*mg/Serving*[2]
Mustard greens, cooked	48	2/3 c	45
Orange juice, frozen	45	1 c	112
Lemon juice	46	1 c	113
Grapefruit	38	½ med	52
Grapefruit juice, canned	38	1 c	84
Cabbage, cooked	33	2/3 c	37
Cantaloupe	33	¼ melon	32
Lime juice	32	1 c	80
Coleslaw	29	2/3 c	23
Spinach, cooked	28	2/3 c	33
Liver, beef, cooked	27	2 oz	15
Tomato, raw	23	1 med	34
Sweet potato, baked	22	1 med	24
Peas, green, cooked	20	2/3 c	22
Potato, baked	20	1 med	20
Tomato juice, canned	16	1 c	39
Pineapple, canned	7	2 sm sl	8
Milk, Whole	1	1 c	2
Beef, ground, cooked	0	3 oz	0
Bread, white, enriched	0	1 sl	tr
Butter	0	1 tbsp	0
Egg 0	1 med	0	
Sugar, white	0	1 tsp	0

Therefore, in order for one to prepare and serve foods which are appetizing as well as nutritious, it is important to know how this vitamin can be conserved.

The environment in which a plant food is grown may have a

significant effect on its ascorbic acid content. Temperature, amount of rainfall, type of soil, and sunlight appear to be factors directly related to the formation of plant ascorbic acid. In addition to these factors, the stage of maturity at which a food is harvested is important in order to obtain the maximum amount of this nutrient. For example, immature Lima beans and white potatoes contain more ascorbic acid than do mature ones, whereas ripe tomatoes have a much higher vitamin content than do green ones.

Food Preservation

The manner which a food is preserved, by freezing, by canning, or by drying (dehydration) has a significant effect on its ascorbic acid content. This fact can be well illustrated by comparing the vitamin values of fresh, frozen, canned, and dried Lima beans, all commonly used foods in the American diet. An edible portion of Lima beans weighing 100 gm has an average ascorbic acid value of 29 mg in the raw form, 22 mg when frozen, 6 mg when canned, and none when dried.

Students often ask if the amount of ascorbic acid in orange juice is in any way affected by the kind of processing it undergoes. This question can best be answered by the report of a study made at the University of Illinois where they compared the reduced ascorbic acid values as well as the cost of an average serving of fresh, frozen concentrated, canned concentrated, and canned juices from organges. All of the orange juice analyzed in this work, regardless of processing procedures, was quite similar in ascorbic acid content, but the frozen concentrate cost less per average serving at the time of that investigation than did either the fresh or the canned.

Effect of Storage

Two other factors that are important in conserving the ascorbic acid content of a food are the time interval between the processing and consumption of a food, and the kind of storage facilities available. From the University of Massachusetts it has been reported that pasteurized milk losses about two-thirds of the reduced ascorbic acid content after four days under refrigeration. However, if the milk was consumed within the first day after processing, about 80 per cent of the original reduced ascorbic acid would have been available. Orange juice and other citrus juices lose negligible amounts of ascorbic acid when stored in the refrigerator; the acid of the juice helps to preserve the vitamin.

The vitamin content of frozen fruits and vegetables has been observed to be affected by changes in storage time and storage temperature. Frozen cherries, peaches, and strawberries were found, by a group at the Pennsylvania State University, to have lost marked amounts of ascorbic acid when stored longer than 4 months at, + 10° F. However, when the same fruits were at a lower temperature, 0° F to -20° F, there was no appreciable loss of the vitamin within the same period of time. Asparagus, green beans, wax beans, broccoli, cauliflower corn, and spinach stored at + 10° F lost 50 per cent of the reduced ascorbic acid content after four months storage, but when stored at 0° F most of the vitamin was retained. It is evident from this work that both frozen fruits and vegetables conserve the ascorbic acid value best by storage at lower temperatures.

Canned foods are also subject to marked losses in ascorbic acid when stored at elevated temperatures for long periods of time. Canned tomato juice after eight months of storage, has been shown to retain only about one-half of the reduced ascorbic acid content when held at room temperature (68° to 77° F) and only about one-third when the temperature was increased to 99° F.

Food Preparation

The amount of ascorbic acid in a food to be eaten is directly related to the kind of food preparation practices used. A raw food that one eats will have more of the vitamin that will one that has been cooked or baked. The removal of the natural skin of a vegetable will increase the loss of the vitamin from the food due to the leaching of the nutrient into the soaking or cooking water. Further losses of the vitamin will occur when a peeled vegetable is sub-divided into smaller sections; for example, when potatoes are diced, the surface area exposed to the soaking or cooking water is increased. The use of copper cooking utensils will decrease the ascorbic acid content in a food because of the destructive nature of the copper iron to the vitamin.

Approximate Contribution of Some Foods to the Day's Recommended Allowances of Ascorbic Acid for College Men and Women over 18 Years

Food	*Measure*	*Per cent/day*
Apple, raw	1 med	4
Banana, raw	1 med	14
Beans, green, cooked	2/3 c	16
Beets, cooked	2/3 c	10
Broccoli, cooked	2/3 c	127
Carbonated beverage	8 oz	0
Carrot, raw	1 med	6
Catsup	1 tbsp	4
Coleslaw	½ c	26
Corn, cooked	2/3 c	13
Grapefruit	½ med	74
Hamburger in roll	1 av	0
Kale, cooked	2/3 c	65
Lettuce	2 lg lv	13
Milk, whole	1 c	3
Orange juice, frozen	½ c	80
Peaches, canned	1 c	10
Peas, green, cooked	2/3 c	31
Pineapple juice, canned	½ c	16
Potatoes, French fried	10 pieces	17
Potatoes, mashed	1 c	27
Pizza, cheese	5½ in. sector	6
Raisins	½ c	1
Tomato juice, canned	1 c	56

Adopted from *Nutritive Value of Foods.* U.S. Dept. Agr., Home Garden Bull. 72, revised 1964.

The method used is cooking of a food will determine the amount of ascorbic acid retained by it. In order to illustrate the amounts of the vitamin that may be lost in vegetable cookery, data from Yale University have been adopted to show the retention of reduced ascorbic acid in fresh green peas cooked by three commonly used home cooking methods (Fig.). The peas retained more of the vitamin when cooking in the waterless cooker or in the pressure saurcepan than when cooked in the tightly covered saucepan. In the tightly covered saurcepan method, reducing the amount of water used from enough water to cover the vegetables to only one-half cup increased the amount of ascorbic acid in the cooked peas by about one-third.

A Guide to the Selection of Foods to Meet the Ascorbic Acid Needs of College Men and Women : Surveys from many areas of the United States have shown that ascorbic acid is one of the nutrients most often found to be deficient in the diets of college men and women. Because of the important role of this vitamin in the maintenance of good health and the fact it is one of the nutrients ingested in sub-optimal amounts, it is important that students know the approximate contribution of the foods eaten in relation to their daily needs.

The contribution of some foods to the day's Recommended Allowances of ascorbic acid for college men and women, over 18 years, has been tabulated (Table). An average serving of a citus fruit or an ascorbic-acid-rich vegetable supplies 50 per cent or more of the daily allowance for the vitamin. Other fruits and vegetables supply lesser but important amounts of ascorbic acid. As can be readily seen from the table, without a serving each day of either a citrus fruit or an ascorbic-acid-rich vegetable , the Recommended Daily Dietary Allowance for ascorbic acid may not be met.

Nutritional Chemistry of Ascorbic Acid (Vitamin C) : Prominent as is the oxdation-reduction behaviour of ascorbic acid *in vitro,* the exact way in which this property functions nutritionally in the animal body has not yet been so fully and clearly worked out as to permit of simple concise statement, though King and co-workers are closely approaching that goal. Meanwhile, the nutritional functions of vitamin C which have thus far been most clearly established are (a) its relationship to the formation and maintenance of intercellular cement substance, (b) its significance in the safeguarding of the body through the blood, and (c) its role in regulating the rate of cellular respiration.

Deficiency of ascorbic acid leads to an increased metabolic rate, as measured by oxygen consumption, both in the intact animal and in animal tissue slices.

Sealock has shown that ascorbic acid is essential for normal metabolism of the amino acid tyrosine in guineapigs. Incompletely oxidized tyrosine was excreted by scorbutic animals, and slices of liver from scorbutic animals showed the same loss in capacity to oxidize tyrosine beyond the quinone stage. When ascorbic acid was added to the tissue slices from scorbutic guineapigs, normal oxidation of tyrosine was resumed. This finding promises to be of far-reaching important as an indication of a major specific function of the vitamin in cellular metabolism and Levine and Cornell Medical College observed a similar relationship in premature infants.

In as much as shortage of vitamin C may have any of such widely distributed results in the body, as indicated in the four preceeding paragraphs, one's individual bodily constitution may determine how and where the penalty will be paid for neglect of vitamin C is one's choice of food.

Those of species (such as the rate and chicken) which thrive on diets containing no demonstrable amount of vitamin C, as well as in species (man, guineapig, and money) which are suspectible to scurvy. The relative distribution in the body is similar in both groups of animals, being highest in glandular tissue (in particular the adrenals, pituitary, corpus luteum, and young thymus) and lowest in blood and muscle; but the concentrations are higher in those species which can synthesize vitamin C (and are therefore presumed to maintain tissue stores at one near optimal level) than in human subjects who have recieved an ordinary or conventional diet. In a series of analyses of human tissues obtained at autopsy, Yavorsky, Almaden, and King (1934) found wide individual variations and concluded that, in approximately one fifth of the people represented, the tissue concentrations of vitamin C were distinctly below what is conductive to optimal health.

The relation of vitamin C to the teeth deserves discriminating consideration. Enthusiastic reports of several years ago, that attributed to this vitamin a specific value in the combat of dental cares, met considerable skepticism. The evidence on this point is conflicting and the question is still debated. The inconclusive state of the caries problem, however, should be distinguished from some other aspects of "the tooth problem" on which evidence is much clearer.

It had earlier been found that vitamin C is highly essential to good internal structure of the teeth; and Boyle, Bessey, and Wolbach (1937) definitely showed the value of this vitamin in the prevention of the systemic type of pyorrhea. Hence the soudness both of the teeth themselves and of their supporting bones and gums is importantly dependent upon the amount of vitamin C supplied by the food.

King and coworkers (1940) found that guineapig teeth, in addition to being very sensitive to structural abnormalities when subjected to mild vitamin C deficiencies, are injured in much greater degree when the animals are injected with diphtheria toxin. To withstand the effect of toxin injections, as shown by retention of normal tooth structure, vitamin C intake levels must be from two to three times higher than are required when the animals are not subjected to the toxin. Thus there was a distinct advantage to the tissues gained from having more than a bare protective allowance of vitamin C.

Concentration levels in the body are influenced both by the rate at which the body receives and the rates at which it destroys and excerts the vitamin. Notwithstanding the solubility of vitamin C, its concentration may vary widely among the tissues and body fluids of the same individual. It has been most frequently determined in the blood plasma, with the concentrations stated in "miligrams per cent" which in this case may, without appreciable error, mean either per 100 grams, or per 100 cubic centimeters or milliters.

An interesting illustration from the work of Farmer and Abt is shown in Fig. From this it will be seen: (1) that one week of feeding a pint of orange juice daily in addition to the previously customary diet brought up into the normal range all of their subjects whose initial blood values were subnormal; and (2) that this more liberal daily allowance of vitamin C in the diet also appreciably increased the level of concentration of vitamin C in the blood plasma (and probably in the body generally) of those whose initial levels were already within the normal range.

That even in the case of so soluble, diffusible, and labile a substance as vitamin C the body habitually carries a higher concentration when it receives a higher intake is a fact of much importance to our growing appreciation of the influence of one's daily food habits upon one's *internal environment.*

This flexibility of the body's internal environment (as contrasted with its previously postulated *fixite)* is very important both theoretically and practically; but of course there are both practicable limits and individual differences.

It will be noted that nine of the ten persons represented in Fig. attained a fairly constant plasma-concentration of between 1 and 15 "mg. per cent" when their intakes of vitamin C were liberal. Similar results have been found in a large majority of the individuals studied by these and other investigators. (In the few exceptional cases the plasma level may be either lower or higher than the general average.)

23

Vitamin D

Children were afflicted with rickets for many centuries before the cause of the disease and the cure for it were known. In Central Europe rickets was known as "The English Disease" because of its prevalence in England. In that country around 1900 it reached almost epidemic proportions when industrialization resulted in both parents working in shops, leaving little time for taking babies out of doors. However, rickets was not just confined to England. It was common in most of the countries of Northern Europe and in the larger cities of the Untied States, where as recently as 35 years ago the disease was still common. The discovery of vitamin D and subsequently its widespread use has spared children the crippling effects of rickets.

A combination of clinical observations of children afflicted with rickets and of experimental animals in whom rickets had been developed under controlled laboratory conditions led to the identification of vitamin D as the *antiricketic vitamin.* An early (1890) significant observation, made by palm an English physician, was that the incidence of rickets is related to the amount of sunshine available; where sunshine was abundant, rickets was rare; where the sun seldom shone, rickets abounded. After a few years the concept developed that rickets, like scurvy and beriberi, may be a dietary deficiency disease. With this possibility in mind, the expected next step was to produce the deficiency in experimental animals. In 1921 Mellanby in England produced rickets in puppies by using a mixture of natural foods; then he induced recovery by giving them cod liver oil. At that time cod liver oil was known to contain only vitamin A, so the cure was attributed to it. Later it was discovered that even after the vitamin A content of cod liver oil was destroyed by oxidation, the anti-ricketic potency still remained. It was then that it became evident that cod liver oil contained two vitamins, vitamin A and the anti-ricketic factor. In the meantime, the association of rickets with calcium and phosphorus metabolism, and the favourable influence of cod liver oil on the utilization of calcium were recognized.

By 1922 the mystery of the effectiveness of sunlight in preventing rickets was on the way to solution. Rats fed a rickets-producing diet failed to develop rickets if exposed to sunlight. More over, rats fed a rickets-producing diet that had been irradiated in the laboratory, given the same effect as exposure to the sun, also failed to develop rickets. After the discovery of the effectiveness of irradiating the diet as a whole, each item was removed, one at a time, and treated separately then incorporated into the diet in an effort to find out which nutrient was responsible for the anti-ricketic effect. Fat was the only nutrient that on irradiation imparted anti-ricketic potency to the diet. With fats established as potential anti-ricketic factors, the next question to be solved was which portion of the fats, the alcohol or fatty acid fractions, became *activated* on exposure to the ultraviolet rays. It was found to be the alcohol portion. Intensive research soon disclosed that numerous alcohols, all of which belonged to the sterol group of alcohols, acquired antiricketic activity on irradiation. The sterols are much more complex in chemical structure than the common medicinal and beverage alcohols.

By the late 1920's it had been established that rickets could be prevented and cured by exposure to direct sunlight, by irradiation with ultraviolet light, by feeding irradiated food, and by feeding cod liver oil. Then followed the identification of the *natural* vitamin D of fish-liver oils as the same substance produced in human skin on irradiation of 7-dehydrocholesterol. The vitamin D formed from

7-dehyrocholesterol is designated as vitamin D_3. The irradiation of ergosterol, a provitamin D in plant foods, produces a different substance, designated as vitamin D_2.

This occurs with vitamin A in fish liver oils and animal fats, but has no connection with carotene.

Chemically it is quite different from vitamin A and belongs to a complex group of compounds called sterols.

VITAMIN D_2

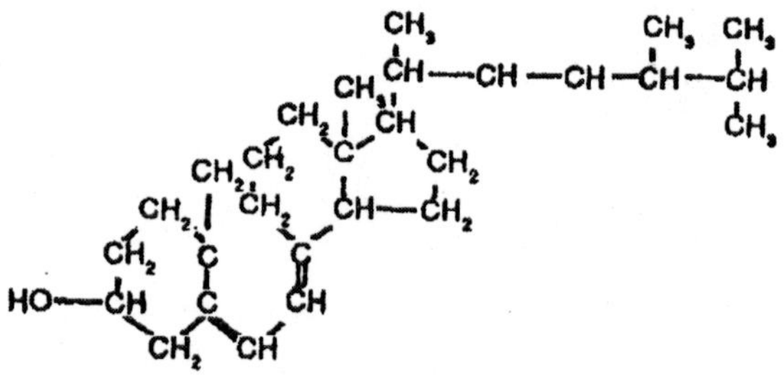

ERGOCALCIFEROL

VITAMIN D_3

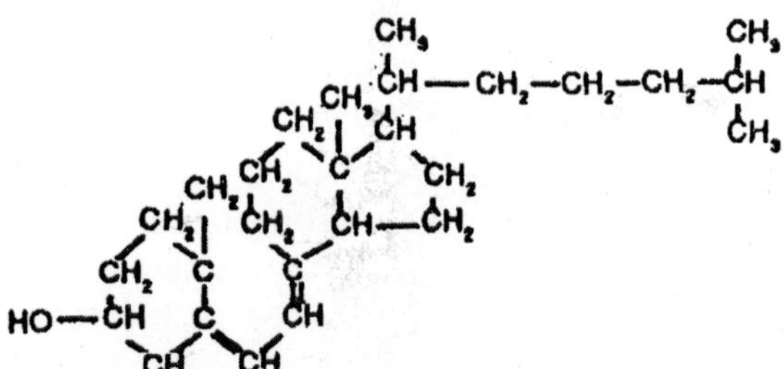

Cholecalciferol

Several of these have, in fact, vitamin D activity. The commonest is that occurring naturally in fish liver oils. This is vitamin D_3. Another is calciferol or vitamin D_2 which is made chemically but the same action for man and mammals as D_3. Birds, however, use D_2 less well (Harris, 1951).

The term as D_1 has been dropped. It was used originally for a compound incorrectly supposed to be the pure vitamin but shown to consist of D_2 associated with an inactive component.

Vitamin D (Ergocalciferol, Cholecalciferol) : It has been said facetiously that civilisation changed vitamin D from a harmone into vitamin. Certainly the beneficial effect of sunlight is the prevention of rickets was recognised long before there was any indication that vitamins existed. More over there is yet some question as to whether a dietary source of vitamin is necessary if the body is regularly exposed to ultraviolet irradiations.

In the mid 1920's Steenbock (1924) demonstrated that irradeation of foods produced antirachitic activity. Later the phenomenas was shown to be due to activation of sterols in foods. The Steenbock irradiation process was to support the winconsin Aluuri research and the rayalties obtained helped to begin and continue Foundation at the university of winconsin.

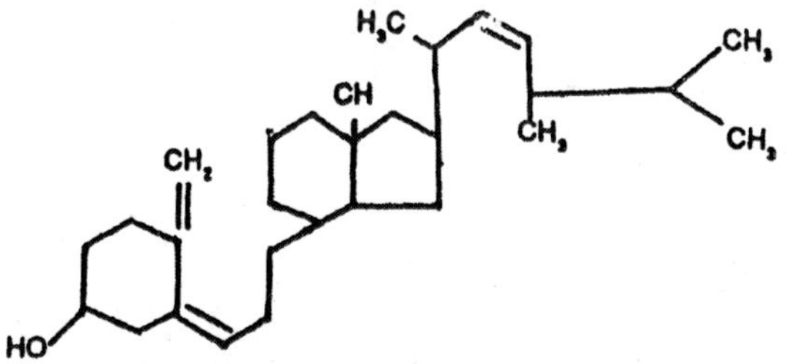

The factor responsible for the preservation of rickets was named vitamin D and was clearly distinguished from vitamin A also present in oils when it was discovered that antirichatic activity remained even after vitamin A activity has been destroyed by oxidation.

Several forms of the vitamin D exist. All arise from irradiation of sterols which are provitamins of precursors of the vitamin. Ergosterol, which occurs in plants, is activated by sunlight to form ergocalciferol (vitamin D_2, calciferol). The form occurring naturally in animal cells is cholecalciferol (vitamin D_3) which is derived from 7-hydroxycholesterol. The similarity of the vitamin structure to cholesterol is evident from the above formula.

Although ultravoilet irradiation produces the vitamin from their sterol precursors, the active forms, ergocalciferol and cholecalciferol are unstable to irradiation. Excessive irradiation will produce some toxic products. The vitamin D are not affected by oxidation or by temperatures below 140°c. In acid media the vitamins are relatively unstable, the instability increasing with increase in temperature. In alkalive solution vitamin D compounds are stable even at elevated temperatures.

The biochemical degradation of cholesterol in the body leads to a host of physiologically important compounds. 7-hydroxycholesterol present in skin undergoes a complex series of photochemical reactions when exposed to sunlight. One of the products is the ring cleaved calciferol or vitamin D_3, which helps to avoid rickets by regulating calcium metabolism.

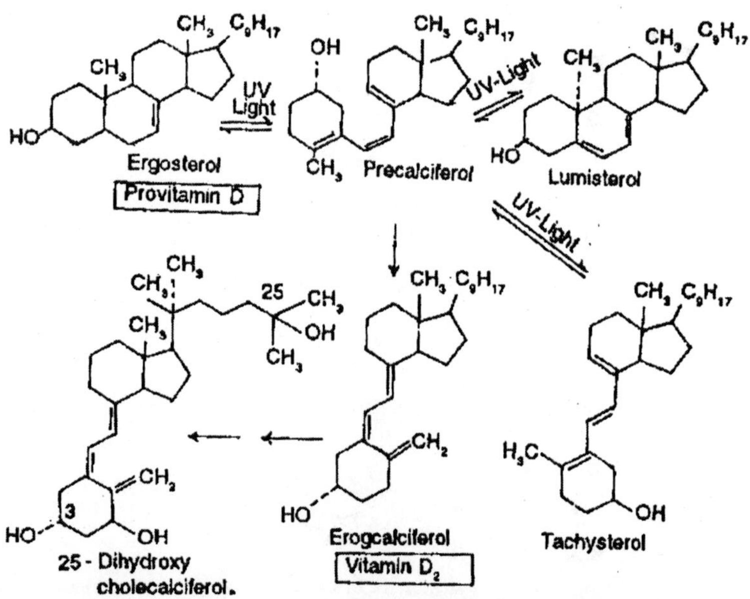

Vitamin D is a group of fat soluble vitamins. Chemically related to the steroids. They are produced from provitamin, 5, 7-unsaturated sterols, by u.v irradiation. In this conversion, ring B of the steroid is opened between C-9 and C-10, forming precalciferol, which acts as a precursor of the vitamin D group (Fig.). Important members are Vitamin D_2 (ergocalciferol), derived from ergosterol, and Vitamin D_3 (cholecalciferol), derived from 7-dehydrocholesterol. Lumisterol and tachysterol are by-products of the synthesis Vitamin D_3 is converted enzymatically in the liver and kidneys to 25-hydroxycholecalciferol, followed by hydroxylation to the highly active la, 25 hydroxycholecalciferol (la, 25- dihydroxy-vitamin D_3);

this compound represents the active form of the vitamin in the human, and its behaviour resembles that of a hormone, rater than a biocatalyst.

Vitamin D_1 is a molecular compound of lumisterol and ergocalciferol.

Vitamin D_4 is 22-dihydroergocalciferol, produced from 22-dihyderoergostrol by UV-irradiation.

Vitamin D_3 is present in cod liver oil in particularly large quantities, and it is also formed in human skin from 7-dehydrocholesterol by the action of sunlight.

The vitamin D-Complex is also present in egg herrings, egg yalk, butter, cheese, intak vig liver and edible fungi.

Antirachitic Factors

Mellanby found experiment with puppies that their skeletal development was influenced by some fat-soluble substance or substances in the food. He especially emphaized the prevention of rickets by virtue of an antirachitic property of certain fats.

Almost simultaneously McCollum and coworkers found this true also of rats. Morever, cod-liver oil, even after destruction of its vitamin A, still has a specific favorable influence in the prevention of rickets and the promotion of good development of the bones and teeth. The substance thus discovered soon came to be called, interchangeably, *antirachitic vitamin* and *vitamin* D. More strictly we say *the vitamins* D, because it has been found that more than one substance has the property of preventing rickets.

As more fully explained by Bills (1939) the term vitamin D (collective singular) is customarily applied to any substance possessing the antirachitic property, which is true of at least ten sterols in some degree, but to an important degree of any two of them: namely, vitamin D_2 and D_3.

Each of these can be produced by the action of ultraviolet light upon its precursor sterol: vitamin D_2 by thus irradiating (activating) ergosterol; and vitamin D_3 by thus activating 7-dehydrocholesterol. This 7-dehydrocholesterol has been shown to occur in the skin and is the chief "animal" precursor of vitamin D_3, just as ergosterol is the chief "plant" precursor of vitamin D_2.

Production by Irradiation

In 1924 it was discovered by Hess and independently by Steen-bock that various foods can be endowed with (or enriched in) antirachitic value by irradiation with certain wavelengths of ultraviolet light.

Ergosterol, concentrated industrially from yeast, was found to be a good starting material for such activation by irradiation. The particular vitamin D formed by activation of ergosterol is the one known as vitamin D_2. Its commercial forms include calciferol and viosteral. It occurs naturally in fish liver oils, where a mixture of vitamins D_2 and D_3 is present, with the former predominating in some varieties, such as tuna, and the latter predominating in cod liver oil.

Vitamin D_3 formed by similar activation of 7-dehy

lrocholesterol, is in some cases more effective nutritionally than vitamin D_2, and it has been identified with the principal natural vitamin D of cod liver oil (and presumably also of milk and eggs). Further activation studies of Bunker, Harris, and Mosher (1940) have also strengthened the evidence that 7-dehydrocholesterol is the substance which is activated, and vitamin D_3 the form which is produced, upon irradiation of the skin.

There is some difference of expert opinion whether vitamin D_3 is preferable to vitamin D_2 in the problem of the more prevention of rickets in infants. In the broader view of child nutrition the fact that vitamin D as obtained in milk and eggs, and in or from fish liver oils, is accompanied by vitamin A is a significant practical reason for preferring such natural sources.

Chicks and rats differ widely from each other in their responses to vitamin D_2: for the rat it is a relatively effective antirachitic, and the potency of its preparation is measured by tests with rats. International and U.S.P. units are therefore rat units. "Rat unit for rat unit," vitamin D_3 is more effective in the nutrition of the chick than is vitamin D_2.

Bills emphasizes the fact that from a given provitamin only one form of vitamin D is produced by irradiation whatever the wavelength or other conditions. The conditions influence the by-products but the vitamin is always the same for the same precursor. He adds that processes in which election bombardment is substituted for ultraviolet irradiation produce the same kind of vitamin.

Prevention of Rickets

According to Park's definition that has now been generally accepted, rickets is a condition where in the mineral metabolism is disturbed in such a way that clarification of the growing bones does not occur normally. We must keep in mind while reading the literature on rickets that the term has been given a narrower definition than Park gives it, older definitions turning more upon the histology of the bone tissue while Park is expressed in terms of calcificaion in the developing bone-an essentially chemical phenomenon.

The failure of normal calcification in the young bone is not generally to be attributed to any initial fault of the bone itself for Shipley found that the rachitic bones placed in normal blood serum began at once to calcify normally; and Kramer and Howland found a subnormal concentration of either calcium or phosphorus (or both) in the blood serum. Thus there are different types of rickets that may be classified according to the nature of the chemical deficiency in the blood.

When the calcium content of the serum is normal but its phosphorus (phosphate ion) content is subnormal then so-called low-phosphorus rickets occurs which corresponds most closely and completely to the classical histological descriptions of the disease. In such cases not only is a characteristic pathological histology but also a tendency to overgrowth of the cartilaginous or osteoid tissue at the ends of the long bones and at the rib junctions. Because the retarded calcification prevents the bones from acquiring normal rigidity at a normal rate, the pressure of the body weight tends further to enlarge the ends of the bones. Together these abnormalities make a typical (and formerly all too familiar) picture recognized in part by the obvious clinical signs of enlarged joints or row of beadlike swellings at the rib junctions and in part by means of the Roentgen ray photograph and the determination of inorganic phosphate in the blood serum; and confirmable at autopsy by histological examination. There has sometimes been a tendency confine the term rickets or "true rickets" to his low-phosphorus type.

When the phosphorus content of the serum is normal but its calcium content is subnormal it results in a similar gross abnormality but a somewhat different histology of the bones. This has been called a "rickets-like condition" or a "second type of rickets," and is now called as *low-calcium rickets.* Low calcium rickets is generally accompanied by tetany; and it appears improbable that in such cases the trouble with the nerves as also with the bones is attributable (at least in part) to the deficiency (subnormal concentration) of calcium in the blood.

When both calcium and phosphorus are reduced below normal concentration in the blood serum, calcification is retarded, but the structural abnormality of the bone differs from that of the two types described above. The condition constitutes *a third type of rickets* under Park's definition, but because of the difference in clinical and histological appearance it is called *an osteoporosis.*

Unless the mineral supply is deficient, any of the three types of rickets can be prevented by vitamin D, which acts to restore to normal the concentrations of calcium and phosphate ions in the blood.

Whence does the vitamin "mobilize" the calcium or phosphate or both? The chief answer seems to be by increasing the *net* absorption of calcium and phosphorus; more being absorbed or less excreted, or both.

This, however may not be quit the whole story of the action of vitamin D in the prevention of rickets.

Sometimes the calcium or phosphorus mobilized through the blood stream and made available to the developing ends of the bones appears to come from stores already in the tissues.

There are also indications that a local factor, affecting, as Hess expressed it, the "anchorage" of the available calcium and phosphate ions in the end of the growing bone, may play a part in particular cases. Morever, Schneider and Steenbock have offered evidence of a specific effect of the vitamin upon phosphorus metabolism such as to favour the deposition of calcium phosphate as bone mineral in the cartilage of the developing bone ends, at the expense of the soft tissues, thus retarding the general growth of the body.

In this connection the report of Zetterstrom that phosphorylated vitamin D activates alkaline phosphatases from intestine, bone, and kidney, thus functioning in the absorption deposition and retention of phosphate, is of interest.

While scientific explanation is still incomplete, the problem of prevention of rickets is now sufficiently solved that few new cases of severe rickets are noticed and the mild rickets now-a-days occurring may in large part be considered as incidental to the fluctuating fortunes with which the soft tissues and the developing bones complete for the phosphate which the blood stream brings.

The prevention of rickets by the exposure of the body to sunshine or ultraviolet irradiation comes about through the production of animal vitamin D" (vitamin D_3) in the skin. Notice, too the efficiency of the "mechanism" which at the same time increases the circulation of blood through the skin, bringing precursor and taking active vitamin into body.

Growth and Development

When severe rickets was more common, its outward signs of bowlegs and knock-knees appeared more noticeable in so far as general growth means increase of soft tissue; and the more active the soft tissue growth the less the supply of phosphorus for the deposition of bone salt which is essential to the normal hardening of the growing bones. If the bones are prevented from adequate hardening and also are made bear an increasing body weight, the bending of the bones and enlargement of the joints will be worse, the greater the weight put them. Under such conditions, the greater the growth, the worse the rickets." But, with the discovery and general use of vitamin D, the situation has greatly changed.

The emphasis at present is on the liberal nutritional intake of calcium, of phosphorus, and of vitamin D. When the intakes of calcium and phosphorus are abundant to the requirements of the growth and development of both soft tissues and bone, liberality of vitamin D tends to promote growth; more particularly liner growth, i.e., increase in the length of the skeleton, perhaps especially of the legs.

The superior height and erectness of the carriage which in some countries has so far been considered "a characteristic of the upper class" is because of long straight legs and is now being largely conferred upon the children of all economic conditions through the discovery of vitamin D and the widespread use of fish live oils.

(This is quite as important a service of chemistry to democracy as was Dr. Slosson's favourite illustration, that through the cheap synthesis of the dyestuff "every working-girl can now wear Royal Purple.")

Like some other of the constructive developments of the newer knowledge of nutrition, the influence of liberal intakes of vitamin D upon growth, when intakes of calcium and phosphorus are also liberal, was first shown by experimentation with laboratory animals of short natural life cycle and then confirmed and extended by clinical experience.

In addition to its role in the prevention and cure of rickets, vitamin D was found (Sherman and Stiebeling. 1929,1930) to promote growth and to facilitate the normal calcification in developing bone even when the mineral metabolism is well above the rickets level.

Along with other factors, vitamin D is of importance in the formation of normal teeth and protection against dental caries.

There is clinical evidence that children who have been protected from rickets cod liver oil are less susceptible to respiratory disease than those who have had rickets , but it remains an open question in how far the stronger respiratory system is due to vitamin D and in how far to the vitamin A which the cod liver oil furnished.

Highly significant is the repeated emphasis laid by Jeans and Stearns upon the fact that while vitamin D is in this respect perhaps even more hopeful to physical development than heretofore realized, ons's confidence in the vitamin must never be allowed to make one less carefully attentive to the importance of liberal calcium intake. Increased calcification means increased retention of calcium in the body, and the body must receive large amounts so as to retain large amounts.

The vitamin must never be expected to decrease the calcium requirement. Rather, it helps the body to make optimal use of a generous calcium intake. The intake of calcium must always be generous is optimal calcification and growth of bone are to be secured.

The Properties

The sterols which are potential sources of vitamin D are known a pro-vitamins. There are many of them, but only two of importance, 7-dehyrocholesterol and ergosterol. The provitamin in higher animals and invertebrates is 7-dehydrocholesterol; ergosterol is present in the plant kingdom, in yeasts, and certain molds.

The pure vitamins D are white, crystalline, odorless substances that are soluble in fats and fat solvents such as ether, chloroform, acetone, and alochol. They are resistant to heat, oxidation, acid, and alkali. The vitamins are comprised of carbon, hydrogen, and oxygen combined in chemical structures that are much alike. Crystals of vitamin D_2 and its chemical formula are shown in Fig. The relative anti-ricketic potency of vitamin D_2 and D_3 appear to be no different for the human being; consequently, in present discussion the term vitamin D *will be* used.

Measurement Unit

Vitamin D potency, like that of vitamin A, is expressed in International and USP units. (Other vitamins are measured in units of weight). The amount of antiricketic activity brought about by 0.025 ug of the International Standard, pure crystalline vitamin D_3 is an International Unit. The unit was adopted by the World Health Organization in 1949 and has been used ever since. The USP unit is the same as the International Unit.

In order to measure the *amount of anti-ricketic activity,* rats are usually employed as the experimental animal. Under standardized conditions of feeding and care, and by using standardized procedures for determining bone change, the potency of vitamin D in terms of International Units is determined.

Expression of Values

Vitamin D values are generally measured either by means of the "line test" or by determination of calcium or total ash in a representative bone.

McCollum and his co-workers have found in their autopsies rats which had been on deficient diets, that a rickets-producing diet of a sufficiently drastic character may result in condition wherever the cartilage and adjacent portions of the metaphyses of the long bones are totally free from visible deposits of calcium phosphate. If, then, such rats are given an effective antirachitic treatment of any kind, then with a few days a line of freshly laid down deposits of calcium phosphate which by proper staining can be made clear to histological examination. This is the *line test.* Bills, Honeywell, et al. (1928,1931) elaborated the technique and gave a very full description of it as used by them. It was also adopted by Steenbock and his coworkers for the routine testing connected with their control of the manufacture of irradiated foods and drugs under the Steenbock patents.

The line test has been adopted as the basis for *the United States Pharmacopoeia* unit for vitamin D potency. Present procedure provides for the comparison of the degree of healing caused by the test material with that caused by a known quantity of a standard reference material in a similar group of animal under identical conditions. In this way the effects of variations in the susceptibility to rickets of different groups of rats, due to environmental and other factors, are minimized.

The International standard reference material has been a crystal line preparation of vitamin D_2; and the International unit has been defined as the antirachitic activity of 0.25 microgram of this standard material. The Expert Committee on Biological Standardization of the World Health Organization offers a crystalline preparation of vitamin D_3, the unit of activity in this case also being defined as that of 0.025 microgram of the reference material. The *United States Pharmacopeia* has adopted this new International standard.

To produce rickets in rats for the line-test method of assay, greatly unbalanced mineral intake are used, the usual rachitogenic diet containing four to five times as much as calcium as phosphorus, that is a very much wider ratio than is likely to be encountered in human experience.

Stiebeling has used a different method wherein the basal diet fed the experimental animals (rats) is adequate in its mineral content as also in all organic factors other than vitamin D, and where in the vitamin D value of the material under investigation is judged by the relative amount essential to allow a standard degree of calcification under fixed experimental conditions.

Young rats receiving the basal diet only (and on irradiation) serves "negative controls"; those receiving the basal diet plus abundant vitamin D (or irradiation) serve as "positive controls." others receive the basal diet plus graded allowances of the food or other material which is being tested for vitamin D value.

At the end of a four or five week experimental period the degree of calcification is found by determining the percentage of ash or of calcium in the freshly removed femur. Within certain limits, this is found to be dependent upon the relative amount of vitamin D supplied. The advantage in calcification shown by the test animals over the negative controls must be because of the material under test provided the experiments are properly conducted and in sufficient numbers to avoid vitiation of results through individual variation. It is suggested that the "end point" for comparison be one half as much improvement in calcification over negative controls as is produced by supplying an abundance of vitamin D (positive controls). The amount of the food required to "reach this end point " (maintain the standard rate of calcification) may be considered as inversely proportional to its richness in vitamin D.

Conversion of Provitamin

The conversion of provitamin D to the vitamin by irradiation is accomplished by a change in the structure of the compound; the second ring in the provitamin D structure opens (note ring B, Fig). The conversion process has been studied in the laboratory with the finding that a series of intermediary products are formed before vitamin D is obtained; further irradiation beyond the vitamin D stage produces toxic products lacking anti-ricketic potency.

Human Requirement

In the opinion of Jeans and Stearns for many persons some vitamin D in addition to that ordinarily obtained by exposure to sunshine is necessary for optimal use of calcium and phosphorus in the body. This may be obtained by special irradiation of the skin, by use of such foods as contain vitamin D, by the taking of the vitamin in some more concentrated form, or in a combination of these ways.

Jeans and Stearns point out that we should recognize three phase; of the vitamin D requirement: (1) that for normal growth and mineralization of the skeleton including the teeth, of infants and children; (2) that for maintenance of these skeletal structures in sound condition throughout adult life; and (3) that to supply the special needs of pregnancy and lactation.

There is still no consenus of scientific opinion as to what constitutes a normal or optimal rate of mineralization of the developing skeleton. Jeans and Stearns state that this is most rapid in early life, and that among the estimates of desirable calcium retention "the higher figures are thought to be a safer."

During infancy the vitamin D requirement depends to some extent upon the rate of growth. The more rapidly the infant grows the more important it is to safeguard him against rickets: also the liberal allowance of vitamin D, gives as such a safeguard, itself tends to increase the rate of growth. Infants whose daily food includes a quart of milk containing 135 units of vitamin D seem to be fully protected from rickets; but additional vitamin D up to a total of 300 to 400 units seems to support a rate of growth and development "nearer the top of the normal range."

The observations of Jeans and Stearns and of Macy and co-workers point out that increasing allowances up to an undetermined level between 135 and 300 units of vitamin D per day tends toward a more uniform ability among infants to assimilate calcium from the food. This appears to be true even of infants fed human milk.

During Childhood there is need of continued emphasis upon liberal intakes of calcium and of vitamin D if optimal development of bones and teeth is to be maintained. According to Jeans and Stearns (1939) the needs of children in this respect do not seem to have been as fully realized. as the needs of infants, probably because it is less easy to make the child's need obvious. Beyond emphasizing the view that there should be no let down of calcium or vitamin D intake as infancy develops into childhood, little can yet be said as to the child's quantitative need.

Jeans and Stearns have studied fifty children aged 1 to 12 years with varying intakes of milk and with and without added vitamin D in the form of cod liver oil to the extent of 300 to 400 added units of vitamin D per day. They found that children varied in their need. If a child utilized calcium efficiently without the added vitamin D, the addition of this to the diet had little effect; while if the original utilization was poor, the ingestion of additional vitamin D increased the retention of calcium.

Whether adults need vitamin D and if so to what extent axe questions upon which there is little evidence beyond the fact, so carefully an repeatedly emphasized by Jeans and Stearns, that vitamin D does not lower the minimum requirement for ingested calcium.

In pregnancy and lactation it is undoubtedly wise to be liberal with the vitamin D allowance.

The National Research Council's Recommended Allowances provide 400 units per day for each child or pregnant or nursing woman.

The Sources

The antirachitic values of fish liver oils, of irradiated ergosterol, and of ultraviolet irradiation of the body, are so great that they have absorbed attention to perhaps a greater extent than is desirable. The presence of important amounts of vitamin D in egg yolk, whole milk, and butterfat has, however, been established. That in some cases the evidence of its presence has failed to appear may be attributed to, in large part, to the use in many experiments of so drastic a rickets-producing diet that symptoms of rickets developed in the experimental animals notwithstanding the presence of amounts of vitamin D (in the food fed for test) which would have been measurable under more nearly normal conditions. Mellanby, working with puppies and using basal diets not so drastically deficient as are the rickets-producing diets used by most of those who work with rats, had no difficulty in showing the vitamin D value of green vegetables and of butterfat. Shipley, Kinney, and McCollum (1924) showed the presence of vitamin D in ether extracts of green leaves. Also in the work of Sherman and Steibeling which demonstrated a considerable vitamin D value in milk, the diets were so arranged and controlled that the effect of the milk in promoting deposition of calcium phosphate in the growing bone must have been due to the vitamin rather than the calcium or phosphorus factor of the food, since the amounts and ratios of calcium and phosphorus in the intake were strictly controlled.

As is true of the vitamin A, so also for vitamin D there seems to be survival value for the species in the capacities that have been developed by the maternal organism to collect the vitamin and pass it on to its young through the fat of the egg yolk or of the milk as the case may be. The concentration is higher in the egg fat, but the total amount of vitamin D obtained through the milk fat is considerable because of the consumption of the latter not only in the form of milk itself but also of cheese, butter, cream, and ice cream. Vitamin D is sufficiently stable, except under irradiation, so that no serious loss need be feared in the making or storing of such foods as these, or in ordinary cooking operations. Direct experimentation has shown it stability at autoclave temperatures.

Vitamin D milk is the term commonly used for market milk of enhanced vitamin D value.

Investigators differ as to the particability of materially increasing the vitamin D value of milk by the irradiation of the cow.

Three methods of enrichment of milk in vitamin D have been on practical use: (1) the addition of a concentrate of natural vitamin D; (2) feeding irradiated material to the milch cow; (3) irradiation of the milk itself.

The choice between these three methods depends partly on economic factors, and partly upon considerations having to do with the difference in kinds and relative proportions of the vitamin D or vitamin D introduced by the respective methods.

Whatever may be the ultimate consensus of opinion as to the relative merits of these different forms of enrichment, it is already established that vitamin D milk is a highly effective means of administering vitamin D.

With milk of such greatly enriched vitamin D value now generally available, and with the wide general distribution of cod liver oil, of the more potent liver oil of the halibut, and of the still more potent *percomorph liver oil,* it is proper enough to place chief dependence upon these highly potent and now well standardized sources. But the natural vitamin D values of foods are scientifically significant to our understanding of the process of nature and how children got along even as well as they did before vitamin D was discovered.

The Storage

Both carefully controlled laboratory experiments with rats and clinical experience with mothers and infants have shown that the antirachitic vitamin can be stored in the body, and to a very important extent. The nutritional condition of the mother influences the storage of vitamin D in the body of the baby before its birth, and her bodily store of this vitamin (or lack of it) affects the antirachitic potency of her milk.

Vitamin D taken by a pregnant or nursing mother is therefore, a nutritional asset both to herself and to the child.

Hess and coworkers encountered cases in which young rats which they had intended using as test animals proved "refractory to rickets," because of the highly antirachitic character of the mother's diet and the fact that the litters were small. This, of course, is a converse demonstration of the clinical experience that severe rickets is more frequent among the children of the poor whose mothers are living on food of inferior mineral and vitamin content, often in dark tenements, and also have borne and suckled many children.

The Functions

Bone Formation : One of the major function of vitamin D is the maintenance of the normal processes of bone formation. A deficiency of the vitamin in the growing young results in rickets. The disease will develop in infants even though the diet provides an ample intake of calcium and phosphorus unless vitamin D is provided in the diet or adequate exposure to sunlight is possible. Late rickets (osteomalacia), a condition similar to rickets in children, occurs in adults.

In rickets the bones are susceptible to deformity when exposed to ordinary stresses. A common and severe change in the skeletal structure is enlargement of the ends of the long bones in the legs and arms. If mineral is not deposited in the developing bone cells, the weight of the body causes the non-calcified portion to flatten and mushroom outward, giving the appearance of enlarged joints. Other bone changes include the *ricketic rosary*, a row of beadlike protuberances on each side of the chest at the function of the rib bones and joining (costal) cartilage (Fig.), bowed legs, and knocked knees—all caused by failure of the bones to calcify normally.

The mode of action of vitamin D in relation to bone calcification remains relatively obscure. Intramuscularly injected radioactive calcium was deposited in the bones of rickets chicks with no vitamin D in the diet. This indicated to the investigators that the primary function of vitamin D might be that of facilitating absorption of calcium from the intestine since calcium deposition in bones could proceed without it.

Calcium Absorption and Metabolism : That vitamin D exerts a favorable influence on the absorption of calcium is evidenced in studies with animals and human beings. Figure illustrates the decrease in calcium excretion in children with rickets after the administration of vitamin D; the fecal exertion is markedly reduced. With children 1 to 12 years of age, Jeans and Stearns found that some utilized calcium efficiently without added vitamin D; others required a supplement. Stearns pointed out that there is no way of discerning from outward appearances which children need the additional vitamin, which do not, hence recommended that vitamin D be given to all children. Numerous balance studies with normal adults show no improvement in the utilization of calcium with additions of sources of vitamin D except during pregnancy and lactation.

The absorption of phosphorus seems not to be influenced by vitamin D. For example, vitamin D-deficient animals absorb phosphorus equally well, with and without an additional supplement of the vitamin.

Growth : Rickets causes a slowing of growth. Animals wherein experimental rickets has been induced grow more slowly than their normal controls with an adequate vitamin D intake. Infants also fail to grow normally without a sufficient amount of the vitamin. Infants given 340 to 400 IU daily of vitamin D were found to grow more in length than those receiving 60 to 135 IU. Giving more than 600 IU daily did not increase the growth rate further; increasing the vitamin D intake beyond 1800 IU daily decreased growth.

The Incidence

Clinically detectable rickets is uncommon in the United States today. The widespread preventive (prophylactic) use of sources of vitamin D has all but eradicated this once common scourge of infancy. Fortification of all evaporated milk and much of the fresh milk with vitamin D almost controls rickets automatically in the United States. Osteomalacia, too, is rare in this country. It is reported to be common among women in the Orient and parts of the Near East during pregnancy and lactation, increasing in severity with each reproductive cycle.

The Hypervitaminosis

Massive dosages of vitamin D may have toxic effects if continued for a period of time. Giving only two massive doses a year is used by some physicians as an acceptable means of providing vitamin D. The level at which deleterious effects result continued high dosages varies with individuals. Some show symptoms with an intake of 50,000 IU per day for a few weeks, others tolerate as much as ten times that amount for longer periods of time. An intake of 1800 IU daily may be mildly toxic in children if continued over a period of time. The early symptoms include loss of appetite, thirst, lassitued, followed by nausea, vomiting, diarrhoea, abdominal discomfort, and loss of weight. The blood calcium chronic condition that has been present for some time becoming acute only when a high degree of toxicity is reached.

Vitamin D preparations on the market are of such potency that dosages are expressed in drops, making it easy to give more than is desirable. In commonly used preparations around 50 drops measure one half teaspoon with a potency of 3000 IU. Most multiple vitamin capsules provide 500 IU or more per capsule.

Recommended Allowances

The Recommended Allowance for vitamin D for late adolescence is 400 IU daily, the same as is recommended throughout childhood. Normal adults seem to need little, if any, supplemental vitamin D. It is generally believed, nevertheless, that for adults with little exposure to the direct sunlight some vitamin D in the diet may be advisable, despite the lack of direct proof to substantiate the presumption. During pregnancy and lactation, when the calcium need is increased, the Recommended Allowance is 400 IU daily. The requirements of infants and children.

Appendices

I

Vitamins in Food

	A or carotene i.u.	B_1 mg.	Ribo-flavin mg.	Niacin ug.	B_6 mg.	C mg.	D i.u.	E mg.
Almonds	0	24	120	334	—	0	0	—
Apple, raw	9c	10	0	100	10	4	0	.21
cooked	—	3	—	25	—	.7	0	—
Apricot, raw	196c	3	20	100	—	3	0	—
canned	142c	3	20	100	—	1	0	—
Arrowroot	0	0	0	0	0	0	0	0
Asparagus; boiled	40c	9	40	280	9	2	0	—
Bacon, fried	0	160	125	1,200		0	—	.15
Banana	14c	9	10	200	90	2	0	.14
Barley	0	20	20	700	—	0	0	1
pearl, boiled	0	20	10	700	—	0	0	—
Beans, baked	47c	20	10	200	—	0	0	—
runner, raw	180c	16	16	200	—	1	0	—
Beef, corned	0	0	40	500			0	0
lear roast	14	11	65	—	—	0	0	.15
Beer, mild	0	1	.10	400	—	0	0	0
Beetroot, boiled	0	5	20	30	30	2	0	—
Bemax	—	450	200	1,700	450	0	0	8
Blackberries raw	85c	6	5	60	—	6	0	—
canned	43c	3	—	—	—	2	0	—

	A or carotene *i.u.*	B_1 *mg.*	*Ribo-flavin* *mg.*	*Niacin* ug.	B_6 *mg.*	C *mg.*	D *i.u.*	E *mg.*
Bovril	—	75	480	17,000	—	0	0	—
Bran, processed	—	180	120	5,000	—	0	0	—
Bread white								
70% extn.	0	10	10	200	60	0	0	.07
81% extn.	0	48	17	330	80	0	0	.2
wholemeal 92% extn.	—	90	50	600	240	0	0	.37
white, fortified	0	48	10	336	60	0	0	.07
Broccoli tops,								
boiled	1,500c	—	—	—	—	6	0	—
Brussels sprouts,								
raw	85c	26	48	190	—	38	0	—
Butter	1,136	0	0	0	0	0	17	.5

Vitamin Values

Published values for the vitamin contents of foods have been obtained from different samples and by different methods of assay. Several are therefore available for the same food.

It has not been thought necessary in a book of this kind to quote all the published figures for every food. Those which are available have therefore been studied carefully and a figure chosen for each food which represents an average value.

	A or carotene *i.u.*	B_1 *mg.*	*Ribo-flavin* *mg.*	*Niacin* *μg.*	B_6 *μg.*	C *μg.*	D *mg.*	E *i.u.*
Cabbage, raw	180c	15	20	100	30	14	0	.03
boiled	160c	1	50	75	—	7	0	—
Carrots, raw	3,500c	15	10	200	100	3	0	.13
canned	3,500c	11	7	180	—	1	0	—
boiled	3,000c	13	8	180	—	1	0	—
Cauliflower,								
boiled	0	1	2.	15	—	11	0	—
Celery, raw	0	7	10	70	500	1	0	—
Cheese, Cheddar	370	9	140	100	18	0	4	.14

	A or carotene i.u.	*B_1 mg.*	*Ribo-flavin mg.*	*Niacin μg.*	*B_6 μg.*	*C μg.*	*D mg.*	*E i.u.*
Roquefort	700	8	100	350	—	0	—	0
Cherries, raw	49c	12	40	40	—	1	0	—
Chicken, roast	—	34	45	1,500	50	0	0	.06
Chocolate, plain	21c	26	70	300	—	0	trace	—
Cocoa, powder	14	30	80	300	—	0	—	.87
Cod, raw	14	11	40	600	—	0	0	—
Cod roe, raw	—	400	260	400	800		14-20	2.0
Cooking fat	0	0	0	0	0	0	0	0
Cornflakes	0	0	—	—	—	0	0	—
Cornflour	0	0	—	30	—	0	0	—
Crab, boiled	—	250	4	800	—	0	0	—
Cream, raw	470	0	42	—	—	0	14	—
Cucumber, raw	0	10	14	50	—	2	0	—
Currants, black,								
raw	26c	13	10	100	—	57	0	—
canned	13c	6	26c	13	10	100	—	57
red, raw	7c	13	—	—	—	13	0	—
Custard powder	0	0	—	30	—	0	0	—
Dates	24c	0	10	100	—	0	0	—
Dripping	28	0	0	0	0	0	0	11
Egg, whole,	250	37	170	0	120	0	17	.6
raw boiled	250	30	160	0	—	0	17	.6
Egg yolk, raw	500	70	170	0	300	0	—	1.5
Egg white, raw	0	0	100	0	90	0	0	0

	A or carotene i.u.	*B_1 mg.*	*Ribo-flavin mg.*	*Niacin μg.*	*B_6 μg.*	*C mg.*	*D i.u.*	*E mg.*
Figs, dried, raw	26c	0	80	500	—	0	0	—
Flour, white								
70-72% extn.	0	24	10	300	72	0	0	.34
81% extn.	0	68	23	450	—	0	0	—
92% extn.	0	130	60	800	100	0	0	.63
70-72% fortified	0	68	10	450	72	0	0	.34

	A or carotene i.u.	B_1 mg.	Ribo-flavin mg.	Niacin μg.	B_6 μg.	C mg.	D i.u.	E mg.
Gooseberries	57c	13	—	—	—	11	0	—
canned	28c	6	—	—	—	5	0	—
Grapes, green, raw	13c	8	26	—	—	1	0	—
Grape-fruit raw	3-4	10	7	—	3	6.5	0	—
canned	3-4	10	7	—	3	6.5	0	—
Haddock, raw	14	8	50	260	—	0	0	11
Halibut, cooked	0	40	50	1,000	30	0	0	—
Ham, boiled	0	110	38	500	100	0	0	—
Hazel nut, raw	0	170	0	200	—	0	0	—
Herring, fried	40	0	0	1,000	—	0	250	—
Honey	0	5	20	100	4	1-6	0	—
Ice-cream	420	0	70	—	—	0	0	—
Jam, black currant	6c	1	10	30	—	17	0	—
gooseberry	6c	1	0	0	—	3	0	0
plum	6c	1	0	0	—	1	0	0
raspberry	6c	1	0	0	—	2	0	0
strawberry	6c	1	0	0	—	6	0	0
Kale, boiled	3,000	32	—	—	—	5	0	—
Kidney, sheep, raw	280	70	400	2,000	—	0	0	—
Kipper, raw	51	0	90	1,000	—	0	250	—
Kisra	0	0	12	800	—	0	0	—
Lard	0	0	0	0	0	0	0	0
Leeks, raw	90c	14	190	190	—	3	0	—
Lemon, juice	0	10	0	0	380	12	0	—
Lentils, raw	57c	190	50	800	—	0	0	—
Lentil, flour (Indian)	—	—	—	300	—	0	0	—
Lettuce, raw	900c	17	20	100	600	3	0	0.12
Liver, calf fried	1,100	110	890	3,500	15	0	—	0.4
Loganberry								
raw	20c	6.	—	—	—	9	0	—
canned	10c	3	—	—	—	3	0	—
Macaroni	—	21	—	600	—	0	0	.34
Mackerel, raw	43	10	190	1800	—	0	100	—
Maize (yellow), whole	28c	94	37	43	200	0	0	3.7

	A or carotene i.u.	B_1 *mg.*	*Ribo-flavin mg.*	*Niacin* µg.	B_6 µg.	*C mg.*	*D i.u.*	*E mg.*
Maize flour, 96% extn.	28	14	9	14	100	0	0	—
Margarine	568	0	0	0	0	0	56	15
Marmite	—	900	1,800	17,000	1,200	0	0	—
Marrow, raw	6c	8	—	76	—	1	0	—
Melon, raw	8c	10	10	200	—	3	0	—
Milk								
whole, fresh	30	13	40	20	20		4 .3	.03
pasteurised	30	12	36	20	20	—	.3	—
evaporated	100	18	100	100	—	—	1	.09
condensed (sweet]	100	30	100	100	25			1
dried	300	85	300	200	90			3
skimmed	0	13	30-40	20	—	—	—	—
Millet, ground	trace	130	10	46	9	0	0	—
Molasses, cane	—	14	57	430	—	0	0	—
Mushrooms, fried	0	50	300	2,000	13	—	0	—
Mustard & cress, raw	—	40	—	—	—	23	0	—
Mutton, lean, roast	14	20	40	1,600	60	0	0	.22
Oats, whole	—	200	30	300	—	0	0	.45
Oatmeal	0	128	40	300	70	0	0	.6
Olive salted	140c	—	—	—	—	4	0	—
Onions, raw	160c	25	10	—	18	5	0	.07
Orange, raw	28c	20	10	100	23	16	0	.07
Orange juice, canned	18c	30	8	100	15	14	0	.07
Oyster, raw	120	190	800	250	9	0	1.4	—
Paprika	—	—	—	300	—	0	0	—
Parsnips, raw	0	22	—	330	—	1.5	0	—
Peaches, raw	185c	5	0	0	—	3	0	—
canned	142c	3	0	0	—	1	0	—
Peanuts, raw	0	179	80	2,600	90	0	0	3.0
Pears, raw	2c	4	10	100	—	1	0	—
canned	1 c	3	5	50	—	1	0	—

	A or carotene i.u.	B_1 mg.	Ribo-flavin mg.	Niacin µg.	B_6 µg.	C mg.	D i.u.	E mg.
Peas, fresh								
raw	57c	48	30	200	—	3	0	.6
boiled	50	38	24	160	—	2	0	—
processed	50	34	24	160	—	2	0	—
split, boiled	50	—	—	—	—	0	0	—
Pineapple, canned	18c	14	20	0	—	3	0	—
Plums, raw canned	107	12	10	20	—	1	0	—
Pork, lean, cooked	0	270	31	1,700	170	0	0	.2
Porridge, cooked	0	125	40	300	—	0	0	—
Potatoes, old								
boiled	0	24	16	225	40	1.5-3	0	.03
roast	0	36	—	—	—	2-3	0	—
fried	0	30	10	200	—	2	0	—
new, boiled	0	30	16	225	—	6.5-7	0	—
Prunes, dried,								
Stewed	200c	8	17	—	—	0	0	—
Pumpkin, boiled	4	6	10	—	—	6	0	—
Rabbit, raw	0	7	200	1,900	—	0	0	—
Radish, raw	0	7	10	40	—	4	0	—
Raisins, dried	13c	0	10	100	27	0	0	—
Raspberries								
raw	20c	6	—	100	—	9	0	—
canned	—	2	—	100	—	3	0	—
Rhubarb, stewed	0	1	—	—	—	1	0	—
Rice, polished								
boiled	0	20	20	300	90	0	—	.13
Rye, wholemeal	—	77	30	340	100	0	0	2.9
Sago, boiled	0	0	0	60	—	0	0	—
Salmon, fresh, raw	—	10	60	2,400	—	0	170	—
canned	70	10	60	1,800	36	—	170	—
Sardines, canned	77	10	80	1,300	46	0	280	—
Sausage pork fried	0	70	50	500	—	0	0	—
breakfast	0	60	20	90	—	0	0	—
Semolina, boiled	0	26	0	570	—	0	0	—
Shrimps boiled	—	26	50	220	—	0	43	
Sorghum ground								
meal	0	130	23	700	—	0	0	—

	A or carotene i.u.	*B_1 mg.*	*Riboflavin mg.*	*Niacin µg.*	*B_6 µg.*	*C mg.*	*D i.u.*	*E mg.*
Spinach, boiled	2,400c	14	42	80	20	8	0	.5
Stout	0	1	14	200	—	0	0	0
Strawberries raw	4c	5	50	80	13	16	0	—
canned	1c	3	—	—	—	7	0	—
Sugar	0	0	0	0	0	0	0	0
Sultanas, dried	14c	0	0	0	—	0	0	—
Swedes, boiled	0	7	—	300	—	6.5	0	—
Sweet breads, stewed	0	10	30	700	—	0	0	—
Syrup	0	0	0	0	0	0	0	0
Tangerines	70c	23	10	90	—	13	0	—
Tapioca	0	0	200	300	—	0	0	0
Tomato, raw	300-500	20	10	100	—	7	0	—
canned	300-500	20	7	100	—	4	0	—
fried	9	18	—	—	—	2	0	—
Tongue, ox, pickled	0	12	60	800	30	0	0	—
Turbot, raw	0	10	25	800	—	0	8	—
Turkey, roast	0	14	50	2,000	—	0	0	—
Turnip, boiled	13c	trace	13	200	30	3	0	—
Turnip, tops, boiled	5000	26	100	150	—	12	0	.7
Veal, lean, cooked	11	70	100	2,000	10	0	0	0
Walnut	0	50	40	300	—6	0	—	
Watercress	1,200c	30	20	500	—17	0	—	
Wheat, whole	0	114	50	1,400	120	0	1	.34
flakes	0	10	20	300	—	0	0	—
shredded	0	73	40	1,400	—	0	0	—
Wine	0	0	5	40	—	0	0	0
Yeast, Baker's	0	70	900	3,000-12,000	—	0	0	—
Yeast, Brewer's	0	340	550	3,000-25,000	—	0	0	—

c = present as carotene

o = known to be absent

— = unknown.

II

Dietary Survey

1. *Food Habits*

   (a) Vegetarian/non-vegetarian/non-vegetarian but usually takes vegetarian food.

   (b) Common dietary pattern:

Morning	*1st day*	*2nd day*	*3rd day*	*4th day*	*5th day*	*6th day*	*7th day*
Mid-day							
Afternoon							
Evening							
Others							

2. *Special Foods*

   (a) For different groups

Groups	*Foods recommended*	*Foods avoided*
1. Adults		
2. Infants		
3. Pregnant women		
4. Lactating women		
5. Invalids		

   (b) During illness

Illness	*Foods recommended*	*Foods avoided*
1. Cold		
2. Fever		
3. Diarrhoea		
4. Dysentery		
5. Other conditions		

(c) During different seasons

Seasons	*Foods recommended*	*Foods avoided*
Summer		
Winter		
Monsoon		

(d) For religious or other reasons (include data on 'fasts' under-taken by family members and foods taken during 'fasting')

Occasion	*Foods permitted*	*Foods avoided*

(e) (i) Foods consumed by previous generation and not consumed now:

(ii) Foods consumed now and not consumed by previous generation:

3. Methods of Cooking

Foods	*Boiling*	*Steaming*	*Frying*	*Baking*	*Pressure Cooking*
(a) Roti					
(b) Rice					
(c) Dal					
(d) Vegetables					

4. *Food Sharing Practices:*
5. *Members Eating Out:*

	Member	*Type of food taken outside*
(a) Regularly		
(b) Occasionally		

6. Foods consumed at home

Foodstuff	*Amount consumed (gm)*	*Source from which obtained*
1. *Cereals and millets*		
(a) Rice		
(b) Wheat		
(c) Other grains		
(d) Other cereal or starchy foods		
(e) Processed cereal foods		
2. *Pulses and nuts*		
(a) Dals		
(b) Whole legumes		
(c) Groundnut, til etc.		
(d) Other nuts		
(e) Parched legumes		
3. *Vegetables*		
(a) Leafy vegetables		
(b) Roots and tubers		
(c) Others		
4. *Animal foods*		
(a) Fish, meat and chicken		
(b) Egg		
5. *Milk and milk products*		
(a) Whole milk		
(b) Toned milk		
(c) Milk powder		
(d) Curd		
(e) Baby food		
(f) Cheese		
(g) Mava		
(h) Milk sweets		

(i) Ice-cream

(j) Other milk products

6. *Fats and oils*

(a) Butter

(b) Dalda

(c) Groundnut oil

(d) Ghee

(e) Refined oil

(f) Other oils (specify)

7. *Sugar*

(a) Refined sugar

(b) Jaggery

(c) Others (honey, molasses etc.)

8. *Condiments*

(a) Tamarind

(b) Cocum

(c) Chillies

(d) Mango slices

(e) Other spices

9. *Preserved and processed foods*

(a) Pickles

(b) Papad

(c) Jams and jellies

(d) Other canned or bottled foods

10. *Beverages*

(a) Tea

(b) Coffee

(c) Cocoa

(d) Malted foods

(e) Carbonated drinks

11. *Miscellaneous*
    (a) Betelnut
    (b) Betel leaves
    (c) Chewing tobacco
    (d) Smoking tobacco

	Per family	***Per capita***	***Per consumption unit***
Foodstuffs			
Cereals (rice + other grains)			
Pulses			
Leafy vegetables			
Root vegetable			
Other vegetables			
Meat			
Fats and oils			
Milk			
Sugar and jaggery			
Nutrients			
Calories			
Protein			
Fat			
Carbohydrate			
Calcium (mg)			
Phosphorus (mg)			
Iron (mg)			
Vitamin A as:			
Carotene (i.u.)			
Pre-formed vitamin (i.u.)			
Thiamine (mg)			
Riboflavin (mg)			
Niacin (mg)			
Vitamin C (mg)			

III

Proforma for Use in Diet Surveys

Interviewer Family No.:

Date: Address:

General Information

1. Head of the family:

   Caste: Religion:

2. Composition of family:

Sl. No.	*Name*	*Sex*	*Age*	*Education*	*Occupation**	*Income*	*Relation to head*
1.							
2.							
3.							
4.							
5.							
6.							
7.							
8.							
9.							
10.							

Include subsidiary occupations

3. (a) Information on married women:

   Number of children born:

   Number of children surviving:

   Miscarriages:

   Still births:

   (b)

	Present family	*Father's family*	*Mother's family*
Number of children born			
Number of children surviving			

(c) Record of deceased family members:

Sl. No.	*Name*	*When occurred*	*Relation to head*	*Age at death*	*Cause of death*
1.					
2.					
3.					
4.					
5.					
6.					
7.					
8.					
9.					
10.					

4. Approximate monthly expenditure on:
   (a) Milk and milk products
   (b) Vegetables and fruits
   (c) Groceries
   (d) Fuel
   (e) Rent

Bibliography

Andrews, S.: *Food and Beverage Science,* Tata McGraw Hill, Publishing Co. Ltd., New Delhi, 1980.

Arora, K.: *Theory of Cooking,* K.N. Gupta & Co., Delhi, 1982.

Aurand, L.W. and A.E. Woods : *Food Chemistry,* The AVI Publishing Co., Connecticut, 1973.

Bennion, M.: *Introductory Foods,* Macmillan Publishing Co., New York, 1980.

Cesaerani, V. and R. Kinton : *Practical Cookery,* Edward Arnold, London, 1981.

Chang, R.: *Physical Chemistry with Application to Biological System,* Collier Macmillan, International Edition, 1977.

Charley, H.: *Food Science,* John Wiley and Sons Inc., New York. 1982.

Clydesale, F. : *Food Science and Nutrition,* Prentice-Hall, Inc., New Jersey, 1979.

Desrosier, N.C.: *Food Preservation,* The AVI Publishing Co. Inc., Connecticut, 1970.

Dick, D.A.T. : *Cell Water,* Butterworth, Washington, 1966.

Dickerson, R.E. and I. Geiss : *Proteins: Structure, Function and Evolution,* Menlo Park, California, 1983.

Eley, G.: *Wild Fruits and Nuts,* E.P. Publishing Ltd., Yorkshire, 1976.

Furia, T.E.: *Hand Book of Food Additives,* C.R.C. Press, T.C. Ohio, 1977.

Goodwin, T.W: *The Biochemistry of Carotenoids*, Academic Press, New York, 1980-84.

Gould, R.F. : *Flavour Chemistry*, American Chemical Society, Washington, D.C., 1966.

Greenstein, J.P. and M. Winitz : *Chemistry of Amino Acids*, John Wiley & Sons, New York, 1961.

Gunstone, F.D.: *An Introduction to the Chemistry and Biochemistry of Fatty Acids and their Glycerides*, Chapman and Hall, London, 1967.

Hall, C.W.: *Dictionery of Drying*, Marcel Dekker, Inc., New York, 1979.

Haschemeyer, R. and A.H. Haschemeyer : *Proteins: A Guide to Study of Physical and Chemical Methods*, John Wiley & Sons, New York, 1973.

Hobbs,B.C. : *Food Microbiology*, Arnold-Heine Mann, New Delhi, 1982.

Holman, R.T. : *The Progress in the Chemistry of Fats and other Lipids*, Pergamon Press, New York, 1971.

Hudson, B.J.F. : *Development in Food Proteins*, Applied Science Publisher, London, 1982,

Jasscler, D.K. and M.A. Joslyn : *Fruit and Vegetables Juice*, The AVI Publishing Co. Inc., Connecticut, 1971.

Jull, M.A.: *Poultry Husbandry*, Tata McGraw-Hill Publishing Co. Ltd., New Delhi, 1977.

Kadans, J.M.: *Encyclopedia of Fruits, Vegetables, Nuts and Seeds*, Parker Publishing Co. Inc., New York, 1973.

Less, R. and E.B. Jackson : *Sugar Confectionery and Chocolate Manufacture*, Leonard Hill Books, An Intertext Publisher, Burks, 1973.

Mackinney, G. and A.C. Little : *Colour of Foods*, The AVI Publishing Company, Connecticut, 1972.

Marton, I.D. and A.J.Macbod : *Food Flavours*, Elesvier Scientific Publishing Co., New York, 1982.

Mate, S. A.: *Cereal as Foods and Feeds*, The AVI Publishing Co. Inc., Connecticut, 1959.

Mcwilliams, M.: *Food Fundamentals,* John Wiley & Sons Inc., New York, 1979.

Mellor, J. D.: *Fundamentals of Freete-Drying,* Academic Press, London, 1978.

Nath, P. : *Vegetables for the Tropical Region,* Indian Council of Agricultural Research, New Delhi, 1976.

Palmer, T.: *Understanding Enzymes,* John Wiley & Sons, New York, 1981.

Peckham, G.G. and J.H. Freeland-Graves : *Foundations of Food Preparation,* Macmillan Publishing Co. Inc., New York, 1979.

Peterson, M.S. and A.H. Johnson : *Encyclopedia of Food Science,* The AVI Publishing Co. Inc., Connecticut, 1978.

Peterson, M.S. and D.K. Tressler : *Food Technology the World Over,* The AVI Publishing Co. Inc., Connecticut, 1963.

Puruthi, J.S.: *Spices and Condiments,* National Book Trust of India, New Delhi, 1976.

Ranganna, K.S. and K.T. Achaya : *Indian Dairy Products,* Asia Publishing House, Bangalore, 1974.

Ranganna, S. : *Manual of Analysis of Fruit and Vegetable Products,* Tata McGraw-Hill Publishing Co. Ltd., New Delhi, 1984.

Schultz, G.E. and R.H. Schirmer : *Principles of Protein Structure,* Springer-Verlage, New York, 1979.

Sebrell, W.H. (Jr.) and R.S. Harris : *The Vitamins,* Academic Press Inc., New York, 1967.

Thangam, P.E.: *Modern Cookery for Teaching and Trade,* Orient Longmans Ltd., Bombay, 1983.

Thankamma, J.: *Food Adulteration,* S.G. Wasani, New Delhi, 1976.

Underwood, E.J: *Trace Elements in Human and Animal Nutrition,* Academic Press, New York, 1977.

Whiteley, P.P. : *Biscuit Manufacture,* Elsevier Publishing Co. Ltd., London, 1971.

Williams, A, : *Bread Making,* Hutchinson, Benhan, London, 1975.